Science and Technology of Micromachines

Science and Technology of Micromachines

Edited by **Eve Versuh**

CWILLFORD PRESS

New York

Published by Willford Press,
118-35 Queens Blvd., Suite 400,
Forest Hills, NY 11375, USA
www.willfordpress.com

Science and Technology of Micromachines
Edited by Eve Versuh

International Standard Book Number: 978-1-68285-139-5 (Hardback)

Printed in the United States of America.

Contents

Preface

Micromachines are miniature machines and its components are used in modern manufacturing and engineering. This book with its design-oriented approach towards explaining different applications of micromachines, microcontrollers, microrobotics, microelectromechanical systems (MEMS), applications of micro machines in different fields, etc. helps the reader gain a comprehensive insight into the field. This book aims to equip students and experts with the advanced topics and upcoming concepts in this area. It is an excellent guide for those wanting to acquire advanced knowledge related to micromachines.

Significant researches are present in this book. Intensive efforts have been employed by authors to make this book an outstanding discourse. This book contains the enlightening chapters which have been written on the basis of significant researches done by the experts.

Finally, I would also like to thank all the members involved in this book for being a team and meeting all the deadlines for the submission of their respective works. I would also like to thank my friends and family for being supportive in my efforts.

Editor

Uncalibrated Visual Servo Control of Magnetically Actuated Microrobots in a Fluid Environment

Jenelle Armstrong Piepmeier [1,*], Samara Firebaugh [2] and Caitlin S. Olsen [1]

[1] Systems Engineering, United States Naval Academy, 105 Maryland Avenue, Annapolis, MD 21402, USA; E-Mail: caitlinolsen.co@gmail.com

[2] Electrical and Computer Engineering, United States Naval Academy, 105 Maryland Avenue, Annapolis, MD 21402, USA; E-Mail: firebaug@usna.edu

* Author to whom correspondence should be addressed; E-Mail: piepmeie@usna.edu

External Editor: Joost Lötters

Abstract: Microrobots have a number of potential applications for micromanipulation and assembly, but also offer challenges in power and control. This paper describes an uncalibrated vision-based control system for magnetically actuated microrobots operating untethered at the interface between two immiscible fluids. The microrobots are 20 μm thick and approximately 100–200 μm in lateral dimension. Several different robot shapes are investigated. The robots and fluid are in a 20 × 20 × 15 mm vial placed at the center of four electromagnets. Pulse width modulation of the electromagnet currents is used to control robot speed and direction. Given a desired position, a controller based on recursive least square estimation drives the microrobot to the goal without *a priori* knowledge of system parameters such as drag coefficients or intrinsic and extrinsic camera parameters. Results are verified experimentally using a variety of microrobot shapes and system configurations.

Keywords: microelectromechanical devices; microrobots; uncalibrated visual servoing

1. Introduction

Tetherless microrobots have been proposed for a number of applications including minimally invasive surgery and micromanipulation of micro gels for *in vitro* tissue culture [1]. Such microrobots, which are at dimensions of tens to hundreds of micrometers, are of comparable size to many biological structures, such as cells, and therefore are an enabling technology. A challenge facing microrobots is providing power and control. In particular, adapting macroscale control strategies to the microscale environment is not always straightforward. The physics of microscale operation does not always scale intuitively. For example, surface forces such as friction or viscous drag play a much larger role than volumetric forces such as magnetic attraction or inertial forces. Adding to the difficulty of the problem, drag forces at this scale can be difficult to model for complex microrobot device geometries.

Tetherless microrobot actuation has been achieved using a number of different power delivery methods including optical [2], electrostatic [3], thermal [4], ultrasonic [5] and electromagnetic [6]. A review of propulsion methods specific to swimming microrobots medical applications is given in [7]. Closed-loop control can be achieved with vision-based feedback of the microrobot position. Controller designs typically utilize a system model that characterizes the interaction between the applied forces (magnetic, drag, electrostatic, *etc.*) on the microrobot and the microrobot motion. A selected number of magnetically controlled systems are reviewed here with specific emphasis on the feedback control methods used and the tracking performance achieved.

Using a clinical magnetic resonance imaging (MRI) system, Tamaz *et al.* [8] develop a proportional-integral-derivative (PID) controller capable of navigating a 1500 μm ferromagnetic bead along a predefined path. They conclude that an adaptive controller would significantly decrease complications in the system and allow for more robust uses in the biomedical field.

Belharet *et al.* demonstrate a generalized predictive control (GPC) scheme to actuate a 500 μm neodymium sphere in an endovascular environment using Maxwell and Helmholtz coils in [9] achieving tracking errors on the order of 50–200 μm depending on the fluid composition and flow conditions.

In [10], Pawashe *et al.* demonstrate model-based learning controllers for 210 μm microbead manipulation using side-pushing by a magnetic 480 μm microrobot operating in a fluid. Microbeads are pushed to within one pixel (7.5 μm). Modeling includes flow velocities induced by the microrobot and equations of motion for the microsphere being manipulated.

Diller [11,12] uses multiple magnetic robots of the same design as [13] with geometric dissimilarities that resulted in differing responses to various frequencies. Planar path errors in [11] are on the order of 1000–1500 μm and three-dimensional (3D) control is achieved in [12] with path errors with less than 310 μm for 350 μm robot and 1500 μm robot.

In [14] Marino *et al.* compare an H_∞ controller with a PID controller for a linear uncertain dynamical model for electromagnetic steering control of a 1000 μm microrobot in low viscosity oil using the OctoMag system [15]. Tracking errors on the order of 270–490 μm are reported using the H_∞ controller. In another work by the same research group [16], Bergeles discusses the difficulties in localizing the position of the microdevice in the ocular environment due to complex optics and distortions. They propose a new projection model that allows localization of the microrobot for control with a proportional-derivative (PD) controller.

Using the MiniMag (Aeon Scientific, Zürich, Switzerland) system, Ghanbari [17] proposes time-delay estimation (TDE) control as a superior approach to H_∞ control for handling the many uncertainties present in modeling microrobot systems. Errors less than 200 μm are demonstrated for a microrobot consisting of a NdFeB cylinder permanent magnet of diameter 500 μm and length 1000 μm. The method does require the selection of controller gains for a particular configuration of the system.

Keuning [18] and Khalil [19] use a proportional-integral (PI) controller and waypoints generated by path planning algorithms to achieve planar navigation of paramagnetic 100 μm beads moving in water. Average errors ranged from 4.7 to 7.0 μm with standard deviations on the order of 2.0 μm.

Initial work [20] by the authors of this paper investigate simple linear model and demonstrated a proportional control strategy for electromagnetic actuation of various microrobot devices operating between fluid layers. Each microrobot design responds differently to the actuating magnetic fields required its own set of gains to compute the required duty cycle.

While varying in complexity, all of these works rely on system models of the various subsystem components including the microrobot device, the environment it is operating within, the actuation system, and the visual feedback system. The contribution of this work is the demonstration of an uncalibrated microrobot control scheme that uses uncalibrated visual feedback for an unmodeled system consisting of a microrobot device operating in a fluidic environment observed via a microscope and controlled by four electromagnets. In this paper, we characterize the performance of this strategy by varying a number of system parameters (microrobot device, magnification, target velocity, *etc.*) without changing any terms in the controller or utilizing configuration specific controller gains. Assumptions made based on the system components are presented in Section 2 and the image-based estimation and control are presented in Section 3. Experimental results are given in Section 4 for planar control of various 200 μm microdevices at various magnifications and system configurations. The experimental results include point-to-point motion and trajectory following with path errors ranging from 1.0 to 4.1 pixels in the image plane (4.1–40.5 μm in the workspace).

2. Electromagnetic Actuation for Microrobotic Control

For this work, we consider a ferromagnetic mass suspended in between two fluid layers and surrounded by two electromagnet pairs whose magnetic fields act primarily in the plane created by the fluid boundary as depicted in Figure 1a. This system was initially developed for participation in the Mobile Microrobot Challenge Competition [21]. While the method is extendable to higher degrees of freedom, for the theoretical and experimental results presented here it is assumed that there are two electromagnet pairs controlling a microrobot device that is acting in a planar, fluid environment.

As shown in Figure 1b, forces acting on this mass include electromagnetic forces $\vec{F}_M$, viscous drag $\vec{F}_d$, surface tension at the boundary $\vec{F}_t$, and apparent weight $\vec{F}_w$. Let $\vec{x}_w$ represent the position of the mass, $[x \quad y \quad z]_w^T$ with respect to a fixed world coordinate frame. Then, the equation of motion for the mass is given by:

$$m\ddot{\vec{x}}_w = \vec{F}_d + \vec{F}_t + \vec{F}_w + \vec{F}_M \tag{1}$$

Further expanding the terms to include drag coefficients (α, β, γ) and the magnetic field strength (B_x, B_y, B_z) results in:

$$m\begin{bmatrix} \ddot{x} \\ \ddot{y} \\ \ddot{z} \end{bmatrix}_w = \begin{bmatrix} \alpha & & \\ & \beta & \\ & & \gamma \end{bmatrix} \begin{bmatrix} \dot{x} \\ \dot{y} \\ \dot{z} \end{bmatrix}_w + \begin{bmatrix} 0 \\ 0 \\ \vec{F}_t + \vec{F}_w \end{bmatrix} + \begin{bmatrix} M_x\|\vec{\nabla}B_x\| \\ M_y\|\vec{\nabla}B_y\| \\ M_z\|\vec{\nabla}B_z\| \end{bmatrix} \tag{2}$$

where M_x represents the magnetic moment of the microrobot along the x-axis.

Figure 1. (**a**) Two electromagnet pairs arrayed about a ferromagnetic mass (the microrobot) operating in a fluid environment; and (**b**) the free body diagram of the microrobot in the fluid.

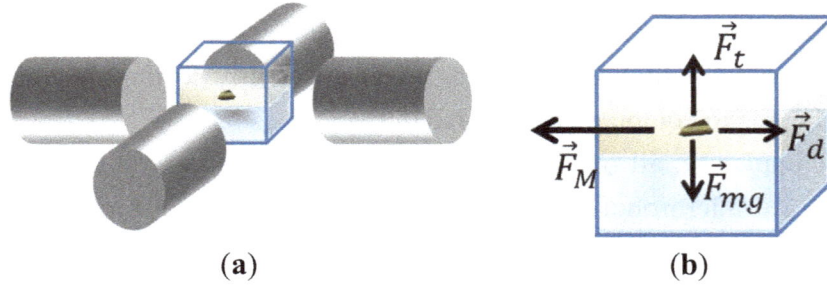

(**a**)   (**b**)

It is important to consider the relative magnitudes of the inertial and viscous forces acting on the body. The Reynolds number is a dimensionless quantity that relates inertial forces to viscous forces in the Navier-Stokes equations for a body moving in an incompressible Newtonian fluid. If the microrobot system is operating in a low Reynolds number regime ($Re \ll 1$), then the inertial term is much smaller than the drag forces as described by Purcell [22].

Furthermore, if we assume that the buoyant and surface tension forces counteract the gravitational forces, we have a microrobot operating in a planar region controlled by orthogonally oriented electromagnets. Thus, Equation (2) can be simplified for planar motion in a low Reynolds fluid environment as:

$$0 \cong \begin{bmatrix} \alpha & \\ & \beta \end{bmatrix} \begin{bmatrix} \dot{x} \\ \dot{y} \end{bmatrix}_w + \begin{bmatrix} M_x \|\vec{\nabla} B_x\| \\ M_y \|\vec{\nabla} B_y\| \end{bmatrix} \tag{3}$$

As the microrobot motion is sensed by a computer vision system, the relationship between pixel coordinate frame and the world coordinate frame can be modeled using projective geometry and homogeneous coordinates. To relate microrobot velocities to pixel velocities, the image Jacobian J_i, sometimes called the interaction matrix, must be computed using partial derivatives where the elements of the matrix will be a function of the intrinsic and extrinsic camera parameters as well as the depth from the camera plane to the microrobot object. If planar robot motion is assumed, the following relationship is given where $J_i \in \mathbb{R}^{2 \times 2}$:

$$\begin{bmatrix} \dot{x} \\ \dot{y} \end{bmatrix}_p = J_i \begin{bmatrix} \dot{x} \\ \dot{y} \end{bmatrix}_w \tag{4}$$

Finally, combining Equations (3) and (4) the observed velocities (in pixels/s) of the microrobot are given by:

$$\begin{bmatrix} \dot{x} \\ \dot{y} \end{bmatrix}_p \cong -J_i \begin{bmatrix} \dfrac{M_x}{\alpha} \|\vec{\nabla} B_x\| \\ \dfrac{M_y}{\beta} \|\vec{\nabla} B_y\| \end{bmatrix} \tag{5}$$

Thus, the fully modeled system would include camera parameters, drag coefficients, and magnetic field properties.

Since the magnetic field strengths are proportional to the applied current (generated by a pulse-width modulated voltage square wave), the system displays a linear relationship between the actuation signal, $\vec{u}(t)$, the microrobot velocity $\dot{\vec{x}}_p(t)$ (observed by the visual system) can be expressed as:

$$\dot{\vec{x}}_p(t) \cong J(x)\,\vec{u}(t) \tag{6}$$

where $J(x)$ incorporates the image Jacobian together with the drag coefficients, magnetic moments, and magnetic field strengths. It is similar to the composite Jacobian matrix used in traditional image based robotic control and will similarly vary as the microrobot moves throughout the workspace. For the 2D system presented in this work, it is a 2×2 matrix that relates the actuation signal to the velocity of the microrobot as seen in the image plane.

Computing a closed form solution for $J(x)$ is possible, but doing so requires accurate and calibrated models of the induced electromagnetic field, drag coefficients, the vision system, *etc.* Any changes to the physical position of the system, components, device geometry, fluid properties, *etc.* require a system calibration step. An alternative is online estimation of the J matrix using iterative methods such as Broyden's method or recursive nonlinear least squares estimation. Such uncalibrated adaptive methods have been successfully implemented in macro-scale manipulators and mobile robots for a variety of applications with more complex nonlinear system models and higher degrees of freedoms (DOF) [23–25]. Experimental results in [20] and [26] show that the J matrix (the relationship between actuation and device velocities) is relatively linear for the experimental system used in this paper. The 2-DOF system described here presents a mathematically tractable problem for online system estimation as presented in the following section.

3. Uncalibrated Visual Servoing

3.1. Recursive Least Squares (RLS) Jacobian Estimation and Control

Consider a microrobot system such as the one described by Equation (6) with an observed state $\vec{x}$ that will vary when the control signal $\vec{u} \in \mathbb{R}^n$, is applied to the system. It is desired that the robot be controlled in such a manner that it is driven towards a goal position or trajectory $\vec{x}^*(t)$. The error $f: \in \mathbb{R}^m$ between the observed and desired or target position, $\vec{x}^*(t)$, is given by:

$$f\big(x(t)\big) = \vec{x}(t) - \vec{x}^*(t) \tag{7}$$

For planar 2-DOF image-based position control, the image data ($\vec{x}_p$ from the previous section) implies that $m = 2$. Similarly, for an electromagnet array consisting of two opposing pairs of magnets, $n = 2$. More complex systems such as those controlling orientation would use higher degrees of freedom.

It is desired to compute a control signal that will minimize the $F(x(t))$ image error squared and drive the microrobot to the target position $x^*(t)$:

$$F\big(x(t)\big) = \frac{1}{2}f\big(x(t)\big)^T f\big(x(t)\big) \tag{8}$$

This can be achieved via a quasi-Newton method utilizing an iteratively estimated Jacobian as developed for various macro scale robotic systems in [23–25]. Here, we use a dynamic recursive least squares (RLS) method presented in [25] for its improved performance in the presence of system noise and ability to follow moving targets.

For a discrete control algorithm updated at iteration k with digital sampling time h_t, let x_k and x_k^* represent the robot position and the target position at the kth iteration as measured in the image plane, respectively. Let Δf_k represent the change in image error $f_k - f_{k-1}$, and let u_k represent the actuation signal for the electromagnets. Then RLS estimate for the Jacobian $\hat{J}_k$ is given by Equation (11) below and the entire iterative control algorithm is given in Algorithm 1. The actuation signal is computed in Equation (13), a quasi-Newton step:

Algorithm 1. Recursive least squares control.

Given: $f \in \mathbb{R}^m$; $u_0, u_1 \in \mathbb{R}^n$, $\hat{J}_0 \in \mathbb{R}^{m \times n}$, $P_0 \in \mathbb{R}^{n \times n}$, $\lambda \in (0,1)$

Do for $k = 1, 2, \ldots$

$$\Delta f_k = f_k - f_{k-1} \tag{9}$$

$$\frac{\partial x_k^*}{\partial t} h_t = x_k^* - x_{k-1}^* \tag{10}$$

$$\hat{J}_k = \hat{J}_{k-1} + \frac{\left(\Delta f_k - \hat{J}_{k-1} u_k + \frac{\partial x_k^*}{\partial t} h_t \right) u_k^T P_{k-1}}{\lambda + u_k^T u_k} \tag{11}$$

$$P_k = \frac{1}{\lambda} \left(P_{k-1} - \frac{P_{k-1} u_k u_k^T P_{k-1}}{\lambda + u_k^T P_{k-1} u_k} \right) \tag{12}$$

$$u_{k+1} = -\left(\hat{J}_k^T \hat{J}_k \right)^{-1} \hat{J}_k^T \left(f_k - \frac{\partial x_k^*}{\partial t} h_t \right) \tag{13}$$

End for

End

Equation (11) is iteratively estimating the relationship between the actuation signal commanded in the previous iteration and the observed change in error. Equation (13) uses this updated estimation of the Jacobian to compute a new command that will drive the microrobot towards the target. The matrix P_k is the estimate of the covariance matrix of the actuation signal, and lambda is a weighting factor that controls the memory of the Jacobian estimation and prevents noise in the term Δf_k (due to system or measurement noise) from resulting in erratic estimation. Values of lambda closer to 1 effect a longer memory; values >0.9 are typical [27]. The result is a control scheme that adaptively learns the relationship between the actuation signal and the robot velocities and drives the robot to the desired position even in the presence of noise. Including the target velocity term $\partial x_k^* / \partial t \, h_t$ in the development allows the controller to follow a moving trajectory. For point to point motion with stationary target positions, this term is simply zero. Figure 2 illustrates the microrobot and target positions used in computing Equations (9) and (10) as well as the control vector computed in Equation (13). No system calibration is required, and the same algorithm will control various microrobot device shapes in at arbitrary optical zoom settings with no system modeling or calibration.

Figure 2. As the microrobot moves, the position error f_k between the target and the robot are monitored in the kth image. The actuation signal u_k is a duty cycle for a square wave sent to each electromagnet pair at the kth iteration of the control loop.

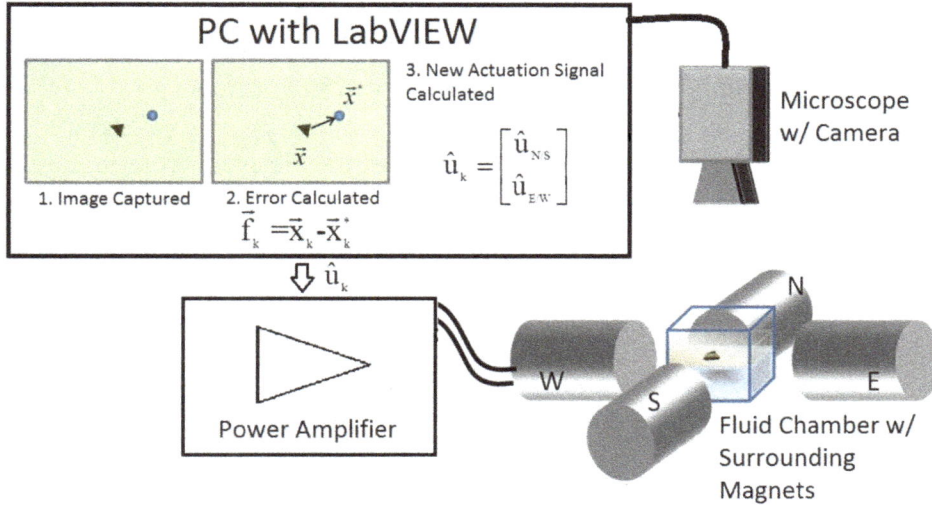

3.2. Practical Implementation

One final consideration regarding the 2D control signal computed in Equation (13) is necessary for implementation on a physical system. First, the magnitude of the computed control signal may be beyond the physical limitations. In this event, the signal may be scaled such that magnitude is within the system's capabilities but that the direction of the vector within the control space is preserved. For a maximum allowable scalar magnitude u_{max}, the scaled actuation signal $\hat{u}_{k+1}$ is used:

$$\hat{u}_{k+1} = \frac{u_{max}}{\max(u_{k+1})} u_{k+1} \tag{14}$$

This is similar to the trust region method employed by Jagersand [24] to prevent large motions outside of the estimated model's current area of validity.

While no modeling is necessary for the algorithmic implementation, it is assumed that the system has been thoughtfully designed such that the magnetic field strength is sufficient to pull the microrobot and overcome viscous drag and other fluid interface reactions.

The algorithm as presented and experimentally verified in this section and the next is for 2D planar control; however, it is plausible to extend the method to three dimensions with additional electromagnets and imaging capabilities to capture three-dimensional position information. Such a system would not utilize a two-fluid interface and gravity would apply a biasing force in the z dimension; however, the Jacobian estimation method would be able to adjust the magnetic field to either work with or against the force of gravity.

4. Experimental Work

4.1. Experimental System

A microrobot system has been implemented comprised of an electroplated nickel slug suspended at the interface between two immiscible fluids, and an electromagnetic actuation system. The microrobot devices are 20 µm thick and fit within a 200 µm diameter circle and were fabricated through the MEMSCAP MetalMUMPS process (Crolles, France) [28]. A variety of device morphologies (developed for an earlier microrobot competition [29]) provide an opportunity to study the effects of different robot shapes (which possess different viscous drag characteristics), and are shown in Figure 3.

The microrobot operates at the interface between vegetable oil, and a solution consisting of sodium chloride and sodium bicarbonate dissolved in water, as shown in Figure 4.

Figure 3. Models of the different microrobot designs derived from the mask layout, rendered in MEMSPro. The designs are referred to as: (**a**) "S", (**b**) bar, (**c**) "wedge", and (**d**) "star". Each device is approximately 200 µm in diameter.

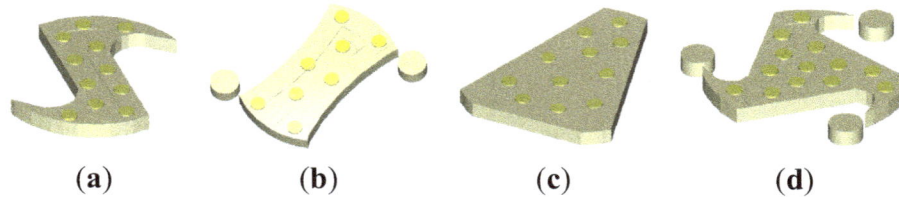

(**a**)   (**b**)   (**c**)   (**d**)

Figure 4. On the left is the fluid chamber (20 × 20 × 15 mm) showing the fluid interface between oil (top layer) and a sodium chloride/sodium bicarbonate solution (bottom layer) that comprises the microrobot's planar work surface. On the right is the concave meniscus that occurs without the oil layer.

interface ⟶

The robots and fluid are in a 20 × 20 × 15 mm vial placed at the center of four cylindrical electromagnets arrayed along the four points of the compass. Each magnet is driven with a pulse of amplitude 11 V and frequency 100 Hz. The duty cycle of the control signal is varied from 0% to 50%. By varying the amplitude or duty cycle of square wave input voltages to each electromagnet, the varying magnetic field imposed on the microrobot imposes varying forces that propel the microrobot through its workspace.

The robots have no permanent magnetization. When placed in a magnetic field they develop an induced magnetization. If the field is non-uniform this leads to motion in the direction of increasing magnetic field strength. We utilize a simple actuation scheme with a magnet for each cardinal direction where only one magnet is actuated at a given time to pull the robot in the desired direction. Visual feedback is used to measure the position of the microrobot device. The vision system consists of a microscope and a 740 × 480 USB camera. System integration is achieved in LabVIEW environment with an 8 Hz vision update rate. Simple thresholding is used to distinguish the microrobot object from the background, and the centroid of the object is used as the robot position. The robustness of binarization is enhanced by backlighting the microrobot beneath the fluid.

4.2. Stationary Target: Point to Point Motion

To demonstrate the robustness of the control a microrobot device is commanded a sequence of point-to-point motions throughout the field of view of system with the following variations:

- Magnification
- Electromagnet position and orientation
- Microrobot device morphology

Figures 5–7 show the variation, trajectory, and error (distance from robot to goal position) for each of these variations, respectively. The figures demonstrate convergent control for a wide variety of system configurations with no *a priori* knowledge, calibration, tuning, or careful fixturing of the components. For each experiment, the initial Jacobian matrix J_0 and the covariance matrix P_0 were set to the identity matrix and the recursive least squares algorithm was used to control the robots from one point to the next using $\lambda = 0.99$.

In Figure 5, results are shown demonstrating point-to-point motion throughout the field of view for three different magnifications {10×, 20×, and 30×}. The corresponding pixel resolutions are approximately 9.8, 6.5, and 3.25 μm/pixel, respectively. Starting in the center of the image, the microrobot devices are commanded to a sequence of points (denoted with A, B, C, and D). This self-intersecting quadrilateral path ensures that all four magnets are utilized for the robot motion and covers a large portion of the workspace. Note that the trajectories in the second column are plotted on axes equivalent to the image resolution shown in the first column.

Figure 5. Point-to-point control for three different magnifications {(**a**) 10×, (**b**) 20×, and (**c**) 30×} showing the image plane trajectory and the error between the goal position and the microrobot over time. The same algorithm and initial parameters were used in each case.

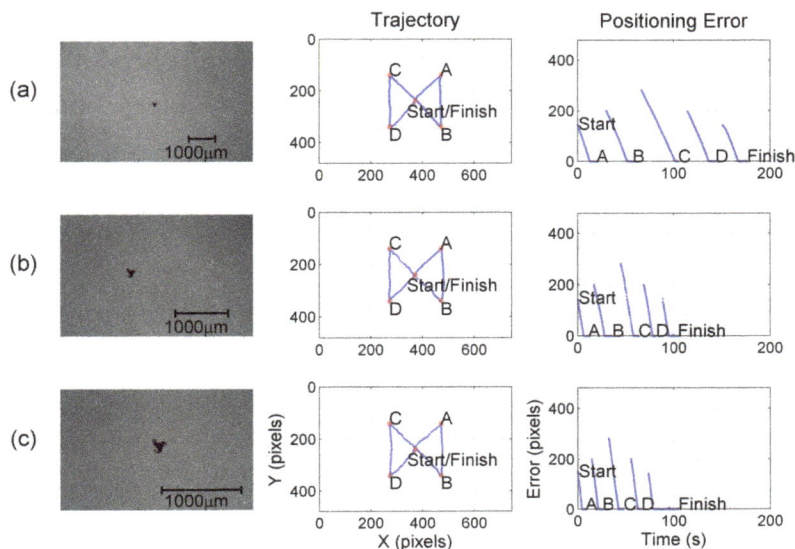

The same experiment is repeated for three different electromagnet configurations at 20× magnification as shown in Figure 6 demonstrating the adaptive ability of the controller to handle vastly different system configurations. If the controller were based on a system model (e.g., H_∞ or PID), these significant alterations to the system configuration would render the controller ineffectual. Rather, the only difference that is demonstrated is found in row (c) where a reduced speed in downward motion is observed. This is due to the weaker magnetic field affected by pulling one magnet away. The system is still convergent and achieves each goal position. Notice that rows (a) and (b) would require significantly different inputs from the electromagnet pairs for up and down motion as compared with row (b) in Figure 5 where up and down motion can be affected with inputs from the north or south magnets.

Figure 6. Trajectory and error for point-to-point motion. In row (**a**), the electromagnets were rotated approximately 45° counter-clockwise from the nominal compass-based orientation; row (**b**) demonstrates results for a in the opposite direction; and row (**c**) gives results when the south magnet was pulled away from the microrobot device several centimeters.

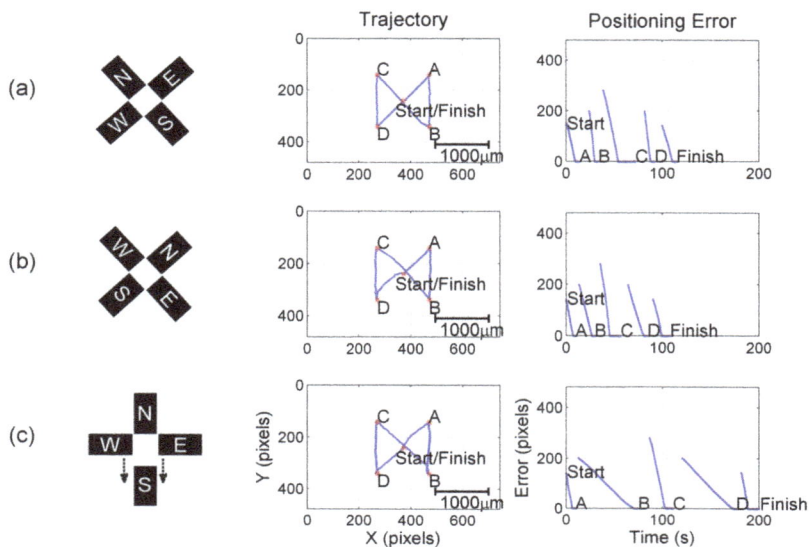

Figure 7. Point-to-point control of three different microrobot devices using the same algorithm and initial conditions: (**a**) "S", (**b**) "bar" and (**c**) "wedge" shaped devices. These were performed at a 20× magnification and may be compared to the "star" device in Figure 5b. Noticeably, the wedge device moves the most quickly, but each microrobot device converges on the target points.

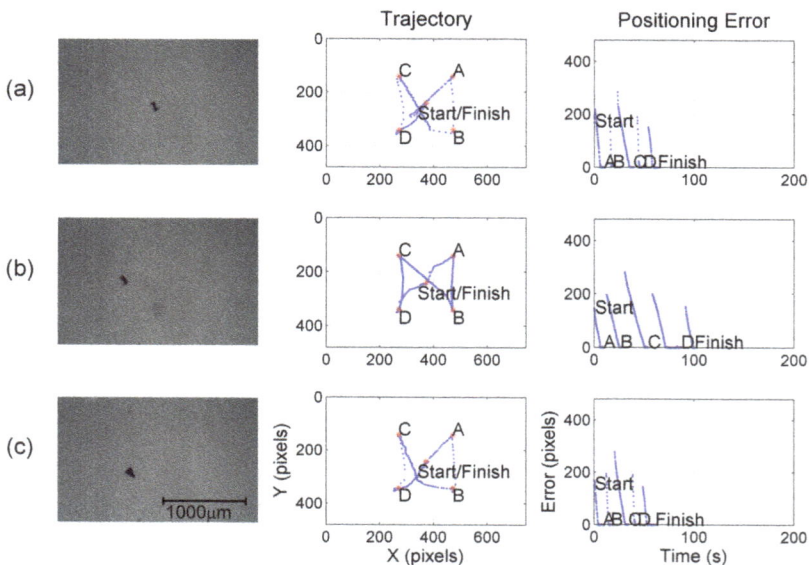

The experiments are repeated again at 20× for three more microrobot device morphologies. This is significant, because at this scale the different shapes will have different viscous drag coefficients. Indeed, this is indicated by the various speeds demonstrated in the positioning error plots; the rows (b) and (c) demonstrate significantly faster systems than row (a) or Figure 5b. A model-based controller would need to be calibrated to each device shape; however, the recursive least squares control is able to learn the system and servo each device to the goal positions. While the motion appears more erratic, it should be observed that the control method is designed to simply converge towards the static target points. The trajectory path is demonstrated in the following section.

All three figures repeatedly demonstrate convergent motion towards the target points under disparate conditions with no initialization or calibration.

4.3. Moving Target: Circular Motion

The inclusion of the target velocity term $\partial x_k^*/\partial t\, h_t$ term in the RLS algorithm given in Section 3 enables the controller to minimize the error even when the target is moving. To demonstrate this, a circular path is given as a desired motion. Again, results in this section demonstrate the ability to actuate a microrobot device with one algorithm (without calibration) while varying the following aspects of the system:

- Magnification
- Target speed
- Electromagnet array orientation
- Microrobot morphologies

As with the stationary experiments, the initial Jacobian matrix J_0 and the covariance matrix P_0 were arbitrarily set to the identity matrix and the RLS algorithm was used to control the robots from one point to the next using $\lambda = 0.99$.

4.3.1. System Performance for Various Magnification and Target Speeds

Here, the steady state tracking error for a moving target prescribed as follows:

$$\vec{x}^*(t) = 100 \begin{bmatrix} \cos(\omega t) \\ \sin(\omega t) \end{bmatrix} \text{pixels} \tag{15}$$

is studied for a range of angular velocities, at three different magnifications {10×, 20×, 30×}. As appropriate for an image-based visual servoing method, the error is presented in pixels at each magnification level.

First, Figure 8 demonstrates the path, tracking error, and control effort made during *one* such experiment with the magnification set at 20×. The target is moving as described in Equation (15) where $\omega = 0.035$ rad/s which results in an average speed (tangential velocity) of 3.5 pixels/s or approximately 22.8 μm/s. The inset in Figure 8 records the path as the microrobot starts at the denoted location, servos towards the desired trajectory and continues around in a counter clockwise manner until it reaches the point where it initially met the desired path. The error between the microrobot and the desired path is shown demonstrating initial convergence and a steady state tracking error of 1.5 pixels.

The normalized control effort $\hat{u}_k$ for the experiment depicted in Figure 8 is presented in Figure 9 with each subplot providing the scaled control signals sent to each electromagnet pair (where a scaled effort of 1 represents a 50% duty cycle signal sent to the electromagnet). This control effort varies a great deal as the controller seeks to make small moves keeping the microrobot on the desired path. The target velocity is such that the change in the goal position is on the order of the resolution of the position measurement and the signal is noise dominated. However, the underlying sinusoidal effort is observed with an expected 90° phase shift between the N/S and E/W electromagnet pairs.

Figure 8. Tracking error in pixels between actual robot position and desired path. The steady state error is 1.5 pixels, and the inset demonstrates the captured path of the robot.

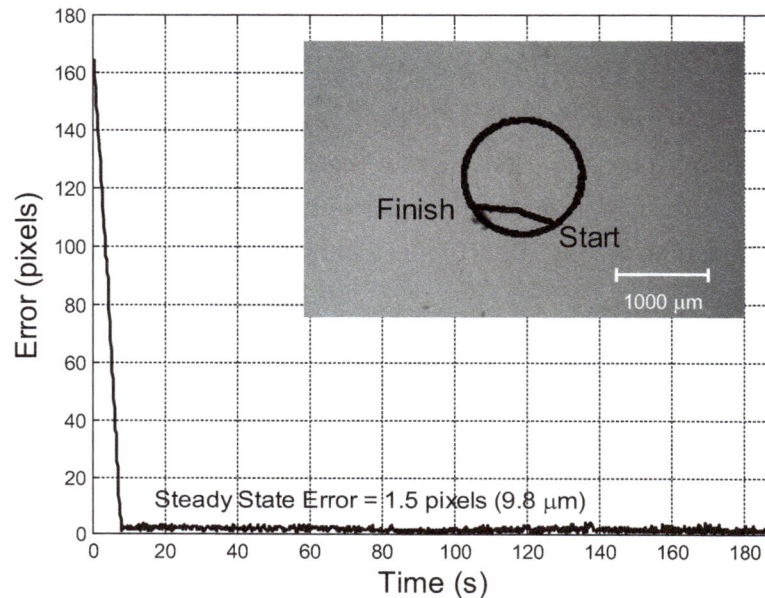

Figure 9. Normalized control effort for the experiment given in Figure 8. The effort is normalized such that a 1 represents the maximum 50% duty cycle applied to a magnet. Positive values are applied to one magnet (E or N) of each magnet pair (N/S, E/W).

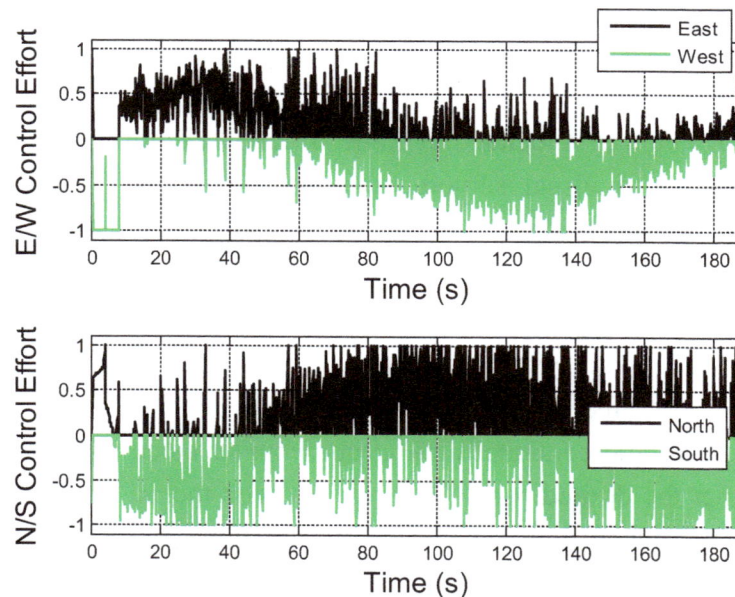

To more thoroughly investigate performance, the steady state tracking error is presented in Figure 10 for a series of experiments conducted at varying target speeds and system magnifications with error bars representing one standard deviation. The results convey both the limitations and strengths of the approach. Clearly the tracking error is greater for the 10× scenario. This makes sense in terms of system signal to noise ratios. At 10× magnification, the device is a smaller blob in the image which can result in more noise in the centroid calculation that is used to determine the position at each step. Furthermore, a given control signal results in a smaller motion (in pixels) than it would at a higher magnification. With greater system noise relative to observed motion, there is increased error in the tracking of the path.

Additionally, it is observed that there is increased tracking error as the target path velocity is increased. To a certain extent, this is to be expected and is similar to the results seen in [25], however the large errors seen at the highest speed for the 10× and 20× magnification are not a failure of the control system but rather due to the limitations of the electromagnets, which are not able to produce the required speeds for the microrobots at these settings. Inspection of the control effort for those two experiments shows saturation at the maximum allowed values.

Overall, Figure 10 demonstrates that when the target is moving at system achievable speeds, we demonstrate stable control for a 200 μm device following a moving target profile with average steady state errors ranging from 1.0 to 4.1 pixels in the image plane which translates to 4.1–40.5 μm in the workspace.

Figure 10. Average steady state tracking error (in pixels) with error bars indicated one standard deviation for a range of velocities and magnifications for a star microrobot following a circular path at three different magnifications (10×, 20×, and 30×). The pixel resolutions are approximately 9.8, 6.5, and 3.2 μm/pixel, respectively.

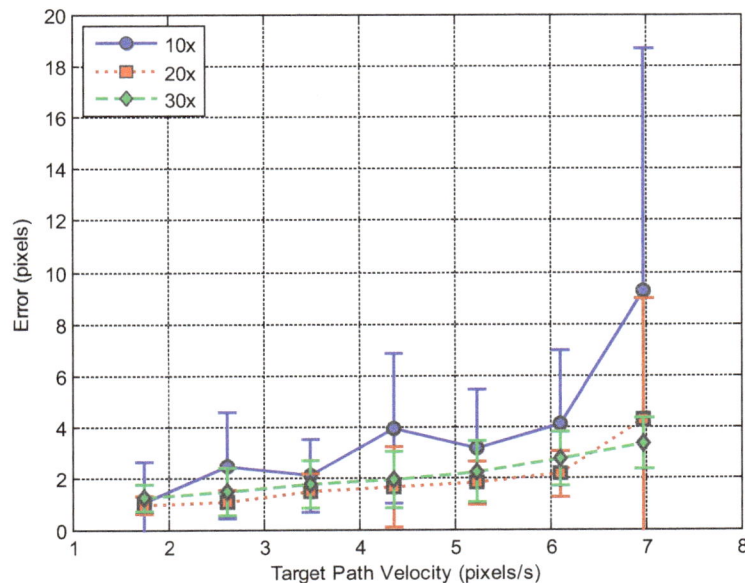

4.3.2. Tracking Performance for Various System Modifications

To demonstrate the versatility of the controller, the same moving target experiment is repeated for two significant system modifications using the same target path described in Equation (15). First, the magnetic array is rotated 45° (the target speed is set at 3.5 pixels/s and the magnification is set at 20×). While this is a simple coordinate frame rotation, it represents completely different system gains since a single magnet pair now induces motion in both the x and y direction of the image plane. In a second experiment, a different microrobot shape, one resembling wedge, is used. At this scale, changes in microrobot morphology can have significant effects on the viscous drag exerted on the device by the surrounding fluids.

If the controller were based on a system model (even a simple proportional controller), these alterations to the system configuration would render the controller ineffectual unless recalibrated. However, with the uncalibrated estimation, the system is still convergent and is able to follow the desired path. Figure 11 shows the Jacobian elements over time ($J_0 = I_2$) for the three different experiments. The values represent the ratio between changes in the robot's position in the image (in both x and y) with respect to changes in the control effort sent to the two electromagnet pairs. In each experiment, the Jacobian values settle into very different numbers for steady state tracking ($t > 5$ s) because fundamental aspects of the system have changed, and the Jacobian estimation scheme is able to learn these changes.

Figure 11. Elements of the Jacobian estimation over time for three experiments: **(a)** a star microrobot with a nominal electromagnet configuration; **(b)** a rotated magnet configuration; and finally, **(c)** a wedge microrobot. In each case, the robot is given a circular path. Low steady state errors indicate effective tracking. Different system configurations result in different Jacobian values necessary for control. Online estimation eliminates the need for *a priori* calculation of these values.

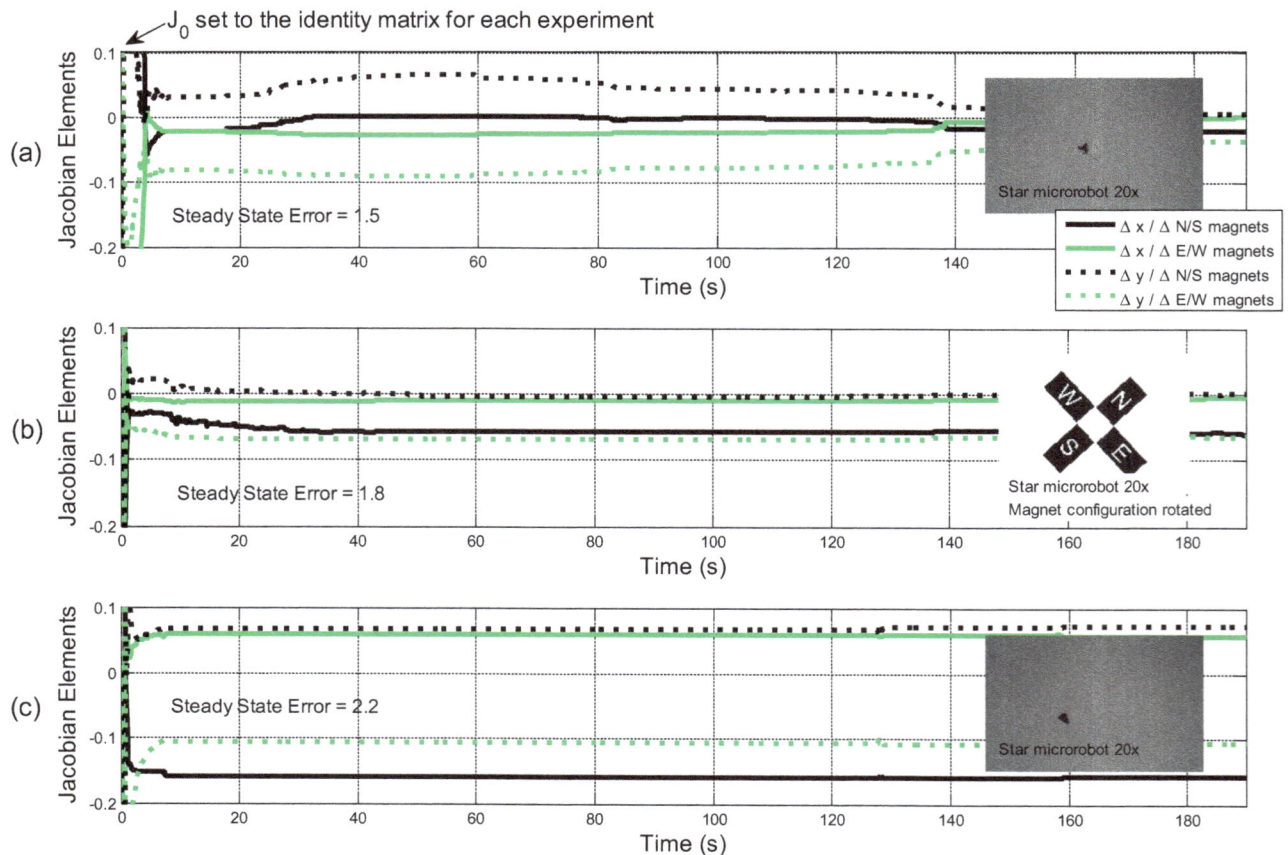

5. Conclusions

This work has experimentally demonstrated an uncalibrated vision-based control method using recursive least squares for a magnetically actuated microrobot system working in a planar fluid environment. The uncalibrated nature of the controller was demonstrated by altering the magnification of the microscope, significantly rotating the electromagnet array, and by testing 200 μm robots with different morphologies. Despite all these changes, the uncalibrated image-based control method converges to stationary target points and demonstrates stable and consistent tracking results for moving target trajectories with average steady state path errors ranging from 1.0 to 4.1 pixels in the image plane (4.1–40.5 μm in the workspace). At a scale where accurate modeling of all system parameters can be difficult, the recursive least squares estimation and control method offers a great deal of flexibility for microrobot control with one algorithm capable of controlling a variety of system configurations.

Acknowledgments

The authors would like to acknowledge the work of Judy Raymond, Norm Tyson, Joe Bradshaw, Daphie Jobe, William Stanton, Erich Keyes, Ken Walsh, and Jerry Ballman, AshleyWessel, Peter Dausman, Louis Henry, Matthew Prevatt, Sean Johnson, Andrew Flora, Justin Aleshire, Kerri Bortz, Jeremy Gray, and Brett Morris.

Author Contributions

Jenelle Armstrong Piepmeier contributed to the writing of the manuscript and the design of the visual feedback control system. Samara Firebaugh contributed to the design and integration of the actuation system. Caitlin S. Olsen conducted a number of the experiments.

Conflicts of Interest

The authors declare no conflict of interest.

References

1. Ishii, K.; Hu, W.I.; Ohta, A. Cooperative micromanipulation using optically controlled bubble microrobots. In Proceedings of IEEE International Conference on Robotics and Automation, Saint Paul, MN, USA, 14–18 May 2012; pp. 3443–3448.
2. Hu, W.; Ishii, K.S.; Ohta, A.T. Micro-assembly using optically controlled bubble microrobots. *Appl. Phys. Lett.* **2011**, *99*, doi:10.1063/1.3631662.
3. Donald, B.R.; Levey, C.G.; McGray, C.D.; Paprotny, I.; Rus, D. An untethered, electrostatic, globally controllable mems micro-robot. *Microelectromech. Syst. J.* **2006**, *15*, 1–15.
4. Sul, O.J.; Falvo, M.R.; Taylor, R.M.; Washburn, S.; Superfine, R. Thermally actuated untethered impact-driven locomotive microdevices. *Appl. Phys. Lett.* **2008**, *89*, doi:10.1063/1.2388135.
5. Denisov, A.; Yeatman, E.M. Micromechanical actuators driven by ultrasonic power transfer. *Microelectromech. Syst. J.* **2014**, *23*, 750–759.

6. Bouchebout, S.; Bolopion, A.; Abrahamians, J.-O.; Régnier, S. An overview of multiple dof magnetic actuated micro-robots. *J. Micro Nano Mechatron.* **2012**, *7*, 97–113.

7. Jian, F.; Cho, S.K. Mini and micro propulsion for medical swimmers. *Micromachines* **2014**, *5*, 97–113.

8. Tamaz, S.; Gourdeau, R.; Mathieu, J.B.; Martel, S. Real-time mri-based control of a ferromagnetic core for endovascular navigation. *IEEE Trans. Biomed. Eng.* **2008**, *55*, 1854–1863.

9. Belharet, K.; Folio, D.; Ferreira, A. Control of a magnetic microrobot navigating in microfluidic arterial bifurcations through pulsatile and viscous flow. In Proceedings of IEEE/RSJ International Conference on Intelligent Robots and Systems (IROS), Vilamoura-Algarve, Portugal, 7–12 October 2012; pp. 2559–2564.

10. Pawashe, C.; Floyd, S.; Sitti, M. Two-dimensional autonomous microparticle manipulation strategies for magnetic microrobots in fluidic environments. *IEEE Trans. Robot.* **2012**, *28*, 467–477.

11. Diller, E.; Floyd, S.; Pawashe, C.; Sitti, M. Control of multiple heterogeneous magnetic microrobots in two dimensions on nonspecialized surfaces. *IEEE Trans. Robot.* **2012**, *28*, 172–182.

12. Diller, E.; Giltinan, J.; Sitti, M. Independent control of multiple magnetic microrobots in three dimensions. *Int. J. Robot. Res.* **2013**, *32*, 614–631.

13. Floyd, S.; Pawashe, C.; Sitti, M. An untethered magnetically actuated micro-robot capable of motion on arbitrary surfaces. In Proceedings of IEEE International Conference on Robotics and Automation, Pasadena, CA, USA, 19–23 May 2008; pp. 419–424.

14. Marino, H.; Bergeles, C.; Nelson, B.J. Robust electromagnetic control of microrobots under force and localization uncertainties. *IEEE Trans. Autom. Sci. Eng.* **2014**, *11*, 310–316.

15. Kummer, M.P.; Abbott, J.J.; Kratochvil, B.E.; Borer, R.; Sengul, A.; Nelson, B.J. Octomag: An electromagnetic system for 5-DoF wireless micromanipulation. *IEEE Trans. Robot.* **2010**, *26*, 1006–1017.

16. Bergeles, C.; Kratochvil, B.E.; Nelson, B.J. Visually servoing magnetic intraocular microdevices. *IEEE Trans. Robot.* **2012**, *28*, 798–809.

17. Ghanbari, A.; Chang, P.H.; Choi, H.; Nelson, B.J. Time delay estimation for control of microrobots under uncertainties. In Proceedings of 2013 IEEE/ASME International Conference on Advanced Intelligent Mechatronics (AIM), Wollongong, Australia, 9–12 July 2013; pp. 862–867.

18. Keuning, J.D.; DeVriessy, J.; Abelmanny, L.; Misra, S. Image-based magnetic control of paramagnetic microparticles in water. In Proceedings of IEEE/RSJ International Conference on Intelligent Robots and Systems (IROS), San Francisco, CA, USA, 25–30 September 2011; pp. 421–426.

19. Khalil, I.S.M.; Keuning, J.D.; Abelmann, L.; Misra, S. Wireless magnetic-based control of paramagnetic microparticles. In Proceedings of IEEE RAS/EMBS International Conference on Biomedical Robotics and Biomechatronics, Rome, Italy, 24–27 June 2012; pp. 460–466.

20. Piepmeier, J.A.; Firebaugh, S.L. Visual servo control of electromagnetic actuation for a family of microrobot devices. In Proceedings of 2013 IEEE Workshop on Robot Vision (WORV), Clearwater Beach, FL, USA, 15–17 January 2013; pp. 209–214.

21. Popa, D. Robust and reliable microtechnology research and education through the mobile microrobotics challenge [competitions]. *IEEE Robot. Autom. Mag.* **2014**, *21*, 8–12.

22. Purcell, E.M. Life at low reynolds number. *Am. J. Phys.* **1977**, *45*, 3–11.

23. Hosoda, K.; Asada, M. Versatile visual servoing without knowledge of true jacobian. In Proceedings of IEEE/RSJ International Conference on Intelligent Robots and Systems (IROS), Munich, Germany, 12–16 September 1994; pp. 186–193.

24. Jagersand, M.; Fuentes, O.; Nelson, R. Experimental evaluation of uncalibrated visual servoing for precision manipulation. In Proceedings of IEEE International Robotics and Automation Conference, Albuquerque, NM, USA, 20–25 April 1997; pp. 2874–2880.

25. Piepmeier, J.A.; McMurray, G.V.; Lipkin, H. Uncalibrated dynamic visual servoing. *IEEE Trans. Robot. Autom.* **2004**, *20*, 143–147.

26. Sakar, M.S.; Steager, E.B.; Cowley, A.; Kumar, V.; Pappas, G.J. Wireless manipulation of single cells using magnetic microtransporters. In Proceedings of IEEE International Robotics and Automation (ICRA) Conference, Shanghai, China, 9–13 May 2011; pp. 2668–2673.

27. Eleftheriou, E.; Falconer, D.D. Tracking properties and steady-state performance of RLS adaptive filter algorithms. *IEEE Trans. Acoust. Speech Signal Proc.* **1986**, *34*, 1097–1110.

28. MetalMUMPs. Availble online: http://www.memscap.com/products/mumps/metalmumps (accessed on 19 September 2014).

29. Firebaugh, S.L.; Piepmeier, J.A. The robocup nanogram league: An opportunity for problem-based undergraduate education in microsystems. *IEEE Trans. Educ.* **2008**, *51*, 394–399.

2

Femtosecond Laser Irradiation of Plasmonic Nanoparticles in Polymer Matrix: Implications for Photothermal and Photochemical Material Alteration

Anton A. Smirnov, Alexander Pikulin, Natalia Sapogova and Nikita Bityurin *

Institute of Applied Physics of Russian Academy of Sciences, 46 Ul'yanov Street, Nizhny Novgorod 603950, Russia; E-Mails: antonsmirnov@ufp.appl.sci-nnov.ru (A.A.S.); pikulin@ufp.appl.sci-nnov.ru (A.P.); ns@ufp.appl.sci-nnov.ru (N.S.)

* Author to whom correspondence should be addressed; E-Mail: bit@ufp.appl.sci-nnov.ru

External Editor: Maria Farsari

Abstract: We analyze the opportunities provided by the plasmonic nanoparticles inserted into the bulk of a transparent medium to modify the material by laser light irradiation. This study is provoked by the advent of photo-induced nano-composites consisting of a typical polymer matrix and metal nanoparticles located in the light-irradiated domains of the initially homogeneous material. The subsequent irradiation of these domains by femtosecond laser pulses promotes a further alteration of the material properties. We separately consider two different mechanisms of material alteration. First, we analyze a photochemical reaction initiated by the two-photon absorption of light near the plasmonic nanoparticle within the matrix. We show that the spatial distribution of the products of such a reaction changes the symmetry of the material, resulting in the appearance of anisotropy in the initially isotropic material or even in the loss of the center of symmetry. Second, we analyze the efficiency of a thermally-activated chemical reaction at the surface of a plasmonic particle and the distribution of the product of such a reaction just near the metal nanoparticle irradiated by an ultrashort laser pulse.

Keywords: plasmonic nanoparticles; gold nanoparticles; laser processing of materials; femtosecond laser pulses; photoinduced nanocomposites; two-photon absorption; photochemical processes; laser heating; thermo-activated chemical processes; polymers

1. Introduction

A plasmonic particle inserted in a transparent material can enhance the effect of laser radiation. The particle absorbs incident radiation and is heated, causing the elevation of the temperature of adjacent material layers. This effect has been considered in numerous papers (for references, see the recent reviews [1,2]). Heating could be provided by irradiation of the materials containing plasmonic particles using sufficiently long laser pulses or continuous waveform (CW) lasers. In this case, the stationary distribution of temperature around the particle is established [3]. Heating of gold nanoparticles within the tumor is considered as a way to kill cancer cells [4]. Heating of metal nanoparticles in water can cause the water to boil. The work in [5] considers the melting of the polymer mediated by laser-irradiated metal particles. It is recognized [5] that the CW irradiation of a separate nanoparticle cannot provide a significant temperature elevation of the surrounding material, whereas the cumulative effect provided by a great number of compactly-situated particles can be dramatic.

The effect of femtosecond laser pulses on the plasmonic particles in a transparent material is reviewed in [2]. Here, a significant heating effect can be provided even by a separate particle.

It should be noted that in both cases of long and short pulse irradiation, the effect of electron emission from the laser-irradiated particle into the surrounding medium could occur. This effect is considered in detail in [1]. It is shown that in the case of long or CW laser irradiation, the electron emission is noticeable for relatively small-sized particles with diameters of about several nanometers.

With femtosecond pulses, the breakdown of the surrounding material is provoked by the laser field enhancement near the plasmonic particles. Here, the effect of electron emission from the metal particle can promote additional seed electrons for the breakdown. This effect provides bubbling of water in a biological tissue, optoporation and transfection of a cell membrane [2].

The effect of the plasmonic particles due to laser heating, field enhancement and electron transfer on some photocatalytic chemical reactions is considered in, e.g., [6] and a recent review [7].

The effect of the plasmonic particles on the surrounding medium is of particular interest concerning the so-called photo-induced nanocomposites [8–10]. Here, the nanoparticles are produced by UV irradiation of the matrix containing dissolved precursor molecules that typically include atoms of a noble metal. After the UV-mediated reduction of those molecules, the extracted metal atoms form an oversaturated solid solution; then, the plasmonic particles appear as nuclei of the new phase [11]. Thus, these nanocomposites are products of the light-material interaction. However, taking into account the ability of the metal nanoparticles to enhance the effect of light, the further irradiation of the formed nanocomposites can provide a further modification of the material, resulting in the alteration of the material properties in a rather sophisticated manner.

The enhanced laser light can modify the properties of the matrix in the vicinity of the nanoparticle, either directly or through specially introduced molecules that serve as initiators of the subsequent transformations.

In [12], photo-induced isomerization in an azobenzene-dye polymer is used for the sub-diffraction imaging of the optical near-field around the plasmonic nanoparticles on the surface of a polymer film irradiated by the laser light.

In this article, we consider a photochemical reaction initiated by the two-photon absorption of light near the plasmonic nanoparticle within the matrix. We also consider the material heating caused by this absorption in comparison with the heating by the light directly absorbed by the particle and analyze the distribution of the product of a thermally-activated chemical reaction just near the metal nanoparticle irradiated by the laser pulse.

This study aims at developing the laser-induced production of 3D nanoplasmonic systems where the formation of photo-induced nanocomposites is only the first step.

2. The Near-Field Effects

Let us consider the irradiation of a metal spherical nanoparticle in a transparent matrix by a series of ultrashort pulses from the laser operated in the visible, near-IR or near-UV ranges (see Figure 1).

Figure 1. The modeling setup. A spherical gold nanoparticle of radius r_p is irradiated by femtosecond laser pulses at the central wavelength λ. The complex refractive index of the particle, its specific heat, density and heat diffusivity are n_p, C_p, ρ_p and χ_p, respectively. The corresponding properties of the matrix are denoted by n_m, C_m, ρ_m and χ_m. For a complete list of notations, see in Appendix A1.

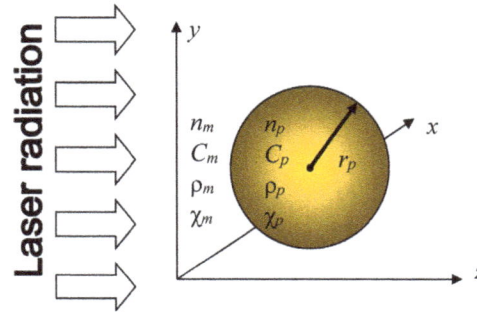

The laser light is significantly enhanced in the near-field domain. We consider a simple photochemical reaction promoted by the two-photon absorption by species X within the matrix. Species X can either belong to the matrix base being, e.g., chromophore groups within the polymer chains or being some dopant molecules dissolved in the matrix.

Assume that the particle radius r_p is on the order of 10 nm and the pulse length τ_{pulse} lies in a range of 10 fs to 1 ps. We approximate the incident laser radiation as a quasi-monochromatic wave at the central wavelength λ.

The evolution of the number density $N_X(\vec{r},t)$ of species X is described by the equation:

$$-\frac{\partial N_X(\vec{r},t)}{\partial t} = N_X \cdot \frac{\eta\,\sigma_X^{(2)}}{2\left(\hbar\omega\right)^2} I^2(\vec{r},t) \tag{1}$$

Here, I is the local laser field intensity, $\sigma_X^{(2)}$ is two-photon absorption cross-section of X, $\hbar\omega = 2\pi\hbar c/\lambda$ is the photon energy and η is the quantum yield. The solution of Equation (1) reads:

$$N_X = N_{X0} \cdot \exp\left(-\int_0^t \frac{\eta\,\sigma_X^{(2)}}{2(\hbar\omega)^2} I^2(\vec{r},t)\,dt\right) \tag{2}$$

where N_{X0} is the initial concentration of X. The fraction $v = (N_{X0} - N_X)/N_{X0}$ of reacted species X is further referred to as the conversion.

In the vicinity of a nanoparticle, the incident field can be approximated as a quasi-monochromatic plane wave with the intensity $I_{inc} = I_0 f(t)$, where I_0 is the peak incident intensity and function $f(t)$ is of a unity amplitude. If the dispersion on the particle size-scale is neglected, then the actual function $I(\vec{r}, t)$ can be calculated by means of Mie theory [13]:

$$I(\vec{r}, t) = I_0 f(t) \cdot \frac{\left| \vec{E}_{Mie}(\vec{r}) \right|^2}{E_0^{\ 2}} \tag{3}$$

Here, $\left| \vec{E}_{Mie}(\vec{r}) \right|^2 / E_0^{\ 2}$ is the enhancement of the electric field squared calculated near the metal nanoparticle irradiated by a plane monochromatic wave. Thus, Equation (2) can be evaluated as:

$$N_X = N_{X0} \cdot \exp\left(- \frac{\eta \, \sigma_X^{(2)} I_0^{\ 2} \tau_{pulse} N_{pulse}}{2(\hbar\omega)^2} \cdot \frac{\left| \vec{E}_{Mie}(\vec{r}) \right|^4}{E_0^{\ 4}} \right) \tag{4}$$

where N_{pulse} is the number of laser pulses applied.

The distribution of $\left| \vec{E}_{Mie}(\vec{r}) \right| / E_0^{\ 2}$ can be calculated by means of different near-field Mie codes; some of them can be found here [14]. The linear absorption of the matrix is assumed to be negligible at the laser wavelength, and the refractive index of the polymer is n_m. The complex refractive index of the metal particle is $n_p = n_p' + i n_p''$.

In Figure 2, we plot typical spatial distributions of the conversion after a series of laser pulses. The case where $r_p = 10$ nm and $\lambda = 800$ nm is presented in Figure 2a,b. Here, two field maxima are formed near the sphere opposite each other, following the dipole approximation. The locations of the maxima are directed by the incident wave polarization and, thus, are the same for all irradiated nanoparticles within the matrix. This gives an opportunity to induce anisotropy into laser-irradiated domains of the material. The results for $r_p = 35$ nm and $\lambda = 400$ nm are presented in Figure 2c,d. Here, the dipole approximation is not applicable, because the thickness of the skin-layer becomes smaller than the diameter of the particle. Thus, the higher-order spherical modes become important, resulting in the shift of the maxima towards the direction of the incident wave propagation. The photochemical reaction in this case results not only in laser-induced anisotropy, but also in the laser-induced loss of the center of symmetry. This is important for the applications in nonlinear optics, in particular for the second harmonic generation. It is notable that such a modification can be performed using reasonable irradiation regimes and materials without an outstanding two-photon sensitivity (see Figure 2 caption). This allows for possible applications of commercially available materials and lasers for a single-step modification of large volumes of about 1 cm^3.

Thermo-sensitive rather than photo-sensitive additives can also be employed. Thermal alteration processes can also be locally induced due to laser irradiation of nanoparticles. In what follows, we consider two main possibilities to heat a transparent matrix near the nanoparticle: either to do it directly via multi-photon absorption or to heat the particle and rely on the heat diffusion. We calculate the temperature rise in the particle and in the surrounding polymer provided by a single laser pulse.

Once an ultra-short laser pulse is absorbed by the gold nanoparticle, the electron subsystem is excited. The electron energy is transferred to the lattice during the electron-phonon coupling time on the order of $\tau_{e-ph} = 1$–3 ps [2]. This is the period during which the particle is effectively heated, even if the actual laser pulse length is shorter. The temperature diffusion time over the particle is $\tau_{diffus} = r_p^2 \chi_p^{-1}$, where χ_p is the heat diffusion coefficient, which is on the order of 1 cm²/s for many metals, including gold and silver. For $r_p = 10$ nm, the value $\tau_{diffus} = 1$ ps, which is on the order of τ_{e-ph}.

Figure 2. The conversion of species X near the gold nanoparticle (indicated by a white circle) irradiated by $N_{pulse} = 10^5$ laser pulses of length $\tau_{pulse} = 100$ fs. The two-photon absorption cross-section is $\sigma_X^{(2)} = 10$ GM (1 GM = 10^{-50} cm⁴·s). The quantum yield is $\eta = 0.8$, and the incident wave is linearly polarized along the x and propagates toward the z direction. For (**a**) and (**b**): $r_p = 10$ nm, $\lambda = 800$ nm; the complex refractive index of gold is $n_p = 0.15 + 4.91i$ [15], the refractive index of polymethyl methacrylate (PMMA) is $n_m = 1.484$ [16]; the incident beam intensity is $I_0 = 5 \times 10^8$ W/cm². For (**c**) and (**d**): $r_p = 35$ nm, $\lambda = 400$ nm, $n_p = 1.47 + 1.954i$ [15], $n_m = 1.503$ [16], $I_0 = 2 \times 10^9$ W/cm².

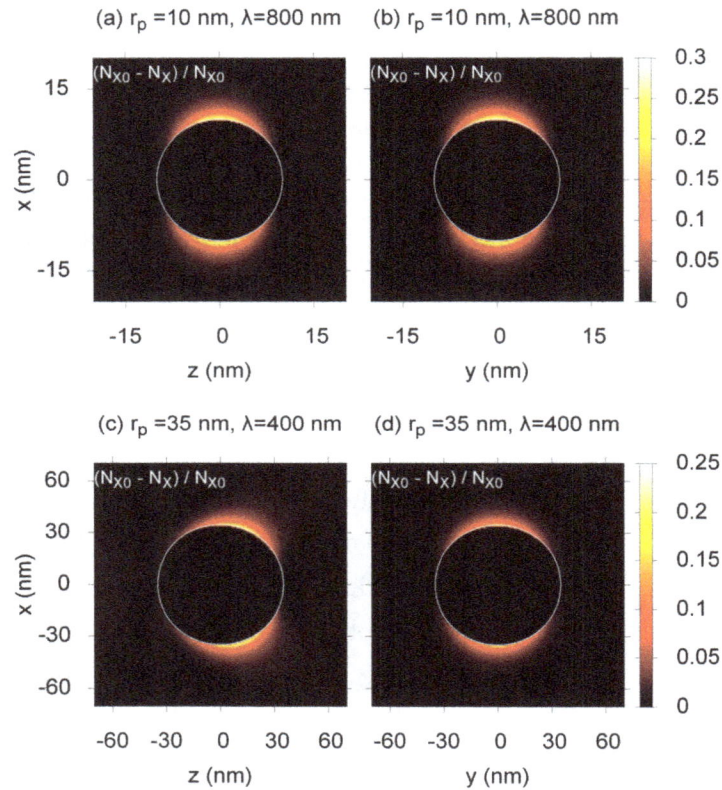

The temperature rise over the particle can be considered uniform if the radius of the particle is well below the thickness of the skin-layer or after the irradiation period t >> τ_{diffus} and t >> τ_{e-ph}.

The heating and thermalization can be considered adiabatic over the time period τ_{adiab} if the fraction of heat that leaks from the particle to the matrix during this period can be neglected. The heat penetration depth from the particle into the matrix by the moment t can be evaluated as $\sqrt{\chi_m t}$. Provided ΔT_p is the temperature rise of the particle due to the laser pulse, the amount of heat transferred to the matrix is evaluated as $Q_m \approx C_m \rho_m \Delta T_p \cdot 4\pi r_p^2 \sqrt{\chi_m t}$. The amount of heat transferred to the particle is $Q_p \approx C_p \rho_p \Delta T_p \cdot 4\pi r_p^3/3$. The ratio is:

$$\frac{Q_m}{Q_p} \approx \frac{C_m \rho_m}{C_p \rho_p} \cdot \frac{3\sqrt{\chi_m t}}{r_p} \tag{5}$$

where C_p and C_m are the specific heat capacities; ρ_p and ρ_m are the densities. Subscripts "p" and "m" denote the particle and the matrix, respectively. It is seen in Equation (5) that $Q_m/Q_p \ll 1$ if:

$$t \ll \tau_{adiab} = \frac{1}{9}\left(\frac{C_p \rho_p}{C_m \rho_m}\right)^2 r_p^{\,2} \chi_m^{\,-1} \tag{6}$$

The ratio of volume-specific heat capacities can be considered $C_p\rho_p/C_m\rho_m \sim 1$ for gold particles in a polymer matrix; thus, $\tau_{adiab} \approx r_p^{\,2}\chi_m^{-1}$. The thermal diffusivity of the polymer, χ_m, is on the order of 10^{-3} cm^2/s. Thus, for a 10-nm particle, $\tau_{adiab} \sim 100$ ps.

It is seen that due to $\chi_m \ll \chi_p$, the heat transfer from the particle to the matrix may be neglected during the time period needed to equalize the temperature inside the particle. The different time scales allow the solution of the problems of the particle heating and the heat diffusion from the particle to the matrix separately. Namely, the laser irradiation causes the uniform temperature rise in the particle by:

$$\Delta T_p = \frac{1}{C_p \rho_p V_p} \iiint_{V_p} d\vec{r} \int_0^{\tau_{pulse}} dt\, w(\vec{r},t) \tag{7}$$

which is the starting condition for the calculations of the heat diffusion within the matrix. Such calculations will be performed in the next section of the paper. In Equation (7), V_p is the volume of the particle and $w(\vec{r},t)$ is the heat power density, which is given by:

$$w(\vec{r},t) = \frac{4\pi}{\lambda} n_p'' I(\vec{r},t) \tag{8}$$

Laser field intensity inside the particle can be calculated by employing the Mie solution again:

$$I(r,t) = \frac{n_p' I_0 f(t)}{n_m} \frac{\left|\vec{E}_{Mie}(\vec{r})\right|^2}{E_0^{\,2}} \tag{9}$$

Finally,

$$\Delta T_p = \frac{F_0}{C_p \rho_p} \frac{4\pi n_p' n_p''}{n_m \lambda} \left\langle \left|\frac{\vec{E}_{Mie}(\vec{r})}{E_0}\right|^2\right\rangle_{V_p} \tag{10}$$

where $<>_{V_p}$ means averaging over the volume of the particle and $F_0 = \int_0^{\tau_{pulse}} I_0 f(t)dt$ is the incident fluence. In the case of Rayleigh scattering, for the small particles, Equation (10) transforms [17] to:

$$\Delta T_p = \frac{F_0}{C_p \rho_p} \cdot \frac{2\pi n_p'}{\lambda} \cdot \frac{9\varepsilon''}{(\varepsilon'+2)^2 + \varepsilon''^2} \tag{11}$$

where ε' and ε'' are the real and imaginary parts of $\varepsilon = n_p^2/n_m^2$.

The multi-photon heating of the matrix directly due to the field maxima near the particles can be treated in a similar way. The temperature rise in the matrix after the electron-phonon thermalization is calculated as:

$$\Delta T_m(\vec{r}) = \frac{1}{C_m \rho_m} \int_0^{\tau_{pulse}} w_{2ph}(\vec{r},t)dt \tag{12}$$

where:

$$w_{2ph}(\vec{r},t) = \frac{\eta_T N_m \sigma_m^{(2)} I^2(\vec{r},t)}{\hbar\omega} \tag{13}$$

is the heat power density due to the two-photon absorption in the matrix. In Equation (13), N_m is the number density of chromophores in the matrix and η_T is the quantum yield. Taking $I(r,t)$ outside the particle from Equation (3), one can evaluate the temperature rise in the matrix as:

$$\Delta T_m(\vec{r}) = \frac{\eta_T N_m \sigma_m^{(2)}}{C_m \rho_m \cdot \hbar\omega} I_0^2 \tau_{pulse} \frac{\left|\vec{E}_{Mie}(\vec{r})\right|^4}{E_0^4} \tag{14}$$

The dependencies of both the temperature rise ΔT_p in the particle and the maximal temperature rise $\max(\Delta T_m)$ in the matrix due to two-photon absorption exhibit a pronounced plasmonic resonance, as is shown in Figure 3.

Figure 3. Temperature rise in the particle (black solid lines, left axis) and the temperature rise in the matrix due to the two-photon absorption (red dashed lines, right axis) after irradiation by a single laser pulse of duration $\tau_{pulse} = 100$ fs. The particle radius is either 10 nm **(a)** or 50 nm **(b)**. The incident laser fluence is $F_0 = 2 \times 10^{-4}$ J/cm^2, which corresponds to the maximum field intensity $I_0 = 2 \times 10^9$ W/cm^2. The incident polarization is circular. The volume-specific heat values for the particle and the matrix are $C_p \rho_p = 2.5$ J·cm^{-3}K^{-1} and $C_m \rho_m = 1.7$ J·cm^{-3}K^{-1}, respectively. The dependencies of $n_p(\lambda)$ and $n_m(\lambda)$ for gold and PMMA are taken from [15,16], respectively. For the two-photon absorption, $\sigma_m^{(2)} = 100$ GM, $N_m = 6 \times 10^{20}$ cm^{-3} and $\eta_T = 1$ are employed.

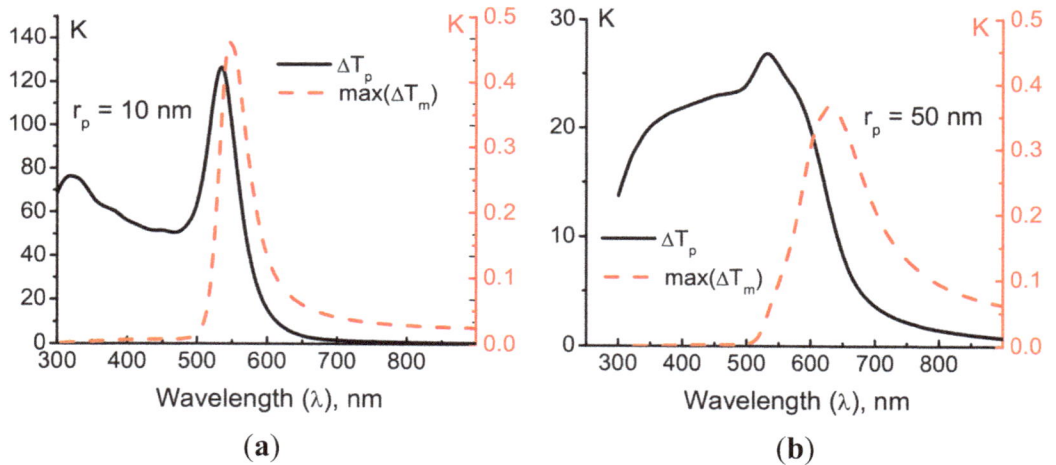

Now, we can compare the contributions of either of the heating mechanisms for typical experimental cases. For that, using Equation (10), we find the incident field intensity needed to achieve a certain temperature rise in the particle (e.g., $\Delta T_p = 100$ K). Then, we calculate the maximum temperature rise $\max(\Delta T_m)$ in the matrix for that intensity. Additionally, we calculate the maximum field intensity that is reached near the particle.

Figure 4 shows the calculation results for two different matrices. The first one has an unrealistically high two-photon sensitivity, $\sigma_m^{(2)} = 100$ GM, and the number density of chromophores $N_m = 6 \times 10^{21}$ cm^{-3}

nearly matches the typical number density of monomer chains in the polymer. The other has realistic parameters: $\sigma_m^{(2)} = 10$ GM and $N_m = 1 \times 10^{20}$ cm^{-3}. It is seen that direct heating of the matrix due to the two-photon absorption becomes advantageous in the longer wavelength part of the visible and near-IR spectrum. However, the field intensity reached near the particle in order to provide a temperature rise comparable to the one in the particle is above 10^{11} W/cm^2 even in the case where the two-photon absorption is overestimated. This intensity is even higher for realistic matrices, overcoming the optical breakdown level.

For the small enough particles (see Figure 4a), the irradiation at wavelengths shorter than the plasmonic resonance (530 nm) causes significant heating of the particle without reaching the breakdown level of intensities near the particle. In this case, direct heating of the matrix due to the multi-photon absorption can also be neglected. This case will be analyzed in the next section of the paper. For the longer wavelengths, the optical breakdown must be taken into account [2].

Figure 4. The maximum temperature rise (dashed lines, right axis) and maximum field intensity (black solid lines, left axis) outside the particle reached when the temperature of the particle is raised by 100 K due to irradiation by a circularly-polarized single laser pulse of duration $\tau_{pulse} = 100$ fs. The particle radius is either 10 nm (**a**) or 50 nm (**b**). The volume-specific heat values for the particle and the matrix are $C_p\rho_p = 2.5$ J·cm^{-3}K^{-1} and $C_m\rho_m = 1.7$ J·cm^{-3}K^{-1}, respectively. The dependencies of $n_p(\lambda)$ and $n_m(\lambda)$ for gold and PMMA are taken from [15,16], respectively. The quantum yield $\eta_T = 1$.

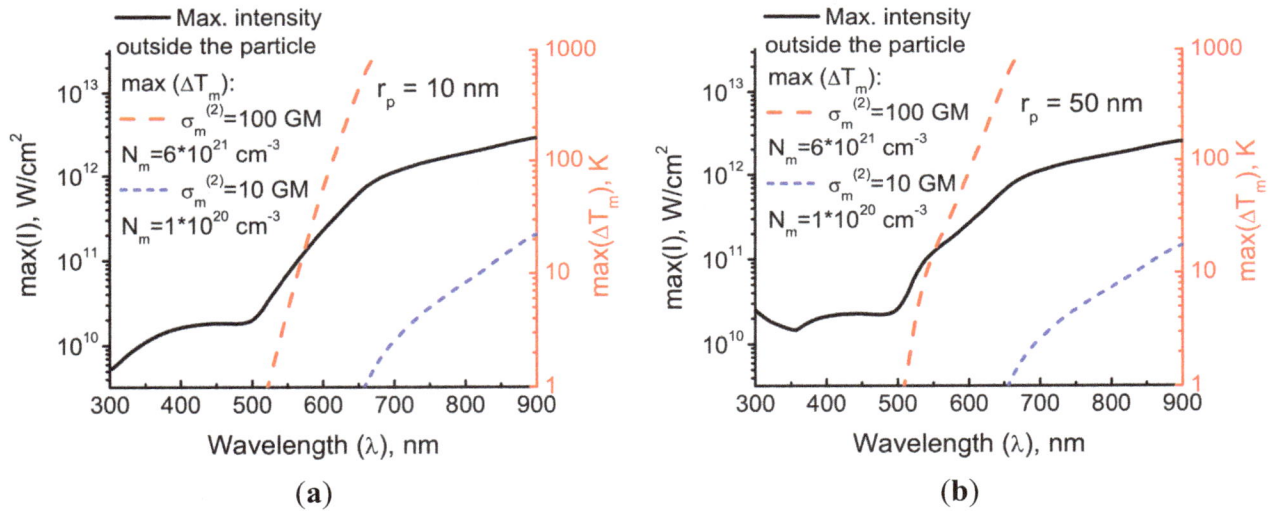

(a)

(b)

3. The Particle Heating Effect

Below, we consider the effect of heat diffusion from the metal nanoparticle uniformly heated by an ultrashort laser pulse. We consider the thermal destruction of some dopant X by a simple, thermally-activated chemical reaction promoted by the temperature elevation resulting from the heat flow from the particle. These molecules are dissolved within the matrix with the starting number density N_{X0}.

The heat diffusion problem is formulated as follows:

$$\begin{cases} \dfrac{\partial T(R,p)}{\partial p} = \dfrac{1}{\beta^2}\dfrac{1}{R}\dfrac{\partial^2 (RT(R,p))}{\partial R^2} \\[2ex] \dfrac{\partial T(R=1,p)}{\partial p} = \dfrac{1}{\beta}\dfrac{\partial T(R,p)}{\partial R}\bigg|_{R=1} \\[2ex] T(R,p=0) = T_r \\[1ex] T(R=1,p=0) = T_r + \Delta T_p \end{cases} \qquad (15)$$

Here, T_r is the "room" temperature (the initial temperature of the surrounding matter or a thermostat), ΔT_p is the initial particle temperature increment after the pulse (above the room temperature), r_p is the radius of the particle, χ_m is the heat diffusivity of the matrix and r is the radial coordinate. Here, we introduce the dimensionless parameter $\beta = (3C_m\rho_m)/(C_p\rho_p)$ (for gold nanoparticles in PMMA matrix, we have $\beta = 1.821$) and use the dimensionless variables: $p = (\beta^2/r_p^2)\chi_m t$ is the dimensionless time and $R = r/r_p$ is the dimensionless coordinate (normalized by the nanoparticle radius). We suppose that only a small amount of dopant molecules are converted due to a single pulse. This allows us to neglect the effect of reaction enthalpy on the heat propagation process.

The way of solving Equation (15) is the Laplace transform. The result is given in [18]:

$$T(R,p) = T_r + \frac{\Delta T_p}{2\sqrt{\pi}R} e^{\frac{\beta(R-1)+p}{2}-\frac{p}{\beta}}\int_0^{\frac{p}{\beta}}\left(\frac{\beta(R-1)+p}{\xi^{3/2}} - \frac{1}{\xi^{1/2}}\right)e^{\left(\frac{1}{\beta}-\frac{1}{4}\right)\xi - \frac{(\beta(R-1)+p)^2}{4\xi}}\,d\xi \qquad (16)$$

This expression can be rewritten via a special function.

We assume the Arrhenius law for the reaction rate:

$$-\frac{\partial N_X(r,t)}{\partial t} = N_X A \exp\left(-\frac{T_A}{T(r,t)}\right)$$

where T_A is the activation temperature and A is the reaction constant. We calculate the part of reacted molecules after one laser pulse. We suppose that this part is much less than unity, so that N_X under the integral can be considered as a constant.

$$\Delta N_X(r) = -N_X(r)A\int_0^\infty \exp\left(-\frac{T_A}{T(r,t)}\right)dt$$

Let us introduce:

$$G(R) = A\int_0^\infty \exp\left(-\frac{T_A}{T(R,t)}\right)dt \qquad (17)$$

Expression (17) represents the fraction of the destroyed species in a pulse. After N_{pulse} pulses, the conversion v reads:

$$v(R) = 1 - (1 - G(R))^{N_{pulse}} \qquad (18)$$

According to (16), we have the distribution of temperature in the matrix in the form $T(R,p) = T_r + \Delta T_p g(R,p)$, where $g(R,p)$ is the normalized function, so that $g(R=1,p=0) = 1$.

Figure 5 shows the time dependence of temperature at different points in the vicinity of the particle. It is seen that the temperature increases rather fast to its maximum and then slowly returns to the initial value. Another situation is on the surface of the particle where the temperature decreases just after the pulse. The most important contribution to the reaction yield occurs during the time interval when the temperature value is close to its maximum.

Figure 5. Evolution of normalized temperature g as a function of dimensionless time p in different points of the polymer matrix surrounding a gold nanoparticle: $R = 1$ (black line), $R = 1.1$ (blue line), $R = 1.5$ (red line) and $R = 2$ (green line).

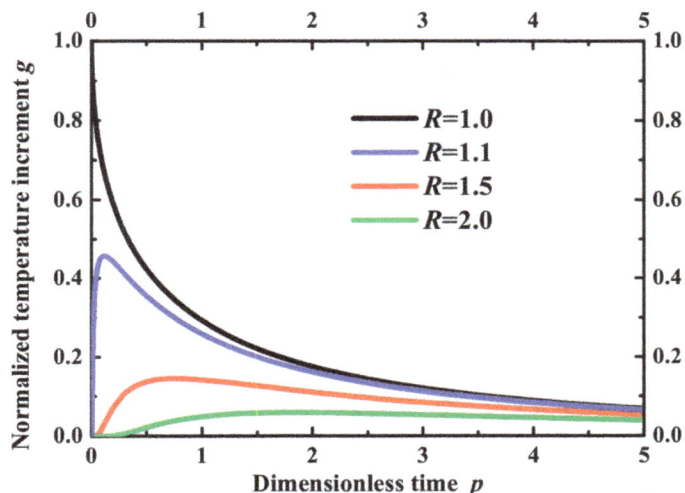

On the surface of the particle, the highest temperature is in the beginning of the process. That is why we can use an approximation for temperature dependence on time on the surface for times close to zero. This approximation follows from Equation (16):

$$T(R = 1, p) \approx \Delta T_p \left(1 - \frac{2}{\sqrt{\pi}} \sqrt{p} \right) + T_r$$

In the problem of the estimation of the integral G, we have large parameter $T_A/(T_r + \Delta T_p)$. It allows one to use the saddle-point method. Here, we can approximately calculate the integral G (see Equation (17)):

$$G(R = 1) \approx \frac{r_p^2}{\beta^2 \chi_m} A \exp\left(-\frac{T_A}{\Delta T_p + T_r} \right) \frac{\pi \left(\Delta T_p + T_r \right)^4}{\left(\Delta T_p T_A \right)^2} \tag{19}$$

This result is in good agreement with numerical calculations (Figure 6).

Having a simple formula, such as Equation (19), for the integral G, we can estimate the realistic conversion of dopant molecules, which can be provided by the particle heating just near the particle surface. We can reach a repetition rate of up to 10^3 pulses per second. Each pulse raises the nanoparticle temperature to ΔT_p. It was mentioned above that for small particles, ΔT_p can be calculated using Equation (11). The laser pulse at the plasmon resonance wavelength ($\lambda = 530$ nm) with the fluence $F_0 = 10^{-4}$ J/cm^2 increases the temperature up to about 100 K. This value of fluence is available from commercial lasers and makes it possible to irradiate simultaneously an area of about 1 cm^2. Thus, for our estimates, we will use $\Delta T_p \sim 100$ K. It should be noted that when energetic femtosecond laser pulses are employed for large-volume processing, the beam power may be significantly higher than the critical value for self-focusing. However, the considered intensities (not higher than 10^9 W/cm^2) are not sufficient to provide filamentation within a sample of reasonable thickness. Indeed, for the filamentation development, the retardation integral $B = \frac{2\pi}{\lambda} \int_0^l I n_2 dz$ should be larger than unity. Here, l is the distance at which filamentation starts to develop and n_2 is the nonlinear refractive index. With the typical value $n_2 \approx 10^{-14}$ cm^2/W and the intensity $I \approx 10^9$ W/cm^2, $B = 1$ will be reached at $l \approx 1.5$ cm. This is an estimate for the possible thickness of the sample. Moreover, our studies [10,19] show that the presence of the plasmonic nanoparticles within the polymer matrix provides a significant negative input in the nonlinear refractive index, which essentially hinders the filamentation.

Figure 6. The integral $G(R = 1)$ as a function of thermostat temperature for $T_A = 8000$ K (black color) and $T_A = 5000$ K (red color). The points show numerical calculations using Equation (16), and the line shows the approximation by Equation (19). The radius of a gold nanoparticle is $r_p = 20$ nm.

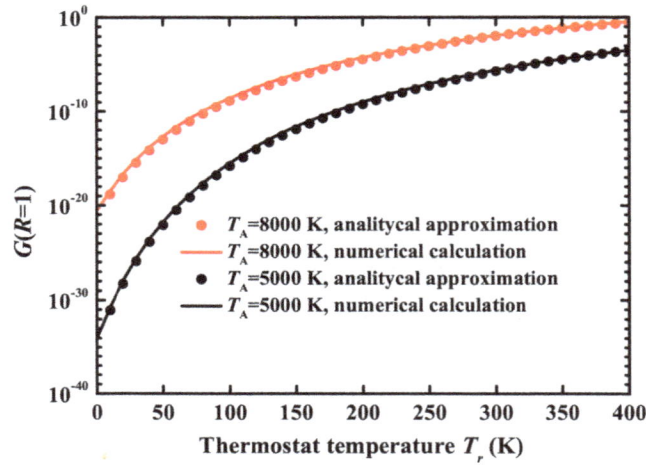

We assume the room temperature value to be about 300 K. However, it should be noted that we can control it by using a thermostat. The reaction constant A is supposed to be less than 10^{13} s^{-1}. An activation temperature range T_A less than 12,000 K is typical of precursor molecules (see, for example, [20–23]). The typical heat diffusivity of polymers is $\chi_m \sim 10^{-3}$ cm^2/s. The radius r_p of a nanoparticle is about 20 nm. The number of pulses N_{pulse} can be up to 10^6 for a few minutes. Thus, the conversion near the particle surface, according to Equations (18) and (19), is:

$$v(R = 1) = 1 - \left(1 - \frac{a^2}{\beta^2 \chi_m} A \exp\left(-\frac{T_A}{\Delta T_p + T_r}\right) \frac{\pi \left(\Delta T_p + T_p\right)^4}{\left(\Delta T_p T_A\right)^2}\right)^{N_{pulse}} \tag{20}$$

Irradiation by a large number of pulses results in overall heating of the sample. It depends on many factors, including the repetition rate, the beam radius, the sample thickness, cooling conditions on the surface, *etc.* The effect of the overall heating on G could be understood from its dependence on T_r (see Equation (19) and Figure 6). The complete analysis of this phenomenon is beyond the scope of this paper.

It can be noted (Figure 7) that for $T_A = 10,000$ K, the increase in temperature by 100 K is not enough for the reaction. $\Delta T_p = 150$ K will transform about 10% of the dopant molecules, and

$\Delta T_p = 200$ K is much more efficient. On the other hand, a large increase in temperature is not desirable, since it can lead to damage of the polymer matrix.

Figure 7. (a) The conversion in the polymer matrix on the surface of a gold nanoparticle (radius $r_p = 20$ nm) as a function of activation temperature, for different temperature increment ΔT_p. **(b)** The conversion in the polymer matrix on the surface of a gold nanoparticle (radius $r_p = 20$ nm) as a function of temperature increment ΔT_p, for different activation temperature T_A. The thermostat temperature $T_r = 300$ K. The number of pulses $N_{pulse} = 10^6$. The curves are calculated using Equation (20).

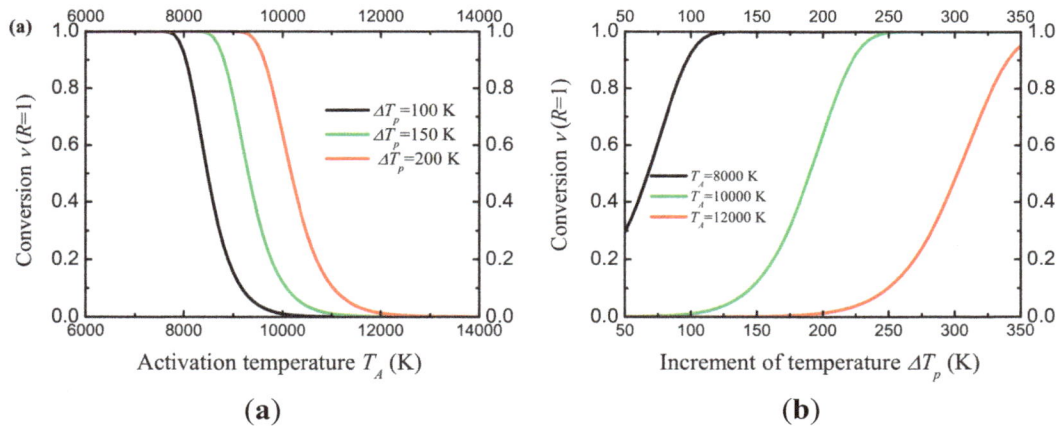

(a)   **(b)**

It has been mentioned above that the maximum temperature at the point is a very important characteristic of the process. It determines the highest rate of reaction. Figure 8 represents spatial distributions of normalized temperature increment g in different moments of time and the envelope of these distributions. The maximum temperature increment can be approximated by the formula $g_{max}(R) = \exp(-a(R-1)^b)$.

This form of temperature confinement was described in [24] for a gold nanoparticle in water with another parameter β. For most of the materials, however, β is of the order of unity. This is the only parameter in the system. The distribution of temperature depends on β. The parameters a and b in the approximation of the maximum temperature can be linearly fitted as functions of β (see inset in Figure 8).

Figure 9 represents the result of the calculation of $G(R)$, introduced above. For a nonzero room temperature, this integral diverges. In fact, at room temperature, the rate of reaction is negligibly small. Since we calculate the laser heating effect, we subtract this background. The resulting expression is:

$$G_1(R) = \frac{r_p^2}{\beta^2 \chi_m} A \int_0^\infty \left(\exp\left(-\frac{T_A}{T_r + \Delta T_p g(R,p)} \right) - \exp\left(-\frac{T_A}{T_r} \right) \right) dp \qquad (21)$$

In order to compare the volume of reaction, we calculate the normalized distribution $G_1(R)/G_1(R=1)$. It is seen that a higher activation energy leads to a smaller volume of reaction. It is seen that the thermally-activated reaction can be effective enough at the very surface of the heated particle. However, the reaction domain is highly localized just near the particle, thus preventing the conversion of a large net amount of molecules. In order to convert remote molecules that are initially located at more significant distances from the plasmonic particles than the particle radius, one should use nonlocal effects, which are beyond the scope of this paper.

Figure 8. The distribution of normalized temperature g *vs.* normalized coordinate R at different moments of time: $p = 0.1$ (black line), $p = 0.2$ (red line), $p = 0.5$ (green line). Blue dots demonstrate the spatial distribution of the maximum temperature. The blue line is its approximation by $g_{max}(R) = \exp(-a(R-1)^b)$. Inset: parameters a and b *vs.* coefficient β and their linear approximation by $a = 2.06 + 0.41\beta$ and $b = 0.61 - 0.034\beta$.

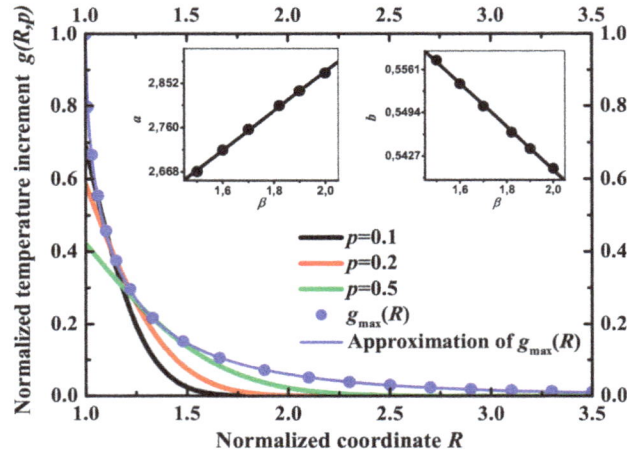

Figure 9. Normalized number of reacted molecules $G_1(R)/G_1$ ($R = 1$) according to Equation (21) for $\Delta T_p = 100$ K, $T_r = 300$ K, $T_A = 10{,}000$ K (green line), $T_A = 8000$ K (black line) and $T_A = 5000$ K (red line).

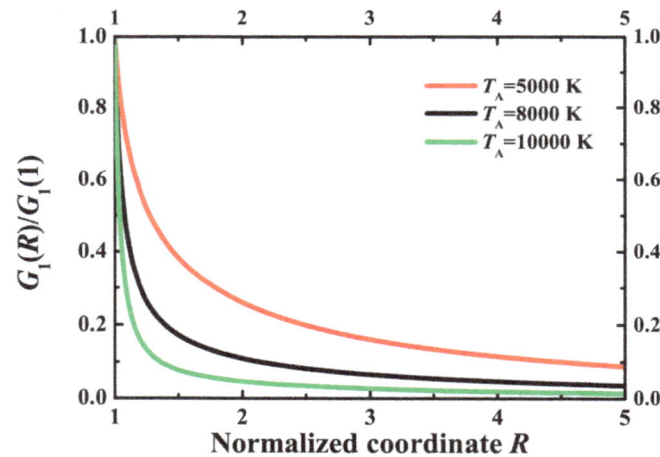

4. Conclusions

We consider the opportunities provided by plasmonic nanoparticles inserted into the bulk of transparent media to modify the material by means of the laser light irradiation.

This is of particular importance for photo-induced nanocomposites. Here, the plasmonic nanoparticles are formed due to laser irradiation of the initially homogeneous material containing precursor molecules by a self-organization process. This process significantly changes the optical property of the medium. The further micro-modification of the material employing previously-formed plasmonic nanoparticles can be considered as the next-step laser alteration of the material properties.

As an example, we discuss a photochemical reaction in the vicinity of the plasmonic nanoparticle initiated by the laser radiation in the near-field domain due to nonlinear absorption. We show that the spatial distribution of the products of such a reaction changes the symmetry of the material, resulting in the appearance of anisotropy in the initially isotropic material or even in the loss of the center of symmetry.

The latter is important for the optical nonlinearity. Photo-induced nanocomposites are highly optically nonlinear (see, e.g., [10]). First of all, this is true for the cubic nonlinearity described by the third-order susceptibility [19]. The second-order susceptibility responsible for the second harmonic generation is zero in the center-symmetric media. The loss of the center of symmetry of the composites due to a photochemical reaction in the near-field of the plasmonic nanoparticles permits one to observe optical nonlinear phenomena of the second order in such materials [25]. Such a modification can be performed using reasonable irradiation regimes with an intensity of about 10^9 W/cm^2 and materials with a not outstanding two-photon sensitivity. This allows for possible applications of commercially available materials and lasers for a single-step modification of large volumes of about 1 cm^3.

It is shown that it is possible to significantly heat small plasmonic nanoparticles by an ultrashort laser pulse without causing optical breakdown in the near field. In this case, the direct material heating by the nonlinear absorption in the vicinity of the particle can be neglected as compared with the heating of the particle itself.

We considered a thermally-activated reaction of the first order initiated in the medium by the femtosecond pulse heating of a plasmonic particle. We analyzed a spherically symmetric solution of the corresponding heat diffusion equation and determined the fraction of reacted molecules (conversion) at the surface of a metal sphere and within the domain surrounding the nanoparticle. We present an analytical formula for the conversion at the very surface and a numerical solution for the conversion distribution within the material bulk. Our estimation shows that a conversion of about unity can be obtained by defocused femtosecond pulses of the mJ energy level for a reasonable number of pulses if the activation energy of the thermally-activated reaction is close to 1 eV or smaller.

It should be noted that the thermoactivated reaction could mean either the thermodestruction of the matrix or the transformation (destruction) of the inserted dopants. An activation energy of about 1 eV is typical of the precursor molecules. Thus, this reaction could start the processes of further complicated changes in the material localized near the plasmonic particles.

Acknowledgments

The authors thank Prof. Dr. Boris Luk'yanchuk for the near-field Mie code and Russian Scientific Foundation (Grant No. 14-19-01702) for financial support.

Author Contributions

Anton A. Smirnov, Alexander Pikulin, and Natalia Sapogova made the calculations; Anton A. Smirnov wrote Section 3 and Appendix A1; Alexander Pikulin wrote Section 2; Nikita Bityurin wrote Introduction and Conclusions; Nikita Bityurin is responsible for the problem formulation, estimations, and overall supervision.

Conflicts of Interest

The authors declare no conflict of interest.

Appendix A1. List of Notations

$\vec{r}$	Radius vector
r	Radial coordinate
x, y, z	Cartesian coordinates
t	Time
r_p	Radius of the particle
V_p	Volume of the particle
λ	Wavelength
$\hbar$	Planck constant
ω	Angular frequency
τ_{pulse}	Pulse length
τ_{e-ph}	Electron-phonon coupling time
τ_{diffus}	Temperature diffusion time
τ_{adiab}	Time period after the pulse, during which the heat transfer from the particle to the matrix can be neglected
T	Temperature
ΔT_p	Temperature rise of the particle after a single laser pulse
$\Delta T_m(\vec{r})$	Temperature rise in the matrix after a single laser pulse due to the multiphoton absorption
T_A	Activation temperature
T_r	Thermostat temperature
C_p, C_m	Heat capacities of the particle and the matrix
ρ_p, ρ_m	Densities of the particle and the matrix
χ_p, χ_m	Heat diffusion coefficients of the particle and the matrix
n_p, n_p', n_p''	Complex refractive index of the particle, its real part (refraction index) and imaginary part (absorption index)
n_m	Refractive index of the matrix
$\varepsilon, \varepsilon', \varepsilon''$	Ratio of dielectric permittivities of the particle and the matrix, its real and imaginary parts
N_X	Number density of species X
N_{X0}	Initial number density of species X
v	Conversion
$\sigma_X^{(2)}$	Two-photon absorption cross section of X
η	Quantum yield of two-photon photochemical destruction of species X
N_m	Number density of chromophores that are involved in heating by two-photon absorption
$\sigma_m^{(2)}$	Two-photon absorption cross section of chromophores
η_T	Quantum yield of two-photon heating
$I(\vec{r}, t)$	laser field intensity
$I_{inc}(t), I_0, f(t)$	Incident field intensity, its amplitude, and normalized temporal shape function
N_{pulse}	Number of laser pulses
F_0	Incident fluence
$\vec{E}_{Mie} / E_0$	Electric field magnification according to the solution of Mie problem for plane monochromatic wave

$w(\vec{r},t)$	Heat power density
$w_{2ph}(\vec{r},t)$	Heat power density due to the two-photon absorption in the matrix
Q_m, Q_p	Amount of heat transferred to the matrix and to the particle due to the laser pulse
β	Dimensionless parameter, $\beta = (3C_m\rho_m)/(C_p\rho_p)$
p	Dimensionless time
R	Dimensionless coordinate
$g(R, p)$	Normalized temperature rise in the matrix due to the laser pulse
A	Reaction constant
$G(R), G_1(R)$	Fraction of destroyed species in a pulse
a, b	Approximation parameters for the maximum temperature
B	Retardation integral
l	Distance at which the filamentation development starts
n_2	Nonlinear refractive index

References

1. Govorov, A.O.; Zhang, H.; Demir, H.V.; Gun'ko, Y.K. Photogeneration of hot plasmonic electrons with metal nanocrystals: Quantum description and potential applications. *Nano Today* **2014**, *9*, 85–101.

2. Boulais, E.; Lachaine, R.; Hatef, A.; Meunier, M. Plasmonics for pulsed-laser cell nanosurgery: Fundamentals and applications. *J. Photochem. Photobiol. C Photochem. Rev.* **2013**, *17*, 26–49.

3. Tribelsky, M.I.; Luk'yanchuk, B.S. Light scattering by small particles and their light heating: New aspects of the old problems. In *Fundamentals of Laser—Assisted Micro- and Nanotechnologies*; Veiko, V.P., Konov, V.I., Eds.; Springer International Publishing: Cham, Switzerland, 2014; pp. 125–149.

4. Liu, X.; Chen, Y.; Li, H.; Huang, N.; Jin, Q.; Ren, K.; Ji, J. Enhanced retention and cellular uptake of nanoparticles in tumors by controlling their aggregation behavior. *ACS Nano* **2013**, *7*, 6244–6257.

5. Govorov, A.O.; Zhang, W.; Skeini, T.; Richardson, H.; Lee, J.; Kotov, N.A. Gold nanoparticle ensembles as heaters and actuators: Melting and collective plasmon resonances. *Nanoscale Res. Lett.* **2006**, *1*, 84–90.

6. Adleman, J.R.; Boyd, D.A.; Goodwin, D.G.; Psaltis, D. Heterogenous catalysis mediated by plasmon heating. *Nano Lett.* **2009**, *9*, 4417–4423.

7. Xiao, M.; Jiang, R.; Wang, F.; Fang, C.; Wang, J.; Yu, J.C. Plasmon-enhanced chemical reactions. *J. Mater. Chem. A* **2013**, *1*, 5790–5805.

8. Alexandrov, A.; Smirnova, L.; Yakimovich, N.; Sapogova, N.; Soustov, L.; Kirsanov, A.; Bityurin, N. UV initiated growth of gold nanoparticles in PMMA matrix. *Appl. Surf. Sci.* **2005**, *248*, 181–184.

9. Athanassiou, A.; Cingolani, R.; Tsiranidou, E.; Fotakis, C.; Laera, A.M.; Piscopiello, E.; Tapfer, L. Photon-induced formation of CdS nanocrystals in selected areas of polymer matrices. *Appl. Phys. Lett.* **2007**, *91*, 153108.

10. Bityurin, N.; Alexandrov, A.; Afanasiev, A.; Agareva, N.; Pikulin, A.; Sapogova, N.; Soustov, L.; Salomatina, E.; Gorshkova, E.; Tsverova, N.; Smirnova, L. Photoinduced nanocomposites—Creation, modification, linear and nonlinear optical properties. *Appl. Phys. A* **2013**, *112*, 135–138.

11. Sapogova, N.; Bityurin, N. Model for UV induced formation of gold nanoparticles in solid polymeric matrices. *Appl. Surf. Sci.* **2009**, *255*, 9613–9616.

12. Hubert, C.; Rumyantseva, A.; Lerondel, G.; Grand, J.; Kostcheev, S.; Billot, L.; Vial, A.; Bachelot, R.; Royer, P.; Chang, S.; Gray, S.K.; Wiederrecht, G.P.; Schatz, G.C. Near-Field photochemical imaging of noble metal nanostructures. *Nano Lett.* **2005**, *5*, 615–619.

13. Born, M.; Wolf, E. *Principles of Optics*, 7th ed.; Cambridge University Press: Cambridge, UK, 1999; pp. 759–789.

14. Mie Type Codes. Available online: http://www.scattport.org/index.php/light-scattering-software/mie-type-codes (accessed on 13 October 2014).

15. Johnson, P.B.; Christy, R.W. Optical constants of the noble metals. *Phys. Rev. B* **1972**, *6*, 4370–4379.

16. Kasarova, S.N.; Sultanova, N.G.; Ivanov, C.D.; Nikolov, I.D. Analysis of the dispersion of optical plastic materials. *Opt. Mater.* **2007**, *29*, 1481–1490.

17. Tribelsky, M.I.; Miroshnichenko, A.E.; Kivshar, Y.S.; Luk'yanchuk, B.S.; Khokhlov, A.R. Laser pulse heating of spherical metal particles. *Phys. Rev. X* **2011**, *1*, 021024.

18. Paterson, S. Conduction of heat from local sources in a medium generating or absorbing heat. *Proc. Glasgow Math. Assoc.* **1953**, *1*, 164–169.

19. Afanas'ev, A.V.; Mochalova, A.E.; Smirnova, L.A.; Aleksandrov, A.P.; Agareva, N.A.; Sapogova, N.V.; Bityurin, N.M. Ultraviolet-induced variation of the optical properties of dielectrics in the infrared region. *J. Opt. Technol.* **2011**, *78*, 537–543.

20. Yang, H.; Hu, Y; Zhang, X; Qiu, G. Mechanochemical synthesis of cobalt oxide nanoparticles. *Mater. Lett.* **2004**, *58*, 387–389.

21. Tada, K.; Sakata, K.; Kitagawa, Y.; Kawakami, T.; Yamanaka, S.; Okumura, M. DFT calculations for chlorine elimination from chlorine-adsorbed gold clusters by hydrogen. *Chem. Phys. Lett.* **2013**, *579*, 94–99.

22. Basavaraja, C.; Kim, J.K.; Huh, D.S. Characterization and temperature-dependent conductivity of polyaniline nanocomposites encapsulating gold nanoparticles on the surface of carboxymethyl cellulose. *Mater. Sci. Eng. B* **2013**, *178*, 167–173.

23. Nair, P.S.; Scholes, G.D. Thermal decomposition of single source precursors and the shape evolution of CdS and CdSe nanocrystals. *J. Mater. Chem.* **2006**, *16*, 467–473.

24. Baffou, G.; Rigneault, H. Femtosecond-pulsed optical heating of gold nanoparticles. *Phys. Rev. B* **2011**, *84*, 035415.

25. Zeng, Y.; Hoyer, W.; Liu, J.; Koch, S.W.; Moloney, J.V. Classical theory for second-harmonic generation from metallic nanoparticles. *Phys. Rev. B.* **2009**, *79*, 235109.

3

Design and Implementation of a Bionic Mimosa Robot with Delicate Leaf Swing Behavior

Chung-Liang Chang * and Jin-Long Shie

Department of Biomechatronics Engineering, National Pingtung University of Science and Technology, No. 1 Shuefu Road, Neipu, Pingtung County 91201, Taiwan; E-Mail: j357753k@yahoo.com.tw

* Author to whom correspondence should be addressed; E-Mail: chungliang@mail.npust.edu.tw

Academic Editor: Miko Elwenspoek

Abstract: This study designed and developed a bionic mimosa robot with delicate leaf swing behaviors. For different swing behaviors, this study developed a variety of situations, in which the bionic mimosa robot would display different postures. The core technologies used were Shape Memory Alloys (SMAs), plastic material, and an intelligent control device. The technology particularly focused on the SMAs memory processing bend mode, directional guidance, and the position of SMAs installed inside the plastic material. Performance analysis and evaluation were conducted using two SMAs for mimosa opening/closing behaviors. Finally, by controlling the mimosa behavior with a micro-controller, the optimal strain swing behavior was realized through fuzzy logic control in order to display the different postures of mimosa under different situations. The proposed method is applicable to micro-bionic robot systems, entertainment robots, biomedical engineering, and architectural aesthetics-related fields in the future.

Keywords: SMAs; fuzzy logic; mimosa robot; micro-controller

1. Introduction

The study field of mimicking the special skills of biological organisms is known as bionics. American scholar, Steele [1] proposed the concept of bionics in 1960, which intended to develop new technology or solve existing technique problems by understanding biological structures and functions. The content of bionics is to observe, study, and simulate a variety of natural biological skills, including biological structures, principles, behaviors, the various functions of organs, *in vivo* physical and chemical processes, as well as the supply of energy, memory, and transmission. Based on these principles, bionics provides new design ideas and system principles for science and technology.

The Mimosa (scientific name of *Mimosa pudica* L.) was the object to be mimicked and designed in this study. It has a palmate compound leaf consisting of four pinnas. Each pinna grows many leaflets, and is composed of pulvinus, petiole, leaflet, and vascular bundle tissue (see Figure 1). The moisture in the pedestal supports the pinna, and when the pinna is stimulated by an external force (such as touch or a bold wind), moisture in the pedestal will quickly flow elsewhere, and the pinnate compound leaf closes. This phenomenon is called turgor movement. The main function of such behavior is to protect itself and frighten away any approaching animal or insect.

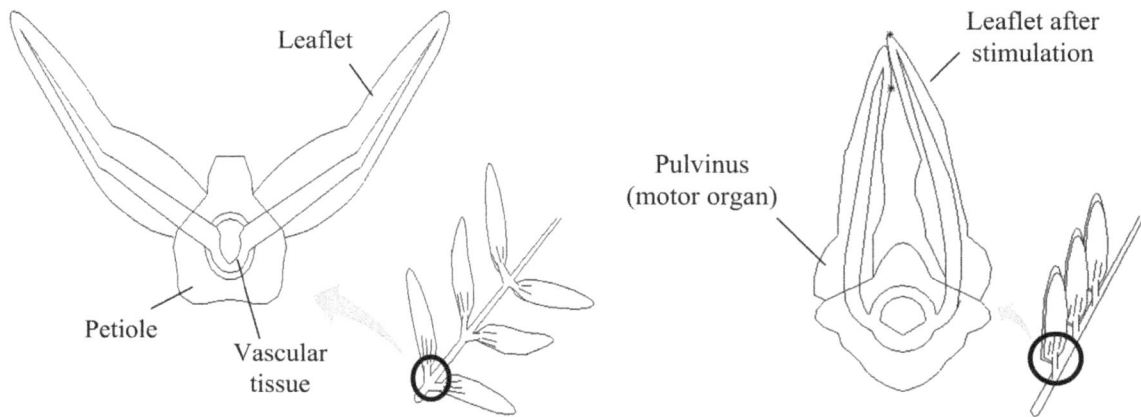

Figure 1. External structure of mimosa (the pinnate compound leaf open (**left**) and close (**right**)).

Most existing bionic products on the market stimulate the external behaviors of biological organisms, which behaviors are realized by the structural assembly, motor, and drive controller [2,3]. However, in order to describe the more delicate operational behaviors, the spatial dimensions of mechanical movements require further improvement, thus resulting in complicated design and high development cost.

Bionic materials have been widely used in various fields to achieve dynamic biological behaviors [4,5], and at present, a variety of advanced materials have been used in bio-sensing and actuator design. Due to material properties, Shape Memory Alloys (SMAs) are popularly used, such as the camera lens in phones, glasses, orthodontic braces, green energy, and medical devices (e.g., stents and endoscopy, ventilation valves of greenhouse cultivation systems, anti-scald valves for water heaters, *etc.*) [6–9].

SMAs have also been used in animal and plant behavior displays, including the micro-robot [10], snake-type robot [11], the fish-like propulsion system [12], Shape Memory Alloy Shirts [13], Self-Actuating Composite Materials [14], and SMAs Flower Robots [15]. The joints of the above robots are constructed of SMAs, as thermal driving can produce the bending and extension of SMAs, which enable bionic robots to present delicate and continuously smooth actions. However, in order to render the above SMS developed robots to display interactive behaviors with its environment, an intelligently controlled algorithm is required. In 2011, Chang *et al.* developed a flower robot, and controlled its growth and blooming behaviors through external light intensity and humidity [16]. Although the robot can control the blooming of its flowers, it cannot control the closing or individual opening behaviors, as SMAs inside the leaves of the robot may result in high temperature due to long time charging.

The joints between the wire and SMAs may be vulnerable to the influence of high temperature, resulting in breakdowns. Therefore, this study proposed a modified SMAs processing method, and applied this technology in the creation of a bionic mimosa leaf. Through the proposed strain and processing processes of SMAs, the swing and control of the leaf can be realized. Moreover, this study applied a fuzzy controller and sensors to realize three mimosa opening/closing behaviors.

In terms of control system design, it is considerably difficult to develop a physiological model of the mimosa, as the behaviors of the mimosa are affected by numerous factors, including internal factors, external factors (environmental factors), biological factors, *etc.* Thus, it is difficult to build a mathematical model of plant physiological behavior. The fuzzy inference approach is well capable of solving and handling the control problems of the imprecise nature existing in all physical systems. However, to solve practical problems, the professional knowledge and skills of experienced experts and scholars were referred to describe the rules and write the program. Hence, the system can infer the results according to the rules, which are used as reference for decision making. This study adopted the decision-making mechanism derived from expert knowledge, which is based on the factors inducing the mimosa's swing behaviors (light beam, sound, *etc.*). Moreover, it applied fuzzy logical inference to adjust the opening and closing rate of the mimosa leaf, thus realizing the robot system. In terms of circuit hardware design, the overall hardware module included the environment sensor module, SMA actuator, power supply, and other peripheral components, which are all integrated in a self-designed embedded board. The micro-controller within the robot system receives the environmental sensing data, which is processed by the internal fuzzy decision-making mechanism, and sends the control signals to the actuator module in the embedded board, so that the mimosa robot can show different swing behaviors. Finally, this study conducted single-pair leaf swinging, strain swinging, situational behavior swinging tests, and displacement analysis of the implemented mimosa robot.

The remainder of this paper is organized as follows: Section 2 describes the bionic mimosa design processes, including the design concept of the proposed system, leaf production, SMAs production, software and hardware design, and implementation; Section 3 introduces the system simulation methods, experimental testing, results, and discussion; Section 4 offers conclusions.

2. Overall Design of the Mimosa Bionic Robot

This section contains the bionic mimosa system design steps and leaf shape production process.

2.1. System Description

The system appearance is shown in Figure 2a. The left top of Figure 2a shows multiple switching groups for users to manually direct leaf swinging, opening, and closing behaviors. Figure 2b illustrates the internal circuit system configuration.

The architecture of the proposed system is as shown in Figure 3, which uses a micro-controller to realize system control. The controller's instructions are encoded by the interpreter prior to being connected to the micro-controller via the transmission interface (USB or RS-232). After the action planning, the micro-controller obtains the values of the environmental factors of Voice and Sunlight through the input and output interface. Finally, the decision-making values are transmitted to the embedded control system for control of the bionic mimosa robot. The micro-controller system is powered by an external 12 V power supply, and the SMA drive module uses a 12 V power supply for the operational experiment.

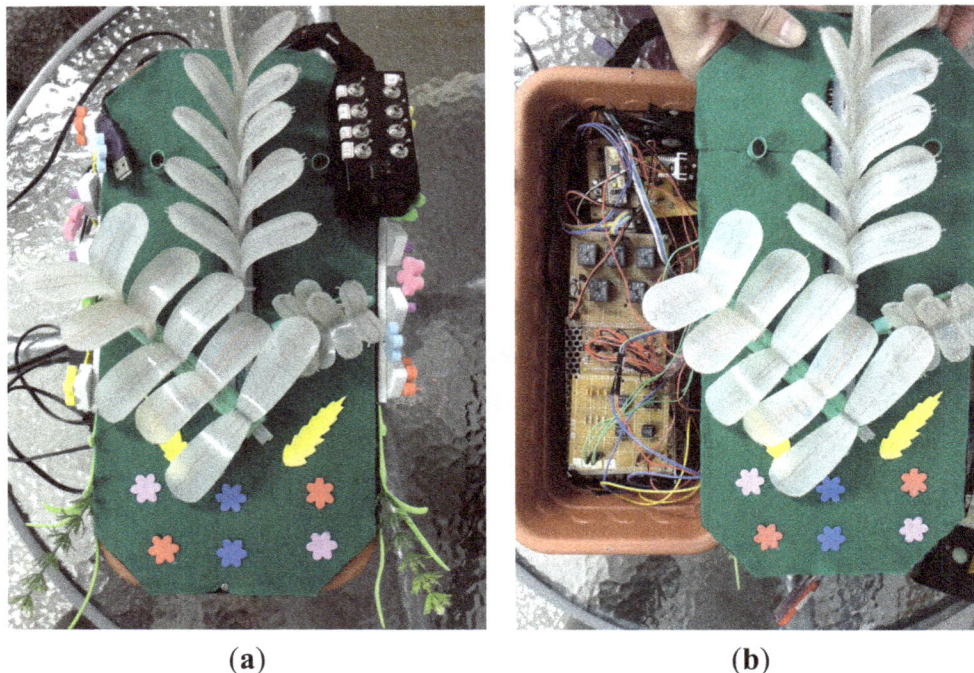

(a) (b)

Figure 2. Mimosa system appearance: (**a**) mimosa robot shape and (**b**) internal circuit board configuration.

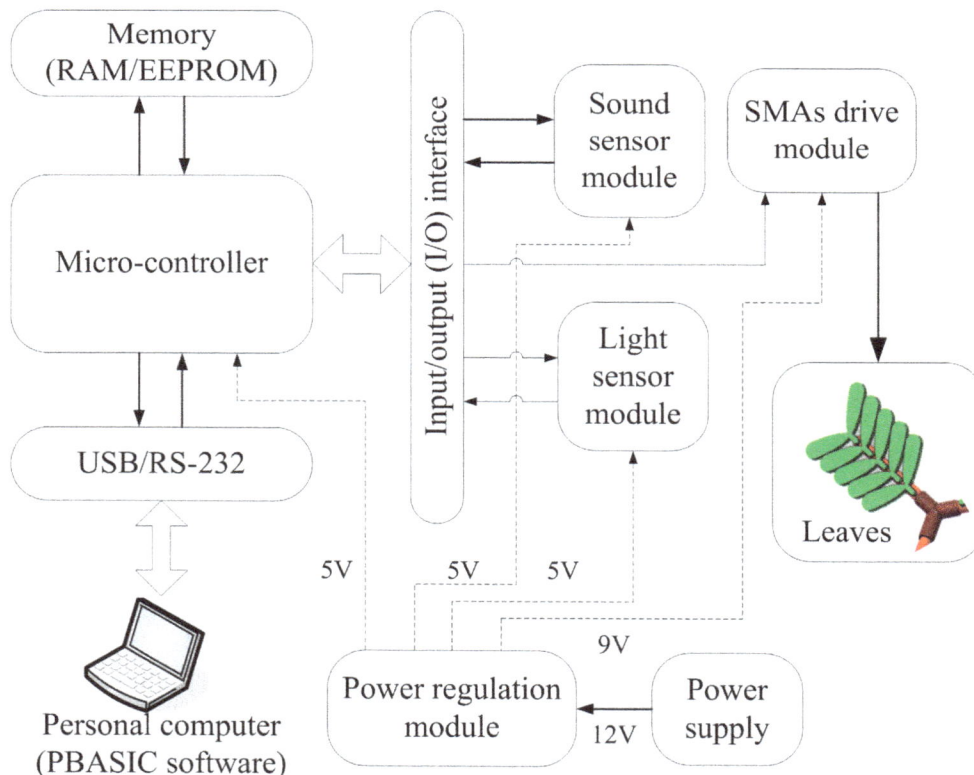

Figure 3. Bionic mimosa robot system architecture.

2.2. Fuzzy Control System

As mimosa opening/closing behaviors are significantly affected by Sunlight and Voice, when light and voice intensities are strong, the opening/closing responses are rapid. The opening/closing of all leaves of the strain are not simultaneously completed in sequence, as it is relatively difficult to establish the mimosa opening/closing model. Therefore, this study applied a fuzzy logic inference system for decision-making regarding mimosa opening/closing behaviors. Three situational modes were developed according to the mimosa behavioral demonstration. Each situational mode had its corresponding fuzzy logic inference, which can be further integrated into a multi-layered fuzzy controller. The framework of the proposed control system is shown in Figure 4.

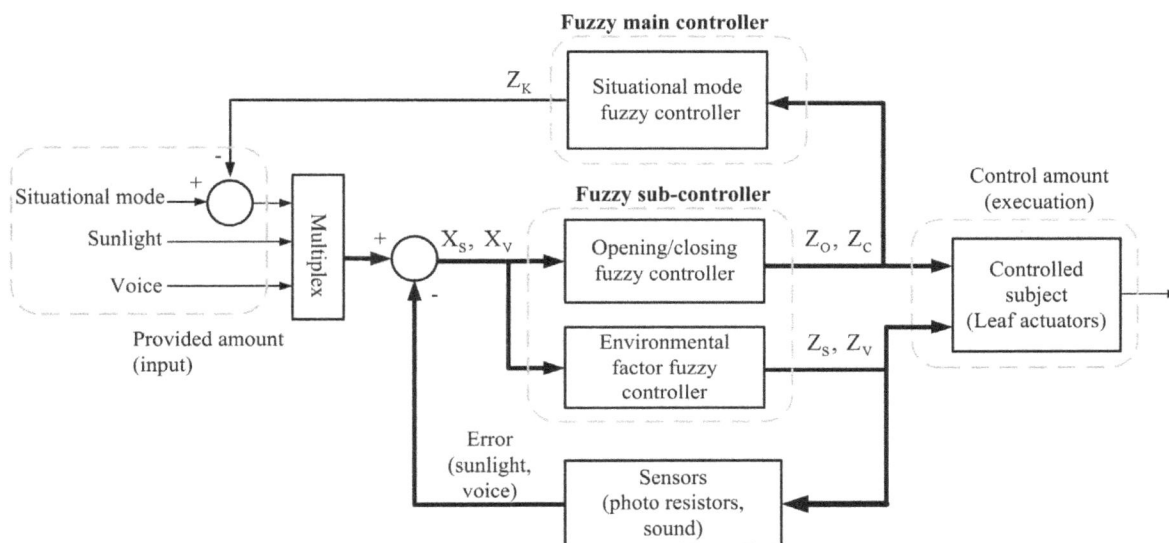

Figure 4. Architecture of the multi-layered fuzzy control system.

The input end of the multi-layered fuzzy controller system receives the values of light sensing and voice, which are transmitted to the environmental factor fuzzy controller and leaf opening/closing controller. The output of the opening/closing controller controls leaf opening/closing time. The output of the environmental factor fuzzy controller controls the leaf behavioral display and swing. Finally, the environmental sensors sense the current environmental factors, and feed the values back to the input end for the next system actuation. During implementation of the system, the output signals of the opening/closing fuzzy controller and environmental factor fuzzy controller can be in the situational mode for the situational mode fuzzy controller. The design steps of the proposed fuzzy control are illustrated as follows [17,18]:

Step 1: Define input and output variables and various language labels databases:

First, this study defined five basic linguistic statements: Very Small (VS), Small (S), Medium (M), Big (B), Very Big (VB), and then defined each controller's input and output variables.

A. Input variables:

(1) Fuzzy sub-controller for environmental factors and opening/closing:

The number of input variables for environmental factors and growth rate are the same.

(1) SUNLIGHT (X_S), where the corresponding linguistic statements are SUNLIGHT VERY SMALL (SVS), SUNLIGHT SMALL (SS), SUNLIGHT MEDIUM (SM), SUNLIGHT BIG (SB), SUNLIGHT VERY BIG (SVB); (2) VOICE (X_V), where the corresponding linguistic statements are VOICE VERY SMALL (VVS), VOICE SMALL (VS), VOICE MEDIUM (VM), VOICE BIG (VB), VOICE VERY BIG (VVB).

(2) Fuzzy main controller:

The input port of the growth stage main controller can be regarded as the output ports of the opening/closing fuzzy controller, meaning the output variables of the opening/closing fuzzy controller become the input variables of the fuzzy main controller. In OPEN TIME (X_O), the input variables include five fuzzy sets, each of which include: OPEN TIME VERY SMALL (OVS), OPEN TIME SMALL (OS), OPEN TIME MEDIUM (OM), OPEN TIME BIG (OB), OPEN TIME VERY BIG (OVB); in CLOSING TIME (X_C), there are also five fuzzy sets, each of which include: CLOSE TIME VERY SMALL (CVS), CLOSE TIME SMALL (CS), CLOSE TIME MEDIUM (CM), CLOSE TIME BIG (CB), CLOSE TIME VERY BIG (CVB).

B. Output variables:

(1) Opening/closing Fuzzy controller:

The output variables for the opening/closing fuzzy controller include opening time (Z_O) and closing time (Z_C). In opening time (Z_O), the five fuzzy sets are OVS, OS, OM, OB, and OVB; in closing time (Z_C), the five fuzzy sets are CVS, CS, CM, CB, and CVB.

(2) Fuzzy main-controller:

The fuzzy main controller acts as the decision-making system, which determines whether the next mode can be reached from the output of the opening/closing fuzzy sub-controller. The main controller has only one situational mode output variable (Z_K), and five output fuzzy sets: Mode 1 (M1), Mode 2 (M2), and Mode 3 (M3).

(3) Environmental factor fuzzy controller:

The output variables of the environmental fuzzy controller are sunlight (Z_S) and voice (Z_V), and their main purpose is to provide compensation for environmental factors. When too much input is provided, supply is reduced, and when input supply falls short, more is provided. Each of the output variables has five fuzzy sets, and the design is the same as that of the input variables.

The input variables, output variables, and fuzzy linguistic labels in the fuzzy sub-controller are organized in Tables 1 and 2, respectively.

Table 1. Input values and linguistic labels.

X_S		X_V	
(as, bs, cs)	Linguistic Labels	(av, bv, cv)	Linguistic Labels
(0, 30, 60)	SVS, SS, SM	(0,0.5,1)	VVS, VS, VM
(30, 60, 90)	SS, SM, SB	(0.5,1,1.5)	VS, VM, VB
(60, 90, 120)	SM, SB, SVB	(1,1.5, 2)	VM, VB, VVB

Table 2. Output values and linguistic labels.

Z_S		Z_V		Z_O		Z_C	
(a_S, b_S, c_S)	Linguistic Labels	(a_V, b_V, c_V)	Linguistic Labels	(a_O, b_O, c_O)	Linguistic Labels	(a_C, b_C, c_C)	Linguistic Labels
(0, 30, 60)	SVS, SS, SM	(0, 0.5, 1)	VVS, VS, VM	(2, 3, 4)	OVS, OS, OM	(2, 3, 4)	CVS, CS, CM
(30, 60, 90)	SS, SM, SB	(0.5, 1, 1.5)	VS, VM, VB	(3, 4, 5)	OS, OM, OB	(3, 4, 5)	CS, CM, CB
(60, 90, 120)	SM, SB, SVB	(1, 1.5, 2)	VM, VB, VVB	(4, 5, 6)	OM, OB, OVB	(4, 5, 6)	CM, CB, CVB

In Tables 1 and 2, a_δ, b_δ, and c_δ illustrate the physical values corresponding to the three vertexes of the triangular membership function, and δ denotes any of the various input/output variables. The design of the fuzzy membership function is described in the next section.

Step 2: determine fuzzification strategy

After defining the input and output variables, the variable values are fuzzified according to the fuzzy set theory. First, the true value range of the input and output values are converted into the fuzzy domain, where membership functions are selected for value fuzzification. Next, the membership degrees of the membership functions are obtained.

Regarding the selection types of membership functions, single tone, trapezoidal, and bell-shaped types are utilized. The triangular type is used as the membership function for the fuzzy sub-controller. The function of $\Lambda_{A_i}(x)$ is shown as follows:

$$\Lambda_{A_i}(x)=\begin{cases} 0, & x < u_i \\ \dfrac{x-u_i}{v_i-u_i}, & u_i \leq x \leq v_i \\ \dfrac{z_i-x}{z_i-v_i}, & v_i \leq x \leq z_i \\ 0, & x > z_i \end{cases} \tag{1}$$

where A_i is the ith fuzzy set, and the values of u_i, v_i, and z_i depend on the linguistic labels. Fuzzification is carried out, which is a crucial procedure where input data are converted into the fuzzy domain using the membership function. When a specific value is entered, it is calculated by the predesigned membership function, which uses an intersection operation to derive the degree of membership, and results in fuzzification. The defined items in the membership function are shown in Figure 5. Figure 6 illustrates the distributions of the input and output membership functions in the fuzzy main controller. It is noteworthy that the single tone type membership function is utilized in the fuzzy main controller.

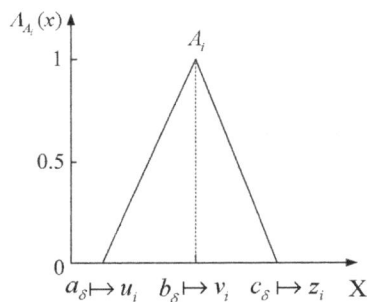

Figure 5. Membership function of fuzzification.

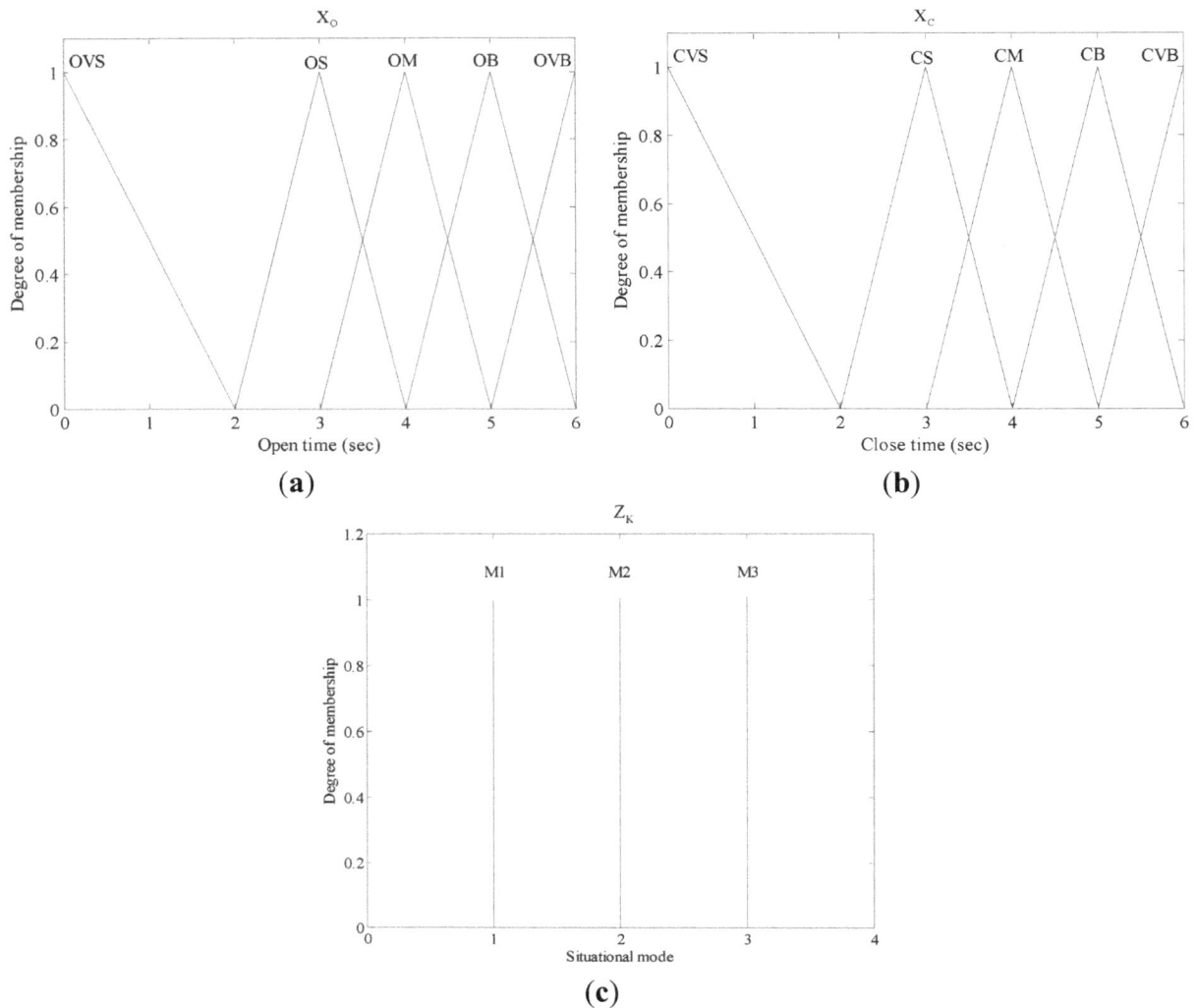

Figure 6. Input and output membership functions of fuzzy main controller: (**a**) membership function of open time fuzzification; (**b**) membership function of close time fuzzification; (**c**) membership function of situational mode controller output.

Step 3: design of fuzzy rule table

The inference rule sentence can be displayed by "*IF ... AND ... THEN ...*" and the example is described, as follows:

If "Sunlight" is "SVS" and "Voice" is "VVS", then "Open Time" is "OVS".

The system has three fuzzy controllers, and the rule base of the situational model controller is as shown in Table 3. The rules are relatively simple, and the situational mode can be judged by the opening/closing signals. The two remaining controls are the opening/closing fuzzy controller and the environmental factor fuzzy controller. The control rule base is as shown in Tables 4 and 5. The rules are defined by judging the changes in the variables of Sunlight and Voice.

Step 4: design fuzzy deduction method

This is the fuzzy inference and analysis decision-making core. In this study, the fuzzy inference method of the AND condition of the same inference rule can be computed using the Min-Dot method [17,18].

Step 5: select method of defuzzification

After inference by fuzzy logic, the output is a fuzzy set. The simplest approach is the bisector method, which first obtains the area of the region enclosed by the membership function curve of the fuzzy sets and vertical coordinates, and then identifies the vertical line that bisects the region into two portions of equal areas [17,18].

Table 3. Situational mode fuzzy rule table.

Z_M		X_O				
		OVS	OS	OM	OB	OVB
	CVS	M1	M1	M1	M1	M2
	CS	M1	M1	M2	M2	M2
Xc	CM	M2	M2	M2	M2	M3
	CB	M2	M2	M2	M3	M3
	CVB	M3	M3	M3	M3	M3

Table 4. Open/Close time fuzzy rule table.

Z_O, Z_C		X_S				
		SVS	SS	SM	SB	SVB
	VVS	OVS, CVB	OVS, CVB	OS, CB	OM, CB	OM, CB
	VS	OVS, CB	OS, CB	OS, CM	OM, CM	OM, CM
X_V	VM	OS, CB	OM, CM	OM, CM	OM, CM	OB, CS
	VB	OM, CM	OM, CM	OM, CS	OB, CS	OB, CVS
	VVB	OB, CM	OB, CM	OB, CS	OVB, CVS	OVB, CVS

Table 5. Rule table of environmental factor fuzzy controller.

Z_S, Z_V		X_s				
		SVS	SS	SM	SB	SVB
	VVS	SM, VB	SB, VM	SVB, VVB	SVB, VM	SVB, VS
	VS	SVS, VS	SS, VM	SB, VB	SVB, VVB	SVB, VVB
X_v	VM	SVB, VM	SB, VM	SM, VM	SS, VM	SVS, VM
	VB	SVS, VVS	SVS, VS	SS, VS	SM, VM	SB, VB
	VVB	SVS, VB	SVS, VM	SVS, VVS	SS, VM	SM, VS

2.3. Leaf and Vein Design

The mimosa leaf production process and vein-leaf integration design are illustrated as follows.

2.3.1. SMAs Processing and Memory

SMA is an alloy that can memorize a shape after thermal processing. When the temperature of the SMA is lower than the deformation temperature, after the plastic deformation of the Martensite Phase, it can restore the pre-deformation shape when the temperature is restored to the range of the Austentile Phase by heating. Such a phenomenon is known as the memory effect [19]. The recovery effect of SMA is related to the sectional area and length, then obtains additional SMAs of different size dimensions requires mask pull plates prior to starting the shape memory. Figure 7 illustrates the SMA processing and memory processes.

A. SMAs pull wire process

First, the wire was annealed to ensure better flexibility, which also increases the wire pull success rate. The wire stiffened slowly after being pulled several times, and it was repetitively annealed and pulled. The annealing flame should not be too hot in order to prevent the wire from melting due to incaution. The fire was quenched when the metal surface turns to white from red, and then the wire was left to cool naturally to room temperature. The wire was calcined by a spirit lamp after it was placed in a stainless steel disc, covered with clay. A stainless steel lid was placed over the disc. In addition, an infra-red thermograph was used to sense the annealing temperature. The wire continued to be calcined for 10 min at over 350 °C, then the lamp was quenched, in order to allow the wire to cool naturally (see Figure 4a).

Second, the punching and grinding operations were conducted. As the front end of the annealed memory wire was threaded through the wire hole, its 2–3 cm long portion was ground fine. The wire was gripped by the gripping head of the carving drill. The drill started to drive the gripping head, and it was held by the wire to rotate. These two files were used to approach the rotating wire in order to grind it fine.

Third, the lubricating operation was performed. Lubricating oil or butter was smeared onto both the wire pulling hole and the wire. The purpose of this step is mainly to reduce pulling friction, increase the pulling success rate, and reduce the probability of alloy cable rupture.

In the final step of wire pulling, the SMA cable diameter was 0.5 mm, and the ideal cable diameter was 0.3 mm for implantation inside a PDMS leaf. The alloy cable was pulled from the larger holes to the smaller holes in order to gradually obtain a thinner alloy cable [20].

B. SMA deformation process

A high temperature heating furnace is commonly used for anneal forming of SMA. The crystal lattices of SMA are rearranged and recombined in the presence of the high temperature of the furnace in order that the SMA can memorize the shape formed at annealing. However, a high temperature heating furnace is very expensive, thus, in this operation, a spirit lamp and clay were used to create the shape forming template of SMA. During calcination, the SMA wire was secured in the clay template, where it underwent even heating and annealing formation (see Figure 7). As the clay template should be made red-hot to provide even heating, the process requires a long time.

This study replaced the clay template heating method with direct heating of the alloy, as this method can shorten heating time. First, SMA was wound on high temperature-resistant objects according to the desired bending degree or method, such as a circular iron bar or copper column. The bending shape was then fixed before placing the object directly above an alcohol lamp for 5–10 min of heating. Since the temperature of alcohol lamp heating was 400–500 °C, it could change the SMA phase to realize the function of memory (see Figure 8).

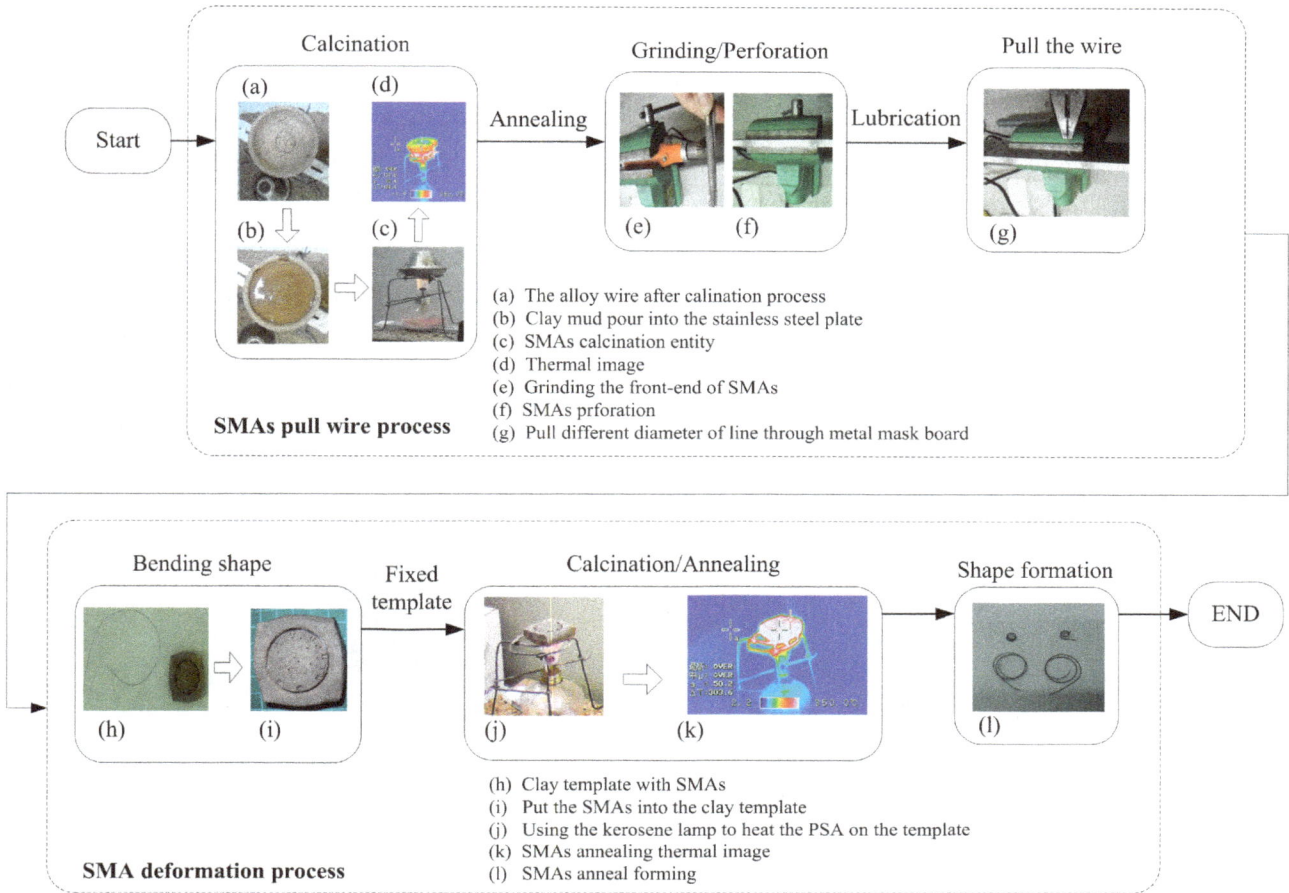

Figure 7. SMA processing and memory process.

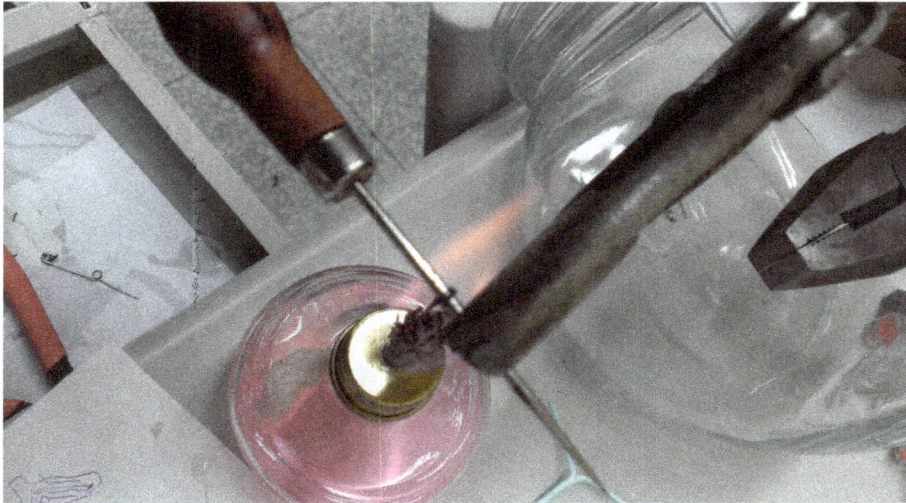

Figure 8. Direct heating of the SMAs.

2.3.2. Leaf and Vein Design and Implementation

This study chose Polydimethylsiloxane (PDMS) as the leaf material [21,22], as cured PDMS can become an elastomer with excellent toughness, high insulation, and water repellency. The major veins are realized by SMAs and wire. The connection points are not realized by soldering, but by the DuPont crimping method, which can avoid breaking the connection point due to solder falling while in the high temperature state, thus, causing bending of unheated SMAs. Next, the crimped wires and SMAs are input into the template. At this step, the memory side of SMAs should be particularly noted, and this study addressed fixing the memory side of the template according to the designed directions of bending SMAs. Afterwards, the prepared PDMS was poured into the template containing the wire and SMAs. In preparation of PDMS, the ratio of PDMS and hardener should be particularly noted. In this study, the ratio was 2.5% (temperature 28 °C), the operation time was 25 min, and the hardening time was 5 h. In the process of preparation and mixing, bubbles were produced inside the PDMS material, which required a vacuum machine for removal in order to avoid structural properties changes caused by any remaining bubbles inside the PDMS during hardening and shaping. Finally, the prepared PDMS was poured into the template to 80% of the volume (see Figure 9a), and placed at room temperature, or in an oven to accelerate curing.

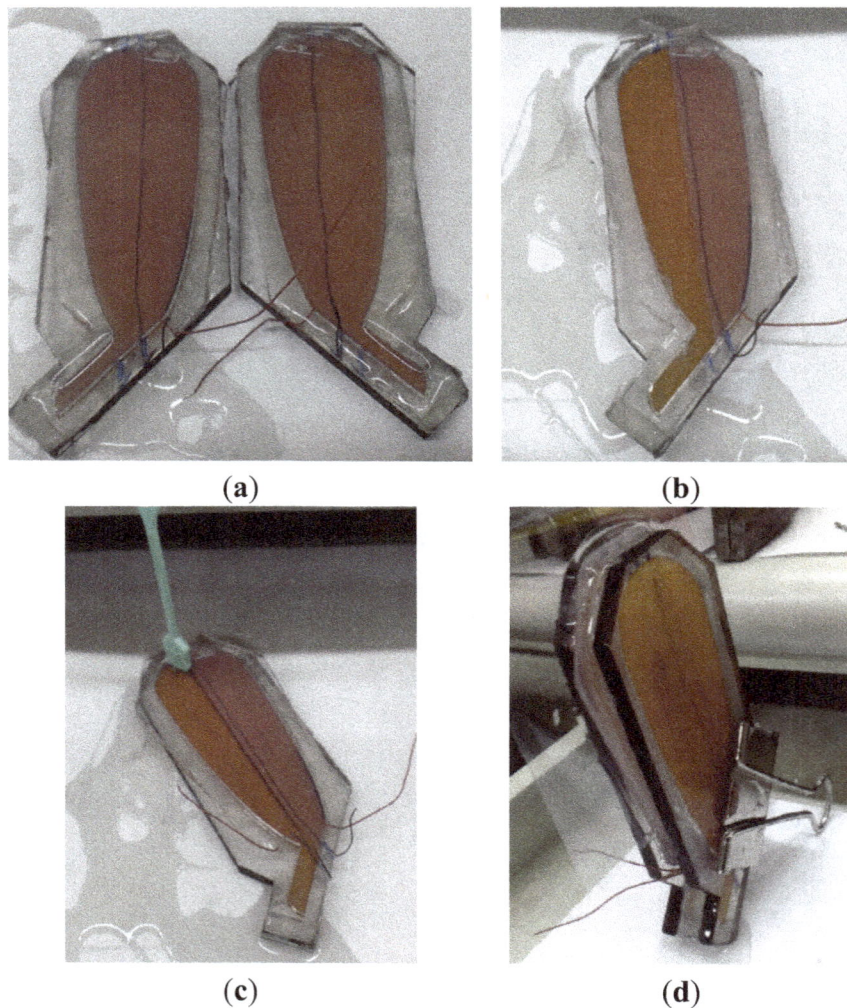

(a) (b)

(c) (d)

Figure 9. Leaf production processes: (**a**) PDMS infusion mode of 80% volume being filled for hardening; (**b**) remove other side of PDMS at the center; (**c**) implant SMA in the position of removal; (**d**) symmetric binding after the second infusion.

The leaf should contain two SMAs of opposite memory sides in order to display the opening and closing directions. However, when making the leaf, if two SMAs were placed in the template before pouring the PDMS, it would be difficult to ensure that the positions of the SMAs are according to the desired directions. Therefore, this study placed an SMA at the center of the vein, and developed a leaf up to 80% of the capacity. After curing and hardening the PDMS, half of the PDMS leaf without the SMA was removed from the center of the template vein (see Figure 9b). Next, another SMA was placed (Figure 9c) to recreate the second half. The finished product was a leaf with dual SMAs of up to 80% in volume. At the early stage, the designed leaf was of single plane production, meaning that the implantation of SMA was placed in the single plane PDMS mode in order to realize the opening/closing of the leaf. However, the opening and closing behaviors were not ideal subjects for plane limitations. Under such a condition, this study applied a different method of creating a mode to fill the void space remaining in the above-mentioned leaf of 80% volume of PDMS, which is the second infusion. Next, the two leaves were bound. Prior to binding, transparent film or slides were used to separate the bonding and non-bonding parts of the leaves at the appropriate positions for molding (see Figure 9d). Figure 10a illustrates completed leaves.

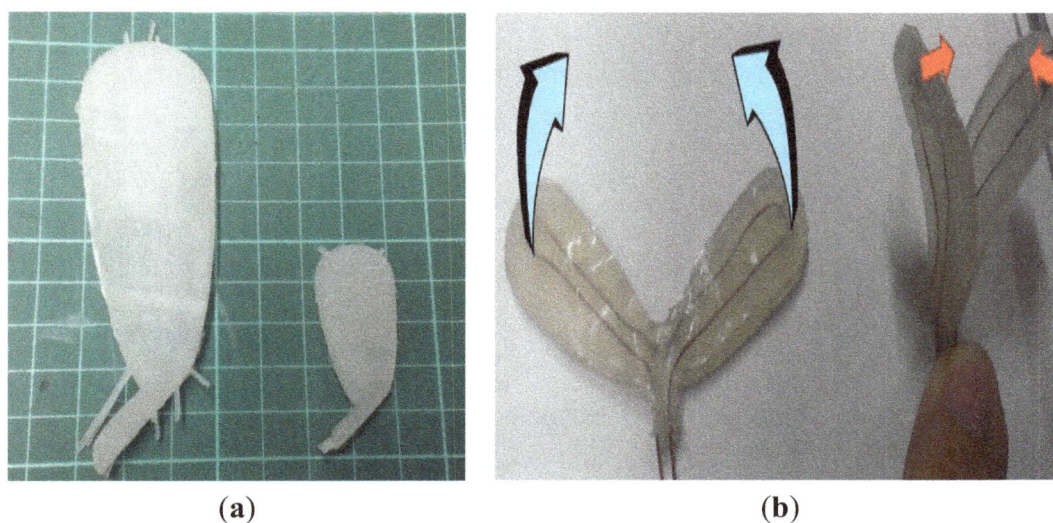

Figure 10. Appearance of completed mimosa leaves: (**a**) completed leaves; (**b**) 2-dimension leaf (**left**), and 3-dimension leaf (**right**).

The leaf production of this study differed from the previous methods of leaf production. In the past, leaf production achieved the two-dimensional state by implanting the SMA in a single-sided PDMS mold in order to realize the opening/closing behaviors of the leaf. However, these behaviors were not desirable due to the limitations of the plane. In such a situation, this study applied another method of creating a mode to fill the void space remaining in the above-mentioned leaf of 80% volume by PDMS, which is the second infusion. Next, the two leaves were bound. Prior to binding, transparent film or slides were used to separate the bonding and non-bonding parts of the leaves at the appropriate positions for molding. Figure 10b illustrates the differences of a two-dimensional leaf and a three-dimensional leaf during opening/closing. As seen, the bending degree of SMAs was greater in the case of the two-dimensional model, thus, the opening/closing actions would be more difficult.

The mimosa branch has numerous pairs of leaves, and the stem should have certain support to facilitate the implantation of a mimosa leaf. Therefore, this study used a plastic cylindrical case to wrap the pairs of leaves. Figure 11 illustrates the prototype of the entire strain of the mimosa.

(a)   (b)

Figure 11. Appearance of strain prototype: (**a**) a single leaf implanted in a plastic cylindrical case; (**b**) numerous pairs of leaves implanted in the prototype.

2.4. Control System Implementation

This section introduces the hardware circuit design, peripheral modules, and software program processes.

2.4.1. Hardware Circuit Design

The bionic mimosa robot hardware devices included a major control module and a sensing and driving module (see Figure 12). The sensing and actuation module consisted of a power stabilization module, Sunlight sensing module, Voice sensing module, and SMA actuation module. The circuit design and functions of the modules are illustrated as follows.

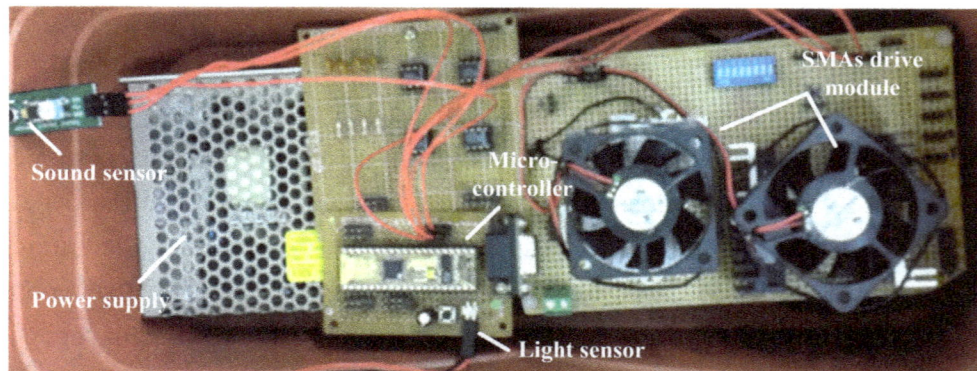

Figure 12. Embedded control board.

A. Major control module

In this study, the control core of the system is a micro-controlling development board. The controller is produced according to Parallax Inc. (Rocklin, CA, USA) [23], and consists of various components, including a micro-controller, electrically-erasable programmable read-only memory (EEPROM) memory chip, serial transmission interface, and power stabilization chip. The system has a built-in BASIC programming language interpreter known as PBASIC, which can provide users with functions for development.

B. USB communication module

The main function of the module is to enable two-way serial communication between the BS240p development board and a PC without a serial communication card.

C. Sunlight sensing module

The Sunlight sensing module consists of a photosensitive resistor, 555 chip vibrator, and a capacity device, which can receive the output values sent by the sensing module through the PULSIN pin of the BS240p development board.

D. Voice sensing module

The Voice sensing module is one of the environmental factors of the proposed system, and is a product of Parallax. The advantage of the sensor is easy operation, which enables it to connect with the BS240p development board.

E. SMA actuation module

The SMA actuation module mainly uses a solid state relay (SSR), a switch, and direct current (DC) power supply to drive the SMAs.

F. Power module

The output voltage of the power supply module can be divided into 5 V, 9 V, and 12 V. The 5 V power supply is mainly for the sensing modules, including BS2, Sunlight sensing, Voice sensing, and SSR relay. In addition, the BS240p development board can provide a power supply of 5 V and 1 A; therefore, it can provide the working voltage for the above modules and electronic devices. The 9 V power supply is mainly for the thermal conduction of the SMA. The power supply module uses MC7809 integrated circuit (IC) to provide a 9 V and 4 A power supply; while the 12 V power supply is for the SMA cooling fan, LED, and other accessory devices.

2.4.2. Software Programming Design

The software programming interface is developed by BASIC Stamp Editor v. 2.5, and consists of 70 groups of simple instruction sets and programming interpreters. The tool set is used for writing the Fuzzy Knowledge Base (FKB) for the mimosa opening and control program in order to realize a fuzzy control system. Figure 13 illustrates the main program control flow. After the program resets the system power supply and BASIC Stamp micro-controller, the micro-controller executes system parameter initialization and sets relevant peripheral parameters. The initial parameter initialization settings include the initial Sunlight and Voice values. After initialization is complete, the PC terminal will display a complete message, and enter into the waiting mode for the user to input the relevant control commands, which determine and implement the commands.

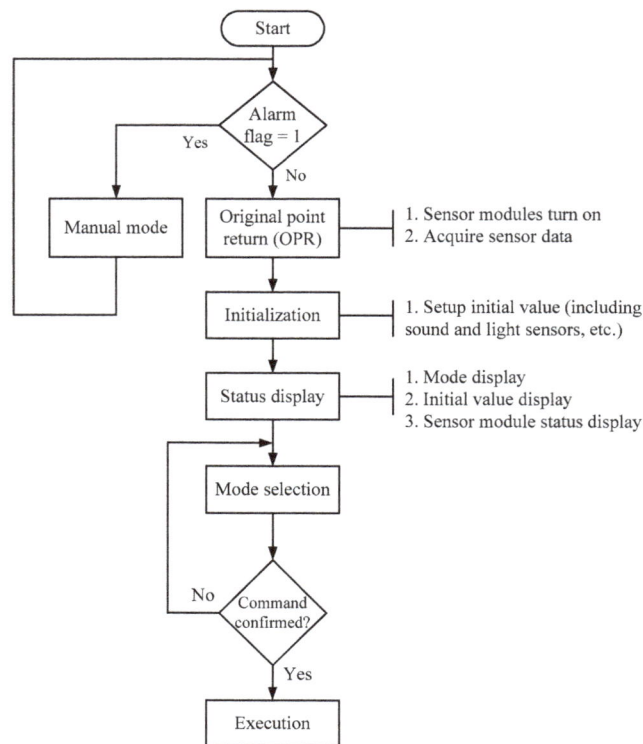

Figure 13. Main program control process.

3. Results and Discussion

3.1. Behavior Mode Simulation and Verification

To analyze the closing/opening expected values and situational modes for different environmental factors (Sunlight and Voice), this study used MATLAB/Simulink version 2010a for simulation and verification of the fuzzy control system. The simulation results are as shown in Figure 14. As seen, opening/closing times are different in the cases of different modes. Mode 1 illustrates leaf opening when the environmental factor stimulation is significant; Mode 2 illustrates the changes and swing of the leaves in the case of environmental factors only; Mode 3 illustrates the unapparent closing and opening of leaves when the external environment stimulation is insignificant.

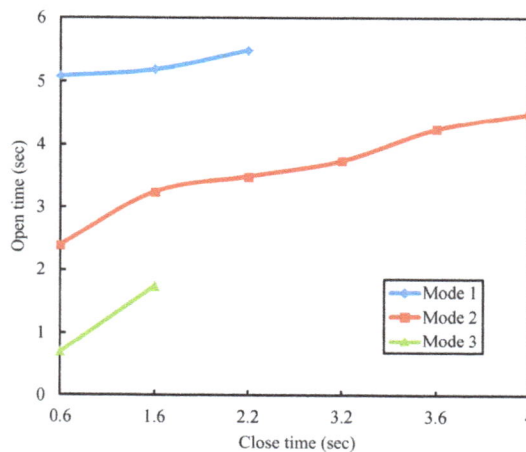

Figure 14. Opening and closing times and situational mode simulation results.

3.2. System Test

This section illustrates single-pair leaf swing, plant strain swing, and bionic mimosa situational behavior swing experimental results, as well as comparative analysis of leaf opening/closing positions.

3.2.1. Single-Pair Leaf Swinging

This experiment observed the coordinated displacement of two leaves over time for opening/closing the leaves, where the leaf top is the point of measurement. Figures 15 and 16 illustrate the single-pair leaf opening/closing experiment curves (actuation time and displacement value). As shown in Figure 16, the horizontal displacement of single-pair leaves is relatively consistent; while two leaves have a difference of 0.3 cm displacement. Regarding the closing experiment, as seen in Figure 16, for the same displacement shown in Figure 15, the horizontal and vertical Close Times are shorter than expected.

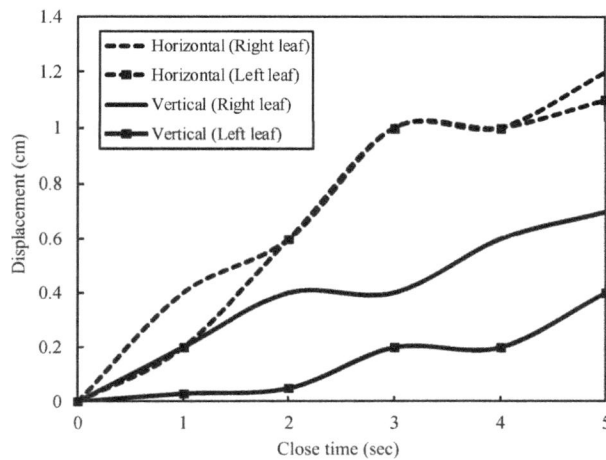

Figure 15. Single-pair leaf opening experiment curve (open time *vs.* displacement value).

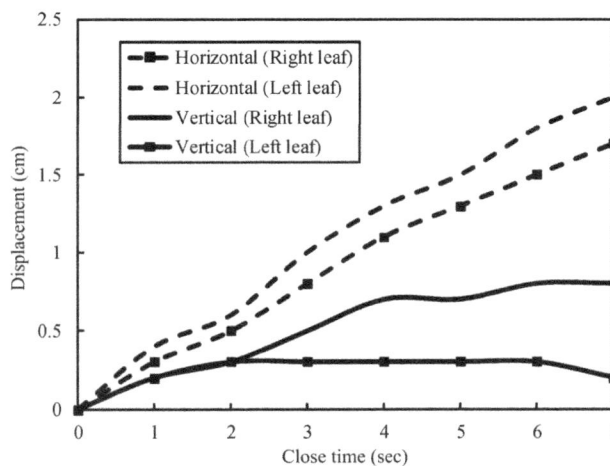

Figure 16. Single-pair leaf closing experiment curve (close time *vs.* displacement value).

3.2.2. Bionic Mimosa Situational Behavior Swinging

This section illustrates the situational behaviors of a bionic mimosa, including Mode 1, Mode 2, and Mode 3. The three situational behaviors are the opening/closing behaviors after introduction of fuzzy control, and according to environmental factors. Figure 17a–c illustrate the shapes of plant strains in the case of Modes 1, 2, and 3, respectively.

According to the values of different environments or adjustments of human factors, the desired closing/opening types are designed. Figure 17d shows the self-designed display.

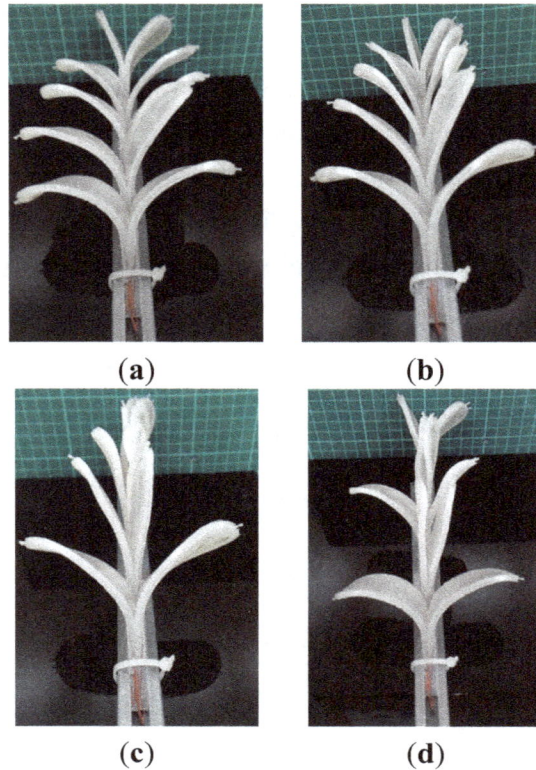

Figure 17. Mimosa system display Modes: (**a**) Mode 1; (**b**) Mode 2; (**c**) Mode 3; and (**d**) self-defined Mode.

3.3. Discussion

In this section, the leaf delicacy process, fuzzy control performance, and strain swing behaviors are analyzed and discussed.

3.3.1. Leaf Delicacy Process

The leaf designed in this study is a completely artificial leaf; thus, due to various factors, such as leaf area, characteristics of materials used, and bending degree of SMA subject to heat, a leaf with the ability to completely close cannot be obtained in the creation process (the statistics shows that the yield rate is about 85%). The cause for such failure is that two SMAs share the same memory direction. In other words, the leaf can swing only in one direction, thus, the arrangement and adjustment of the memory direction of SMA is an area for improvement in future micro-processing approach studies.

3.3.2. Fuzzy Control Performance

The fuzzy logical decision-making rule implemented in this study is derived from several experiments, through which the mimosa can show three swing behaviors. The design can reduce the number of fuzzy subsets and rule base, which decreases the complexity of fuzzy calculation and allow it to be more easily realized in the micro-controller. However, if more mode changes (only three modes now) are required, the controller number should be adjusted, and the psychological mode of the mimosa should be involved.

3.3.3. Strain Swing Behaviors

A total three swing modes are designed for the leaf. The swing position and direction of these three modes are preset and finalized after several adjustments. Our future study will construct a mathematical model based on the psychological mathematical model of a mimosa and the non-linear thermodynamic model of the SMA. More strains and environmental sensors will be added, and movable joints will be designed on leaf trunks and branches. In this way, the mimosa robot will be more lifelike and have better interaction.

4. Conclusions

This study developed SMA as the micro-actuation component for application in a mimosa vein. Regarding the processing and production of SMAs, through wire masking and annealing technology, this study realized the ideal wire diameter of the SMA, where SMA heating was realized by the direct heating memory method. The connection of the wire and SMA was realized by a DuPont joint that replaces the traditional soldering technique. Using two SMAs for different actuation directions can directly result in two-way bending of leaves, and the three-dimensional design of the leaf is more consistent with the actual shape of the mimosa.

Regarding the implementation of a control system, this study established a bionic mimosa situational mode, and determined the amount of environmental factors through fuzzy logic procedures. Finally, plant opening/closing situations were controlled by defuzzification. The proposed method can be applied in analysis of diverse systems and mode establishment. The entire experimental process was demonstrated by three modes, where each pair of leaves displayed different closing/opening postures according to different time durations for opening and closing. Users may choose manual or automatic modes according to different initial environmental settings in order to design different opening/closing display postures. The proposed method can be applied in micro-bionic robot systems, entertainment robots, biomedical engineering, and building aesthetics-related fields.

Acknowledgments

The authors appreciate the comments of the editor and reviewers. This study was financially sponsored by the Ministry of Science and Technology under Grant No. MOST 103-2221-E-020-041.

Author Contributions

The original concept and methods in research were provided by Chung-Liang Chang; he also wrote the paper, designed of experiments and supervised the work at all stages. The model simulation, measurements, and data analysis were performed by Jin-Long Shie. Both authors have read and approved the final manuscript.

Conflict of Interests

The authors declare no conflict of interests.

References

1. Steele, J.E. How do we get there. In *Bionics Symposium: Living Prototypes—The Key to New Technology*; Gray, C.H., Ed.; Routledge: New York, NY, USA, 1995; pp. 55–60.

2. Abbott, J.; Nagy, Z.; Beyeler, F.; Nelson, B. Robotics in the small. *IEEE Robot. Autom. Mag.* **2007**, *14*, 92–103.

3. Firebaugh, S.; Piepmeier, J.; Leckie, E.; Burkhardt, J. Jitterbot: A mobile millirobot using vibration actuation. *Micromachines* **2011**, *2*, 295–305.

4. Bar-Cohen, Y. *Electroactive Polymer (EAP) Actuators as Artificial Muscles: Reality, Potential, and Challenges*, 2nd ed.; SPIE Press: Washington, DC, USA, 2004.

5. Nishida, G.; Sugiura, M.; Yamakita, M.; Maschke, B.; Ikeura, R. Multi-input multi-output integrated ionic polymer-metal composite for energy controls. *Micromachines* **2012**, *3*, 126–136.

6. Callister, W.D., Jr. *Materials Science and Engineering: An Introduction*, 8th ed.; John Wiley and Sons: Hoboken, NJ, USA, 2009.

7. Jani, J.M.; Leary, M.; Subic, A.; Gibson, M.A. A review of shape memory alloy research, applications and opportunities. *Mater. Des.* **2014**, *56*, 1078–1113.

8. Son, H.M.; Lee, Y.J. New variable focal liquid lens system using antagonistic-type SMA actuator. In Proceedings of the 4th International Conference on Autonomous Robot and Agents, Wellinfton, New Zealand, 10–12 February 2009.

9. Makishi, W.; Matunaga, T.; Haga, Y.; Esashi, M. Active bending electric endoscope using shape memory alloy coil actuator. In Proceedings of the First IEEE/RAS-EMBS International Conference on Biomedical Robotics and Biomechatronics, Pisa, Italy, 20–22 February 2006; pp. 217–219.

10. Lee, Y.P.; Kim, B.; Lee, M.G.; Park, J.O. Locomotive mechanism design and fabrication of biomimetic micro robot using shape memory alloy. In Proceedings of the IEEE International Conference on Robotics and Automation, Seoul, Korea, 26 April–1 May 2004.

11. Liu, C.Y.; Liao, W.H. A snake robot using shape memory alloys. In Proceedings of the 2004 IEEE International Conference on Robotics and Biomimetics, Shenyang, China, 22–26 August 2004.

12. Zhang, Y.; Li, S.; Ma, J.; Yang, J. Development of an Underwater Oscillatory Propulsion System Using Shape Memory Alloy. In Proceedings of the IEEE International Conference on Mechatronics & Automation, Niagara Falls, NY, USA, 29 July–1 August 2005; pp. 1878–1883.

13. Liu, Y.L.; Chung, A.; Hu, J.; Lv, J. Shape memory behavior of SMPU knitted fabric. *J. Zhejiang Univ. Sci. A* **2007**, *8*, 830–834.

14. Mingallon, M.; Ramaswamy, S. Bio-inspired self-actuating composite materials. In *Composites and Their Applications*; Ning, H., Ed.; InTech: Rijeka, Croatia, 2012.

15. Huang, H.L.; Park, S.H.; Park, J.O.; Yun, C.H. Development of Stem Structure for Flower Robot using SMA Actuators. In Proceedings of the IEEE International Conference on Robotics and Biomimetics, Sanya, China, 15–18 December 2007; pp. 1580–1585.

16. Chang, C.L.; Sie, M.F.; Shie, J.L. Development of a multisensor-based bio-botanic robot and its implementation using a self-designed embedded board. *Sensors* **2011**, *11*, 11629–11648.

17. Lee, C.C. Fuzzy logic in control system: Fuzzy logic controller. I. *IEEE Trans. Syst. Man Cybern.* **1990**, *20*, 404–418.

18. Lee, C.C. Fuzzy logic in control system: Fuzzy logic controller. II. *IEEE Trans. Syst. Man Cybern.* **1990**, *20*, 419–435.

19. Otsuka, K.; Wayman, C.M. *Shape Memory Materials*; Cambridge University Press: Cambridge, UK, 1998.

20. Chang, C.L.; Shie, J.L. Design and implementation of actuator for the swing mechanism of interactive bio-mimosa robot. *J. Chin. Soc. Mech. Eng.* **2013**, *34*, 137–142.

21. Lotters, J.C.; Olthuis, W.; Veltink, P.H.; Bergveld, P. The mechanical properties of the rubber elastic polymer polydimethylsiloxane for sensor applications. *J. Micromech. Microeng.* **1997**, *7*, 145–147.

22. McDonald, J.C.; Duffy, D.C.; Anderson, J.R.; Chiu, D.T.; Wu, H.; Schueller, O.J.A.; Whitesides, G.M. Fabrication of microfluidic devices using poly(dimethylsiloxane). *Electrophoresis* **2000**, *21*, 27–40.

23. Parallax Inc. BASIC Stamp. Available online: http://www.parallax.com/microcontrollers/basic-stamp (access on 5 June 2014).

4

High Resolution Cell Positioning Based on a Flow Reduction Mechanism for Enhancing Deformability Mapping

Shinya Sakuma [1,†,*], Keisuke Kuroda [1], Fumihito Arai [2], Tatsunori Taniguchi [3], Tomohito Ohtani [3], Yasushi Sakata [3] and Makoto Kaneko [1]

[1] Department of Mechanical Engineering, Osaka University, 2-1 Yamadaoka, Suita, Osaka 565-0871, Japan; E-Mails: kuroda@hh.mech.eng.osaka-u.ac.jp (K.K.); mk@mech.eng.osaka-u.ac.jp (M.K.)

[2] Department of Micro-Nano Systems Engineering, Nagoya University, Furo-Cho, Chikusa-Ku, Nagoya 464-8601, Japan; E-Mail: arai@mech.nagoya-u.ac.jp

[3] Department of Cardiovascular Medicine, Osaka University, 2-1 Yamadaoka, Suita, Osaka 565-0871, Japan; E-Mails: t-taniguchi@outlook.com (T.T.); ohtani@cardiology.med.osaka-u.ac.jp (T.O.); yasushisk@cardiology.med.osaka-u.ac.jp (Y.S.)

† Present Address: Department of Micro-Nano Systems Engineering, Nagoya University, Furo-cho, Chikusa-Ku, Nagoya, Aichi 464-8603, Japan.

* Author to whom correspondence should be addressed; E-Mail: sakuma@mech.nagoya-u.ac.jp

External Editor: Jeong-Bong Lee

Abstract: The dispersion of cell deformability mapping is affected not only by the resolution of the sensing system, but also by cell deformability itself. In order to extract the pure deformability characteristics of cells, it is necessary to improve the resolution of cell actuation in the sensing system, particularly in the case of active sensing, where an actuator is essential. This paper proposes a novel concept, a "flow reduction mechanism", where a flow is generated by a macroactuator placed outside of a microfluidic chip. The flow can be drastically reduced at the cell manipulation point in a microchannel due to the elasticity embedded into the fluid circuit of the microfluidic system. The great advantage of this approach is that we can easily construct a high resolution cell manipulation system by combining a macro-scale actuator and a macro-scale position sensor, even though the resolution of the actuator is larger than the desired resolution for cell manipulation. Focusing on this characteristic, we successfully achieved the cell positioning based on a visual feedback control with a resolution of 240 nm, corresponding to one pixel of the vision system. We show that the utilization of this positioning system contributes to reducing the dispersion coming from the positioning resolution in the cell deformability mapping.

Keywords: flow reduction mechanism, cell positioning; high resolution; deformability; red blood cell

---

1. Introduction

There have been a number of works measuring the cell mechanical characteristics, such as cell mechanical impedance [1–3], cell deformability [4–7] and cell property under reciprocated mechanical stress [8,9], by utilizing a microfluidic chip whose cross-sectional area is close to that of the minimum size of the human blood pipe. The measurement of cell mechanical impedance or cell deformability needs an appropriate actuator for manipulating the cell, as shown in Figure 1a. The dispersion of the obtained data is influenced by both the resolution of the actuator and the property of the cell, as shown in Figure 1b,c. Figure 1b,c shows the heat map on the frequency of the data, where the blue and red color show the low frequency and high frequency, respectively. Suppose cellular properties are the same for both Figures 1b,c. Furthermore, suppose an extreme case where the system holds infinitely small resolution and no sensing error. In such a case, the measured property exactly coincides with that of cellular one, as shown in Figure 1b. In a general case, however, the system has a limitation of resolution and includes some sensing errors. Eventually, the measured property has a wide dispersion depending on the one coming from the sensing system, as shown in Figure 1c. Therefore, in order to obtain the pure dispersion characteristics of the cell group, it is necessary to increase the resolution of cell manipulation. This is the reason why we focus on high resolution cell positioning in this paper.

Under these circumstances, the goal of this work is to provide a cell positioning system with an extremely high resolution, such as less than 250 nm. There are two approaches for fluid-based cell actuation in a microfluidic chip. One is the internal actuation, where the actuator component for producing a propelling force is installed within the microfluidic chip, and the other is the external actuation, when such an actuator component is placed outside of the microfluidic chip [10–13]. An electrically-driven micropump installed inside of a microfluidic chip is a good example of the internal actuation [13]. Since the resolution of the actuator directly affects the resolution for the cell positioning, the internal actuation has an advantage for achieving a high resolution. However, this approach usually requires a complicated actuation mechanism due to the necessity of mechanical and/or electrical components in the chip, and results in a lack of robustness of the actuation system. Furthermore, when the microchannel is fully blocked by the cell, we are obliged to throw away the microfluidic chip together with the micropump. On the other hand, a syringe pump [14–17], a peristaltic pump [18] and a pneumatic pump [19] are good examples of external actuation. These external pumps enable us to achieve the simple set up, where they are connected to the microchip through a hard tube. This mechanical configuration provides a big advantage for disposable applications using microfluidic devices. Among them, the syringe pump has a certain advantage due to its quick response time based on the powerful actuation. For these reasons, this work focuses on using a syringe pump as the actuator, as shown in Figure 1a. The cell positioning system is composed of a vision sensor, a syringe pump and a microfluidic chip. Although the system has a simple configuration, there is an issue where the cell velocity is geometrically amplified in the microchannel. For example, suppose that we combine a syringe pump with the cross-sectional area of A_1 with a diameter of 10 mm and connected to the microchannel with a cross-sectional area of A_2 with a diameter of 10 μm. In this case, the area ratio $R_{Incom} = A_1/A_2 \simeq 10^6$, which results in increasing the velocity in the microchannel with a huge amplification ratio. Even if the syringe pump has a high resolution, such as $\delta_1 = 10$ nm, the equivalent resolution $\delta_2 = R_{Incom}\delta_1$ in the michrochannel becomes in the order of 1 mm, which is far from ensuring a high resolution in the order of 100 nm. Therefore, an appropriate flow reduction mechanism is very much required for the cell positioning with high resolution.

Figure 1. Cell positioning system. (**a**) The overview of the system and an illustration of cell movement in a microchannel and the contribution of system dispersion to the deformability map by using (**b**) a high resolution system and (**c**) a low resolution system.

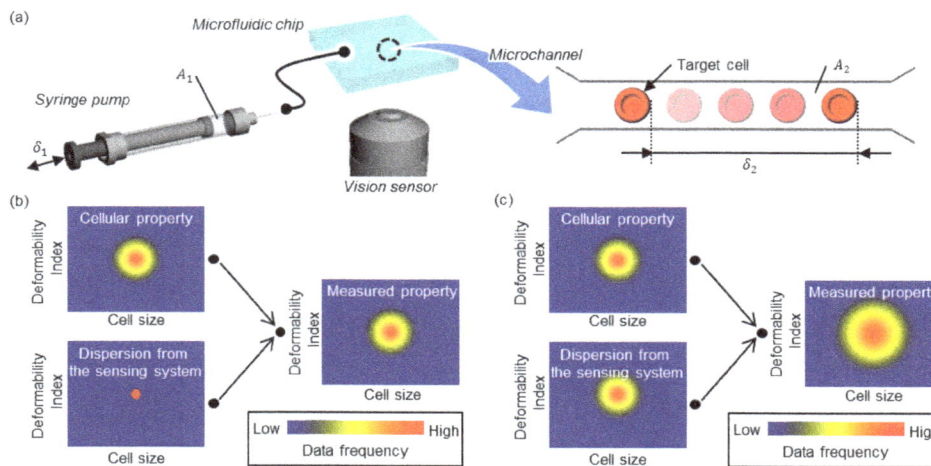

To overcome this serious issue of the geometrically amplification of the flow in a microchannel, we propose to utilize the flow reduction mechanism embedded in the microfluidic system itself. Generally, a microfluidic chip made of polydimethylsiloxane (PDMS) has elasticity and produces a volumetric change of the microfluidic system when the pressure in the microchannel increases stepwise. While the volumetric change is usually considered as an issue for cell position control, we can regard it as the function of the flow reduction embedded in the microfluidic system. This characteristic contributes to drastically improving the positioning resolution of the cell at the manipulation point. To confirm this characteristic, we constructed a full experimental system equipped with online high speed vision and succeeded in cell position control with a resolution of less than 250 nm by using visual feedback control. This is in the same order of the resolution of a standard optical microscope.

This paper is organized as follows. After explaining the elasticity of the microfluidic chip in the Introduction, we discuss how to achieve high resolution in the microfluidic channel in detail in Section 2. In Section 3, we explain the experimental system and the results for confirming the basic principle. In Section 4, we show an example of the deformability map of blood transfusion, where there are several groups, while there is only one before blood transfusion. We give also a discussion in Section 5 before concluding remarks in Section 6.

2. Flow Reduction Mechanism

2.1. Under Incompressibility of the Microfluidic Chip

While implementing mechatronics devices, we generally combine a motor and a reduction gear, so that we can increase the force (or torque) and position resolution (or angle resolution) at the output shaft. As mentioned in the Introduction, a syringe pump is directly connected to a microfluidic channel through a tube. This is a typical example of a velocity amplification mechanism, where the flow velocity at the piston of the syringe pump increases with $R_{Incom} = A_1/A_2$; because A_1 is usually even larger than A_2 and δ_2 is the order of 1 mm, as shown in the Introduction. Most of the conventional works utilizing a syringe pump together with a microfluidic channel are categorized into this group. However, so far, this velocity amplification mechanism has not been treated as a serious issue, since precise cell manipulation with high resolution has not been discussed precisely. Figure 2a shows the behavior of cell motion under the incompressibility of the microfluidic chip when the piston of the syringe is actuated with δ_1. The target cell is flowing in the microchannel without any time lag, and it maintains the position until the next command is given to the syringe pump. In this case, the displacement of cell δ_2 results in $\delta_2 = R_{Incom}\delta_1$ instantaneously, irrespective of each observation time t_1, t_2, t_3 and t_∞, as shown in Figure 2a. We note that $\delta_2 = R_{Incom}\delta_1$ occurs instantaneously.

Figure 2. Analytical model of the flow reduction mechanism and the target cell motion in the microchannel at each observation time t_1, t_2, t_3 and t_∞ when the syringe pump is actuated with a step input of δ_1. (**a**) Without elasticity and (**b**) with elasticity.

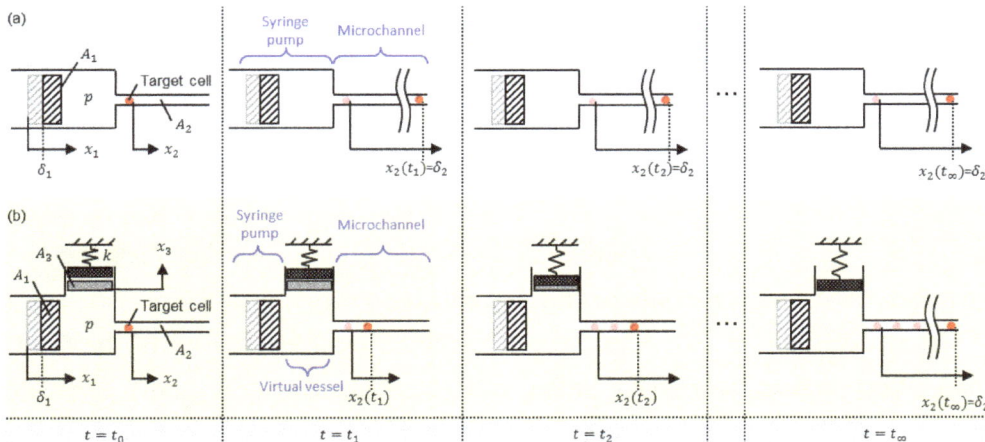

2.2. Under Compressibility of the Microfluidic Chip

Figure 2b shows a model of how the target cell moves when the piston of a syringe is actuated under a compressible vessel with a microchannel. For simplicity, we suppose that all other parts, except the vessel, are rigid, and therefore, there is no deformation from the pressure increase. In other words, all compressive elements are integrated in the virtual vessel in Figure 2b, where A_3 is the equivalent cross-sectional area of the elastic vessel. For further simplicity, we suppose that the mass of the piston is neglected. Now, suppose that a piston is pushed with a step input δ_1, and then the position is maintained. In the case of Figure 2a, the pressure in the vessel sharply increases and decreases with the interval of the liquid flow in the microchannel, which is the basic characteristic of the incompressible liquid/structure system. On the other hand, in the case of Figure 2b, the pressure in the vessel sharply increases with a step input, and the liquid in the microchannel starts to flow. Since there is a compressible element, the liquid in the microchannel is continuously flowing, and after the pressure increases, it gradually decreases with an appropriate time constant, which is determined by the stiffness, the damping of the virtual vessel and the flow in the microchannel. We note that A_3 is the cross-sectional area of the virtual vessel and is unknown. We can observe that the position of the target cell δ_2 gradually changes with respect to each observation time t_1 through t_∞, as shown in Figure 2b. The compressibility comes from both the tube and the PDMS microfluidic chip, and therefore, we cannot confirm it, until we build up the total system. However, the advantage of using the combination of a macroactuator and a microfluidic channel is that we can keep an extremely high area ratio $R_{Incom} = A_1/A_2 \simeq 10^6$, by which we can keep the pressure almost constant with respect to time, since the pressure drop due to the liquid flow is extremely small and negligible. This nature brings us two advantages, where one is to produce a flow reduction in the microchannel and the other is to contribute to an easy control scheme. This is what we call the "flow reduction mechanism" characteristic embedded in a microfluidic system.

2.3. Analysis of the Flow Reduction Mechanism

In order to provide a more general discussion, we consider a model, as shown in Figure 2b, where A_i, x_i, $\dot{x}_i$ ($i = 1, 2, 3$) and p are the i-th equivalent cross-sectional area, the displacement, the velocity and the pressure in a vessel, respectively. We note that A_3, x_3, $\dot{x}_3$ are unknown parameters of the virtual vessel, which is imaginarily integrated into all compressive elements of the fluid circuit. For incompressible fluid and an elastic vessel, we have the following Equations (1)–(3) as a function of time t.

$$A_1\dot{x}_1(t) = A_2\dot{x}_2(t) + A_3\dot{x}_3(t) \tag{1}$$

$$A_2 p(t) = c\dot{x}_2(t) \tag{2}$$

$$A_3 p(t) = kx_3(t) \tag{3}$$

where Equations (1)–(3) denote the continuity of liquid, the relationship between the pressure and the cell velocity and the force balance equation for the virtual plate, respectively, where k and c are the stiffness of the vessel and the constant parameter determined by the viscosity of the fluid, respectively. We note that the cell is supposed to move together with the fluid. By using Laplace transform under the condition of $x_1(0) = x_2(0) = x_3(0) = 0$, we have the following Equations (4)–(6).

$$A_1 X_1(s) = A_2 X_2(s) + A_3 X_3(s) \tag{4}$$

$$A_2 P(s) = csX_2(s) \tag{5}$$

$$A_3 P(s) = kX_3(s) \tag{6}$$

From Equations (4)–(6), we can obtain the following relationship between $X_1(s)$ and $X_2(s)$,

$$X_2(s) = \frac{kA_1A_2}{kA_2^2 + csA_3^2}X_1(s) \tag{7}$$

Now, suppose that a step input is given as in Equation (8).

$$X_1(s) = \frac{\delta_1}{s} \tag{8}$$

where δ_1 is the amplitude of the step input. By inputting Equation (8) into Equation (7), we can obtain the substituting Equations (9)–(11):

$$x_2(t) = R_{Incom}\delta_1 \left(1 - \exp\left(-\frac{t}{\tau}\right)\right) \tag{9}$$

$$\tau = \frac{1}{r_A}\frac{c}{k} \tag{10}$$

$$r_A = \frac{A_2^2}{A_3^2} \tag{11}$$

where r_A and τ are the design parameter of the microchannel and the time constant during the relaxation of the pressure due to the flow in the microchannel, respectively. Eventually, we can obtain the following Equation (12) about the flow rate in the microchannel $\dot{x}_2(t)$,

$$\dot{x}_2(t) = \frac{R_{Incom}\delta_1}{\tau}\exp\left(-\frac{t}{\tau}\right) \tag{12}$$

Now, we define the reduction ratio R_{Com} by Equation (13):

$$R_{Com} \overset{\text{def}}{=} \frac{\delta_2}{\delta_1} \tag{13}$$

By supposing that the initial velocity can be kept constant during the sampling interval Δt, the displacement of the cell together with the fluid is given by:

$$\delta_2 = \dot{x}_2(0)\Delta t = \frac{R_{Incom}\delta_1\Delta t}{\tau} \tag{14}$$

By the definition, R_{Con} can be rewritten by:

$$R_{Com} = \frac{\Delta t}{\tau}R_{Incom} \tag{15}$$

Equation (15) means that we can improve the positioning resolution of the target cell by utilizing the compressibility of the microfluidic system under the condition of $\Delta t/\tau < 1$.

3. Experiments

3.1. Experimental System

Figure 3 shows the experimental system, where it is composed of a syringe pump utilizing the piezoelectric actuator (MESS-TEK Co., Ltd. Saitama, Japan), online high speed vision (I-I-LAB. Co., Ltd. Hiroshima, Japan) and a microfluidic chip. The cross-sectional diameter of the piston of the syringe is 7.29 mm ($A_1 = 41.7$ mm^2). The piezoelectric actuator is directly connected to the piston through a bolt for the purpose of high speed response. Two tubes (PTFE, ID: 0.96 mm, OD: 1.56 mm, Chukoh Chemical Industries Ltd., Tokyo, Japan) are connected to the inlet and outlet of the microfluidic chip.

Cells are injected into the microchannel through the tube, and then, the position is detected by online high speed vision, where 1 pixel corresponds to 240 nm. The maximum sampling rate of the total system Δt is 1 ms.

Figure 3. The developed cell positioning system. (**a**) An overview of the system and (**b**) the developed syringe pump.

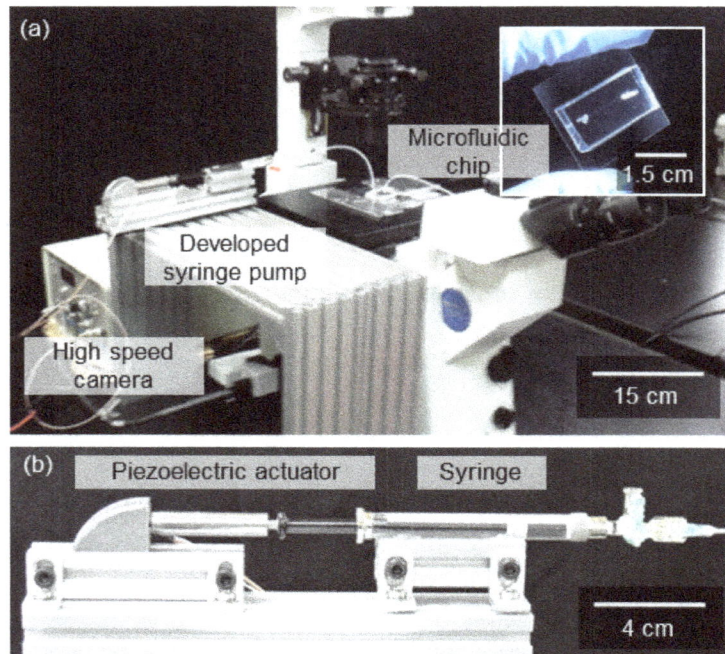

3.2. Fabrication Process of the Microfluidic Chip

The microfluidic chip is fabricated by standard soft-lithography using PDMS for a disposable application.

(i) Spin-coated SU-8 (KAYAKU, Co., Ltd. Tokyo, Japana) is patterned by laser lithography.

(ii) The fabricated pattern is transcribed to PDMS (SILPOT 184 W/C, DOW CORNING TORAY, Co., Ltd. Tokyo, Japan). The thickness of the molded PDMS is 5 mm.

(iii) The molded and punched PDMS is bonded with the glass substrate by heating after O_2 plasma treatment. The size of the inlet and outlet hole is 1.5 mm.

We use a red blood cell (RBC) as the target cell. By considering that the height and the diameter of an RBC typically range from 2 through 3 μm and 6 through 8 μm, respectively, the height and the width of the microchannel are designed as 3 μm and 10 μm, respectively.

3.3. Sample Preparation

To achieve single RBC evaluation from whole blood, we dilute blood with standard saline solution. RBCs were obtained from a volunteer subject who read and signed the donor consent for the evaluation. The blood was withdrawn by a licensed medical doctor. The procedures of the preparations are described as follows:

(i) The microchannel is filled with standard saline solution.

(ii) The blood is diluted by saline with a density of 2%.

(iii) The blood-saline mixture is injected into the microchannel from the sample inlet.

Equation (15) says that the reduction ratio under compressibility R_{Com} varies depending on $\Delta t / \tau$ and incompressibility R_{Incom}. We first estimate τ by giving the step input δ_1 to the pump. Figures 4 shows the measured position of RBCs in the microchannel with respect to time, where the step input is $\delta_1 = 5.0\,\mu$m. The blue, black and red line show the results of the cases using the tubes whose lengths are 10, 25 and 50 cm, respectively. From these results, we can evaluate τ by the elasticity of the fluid circuit. From Equation (15), we can see that R_{Com} is a function of R_{Incom}, Δt and τ, where R_{Incom} and Δt are given parameters, and τ is an unknown parameter.

Figure 4. The measured cell position when a step input of $\delta_1 = 5\,\mu$ m is given to the syringe piston.

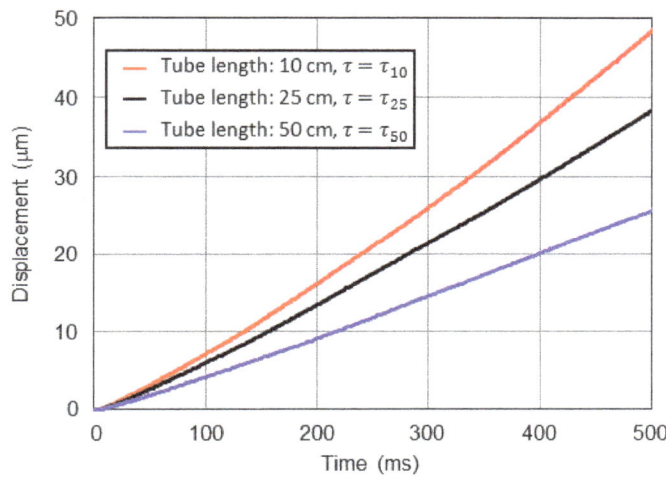

3.4. Without Visual Feedback Control

From Equation (14), we can obtain the following Equation (16),

$$\tau = \frac{R_{Incom}\delta_1 \Delta t}{\delta_2} \qquad (16)$$

By using Equation (16), we can evaluate τ without measuring unknown parameters A_3, c and k. For example, R_{Incom} and δ_1 are measured as 1.4×10^6 and $5\,\mu$m in the developed system, respectively. Supposing $\Delta t = 100$ ms, we can compute τ, and τ_{10}, τ_{25} and τ_{50} are calculated as 2.3×10^7, 2.8×10^7 and 3.9×10^7 ms, respectively, where the subscript of τ shows the length of the tube.

3.5. With Visual Feedback Control

In this section, we discuss the resolution of cell positioning under visual feedback control. In order to make the sampling interval Δt small, we utilize online high speed vision. Figure 5a shows a series of photos when the position of the RBC is controlled under the linear input of $10\,\mu$m/s. Figure 5b shows the response of position control under the given target position of the RBC with respect to time, where the each position command is given as the step input, a linear input of 10, 25 and 50 μm/s, respectively. The blue, red and green curves show the target position, the measured position and the error between them, respectively. The amplitude of each input is 208 pixels, which corresponds to 50 μm. The error of cell positioning is within $\pm$240 nm ($\pm$1 pixel) in an equilibrium state. From these results, we could confirm that the embedded flow reduction mechanism can drastically contribute to reducing the flow rate in the microchannel generated by a macro pump system placed outside of a microfluidic chip. Moreover, we can control the cell position with the extremely high resolution of $\pm$240 nm, corresponding to the limitation of the pixel of the online high speed vision sensor.

Figure 5. Demonstration of cell positioning. (**a**) An example of cell position control and (**b**) the measured cell position with respect to time.

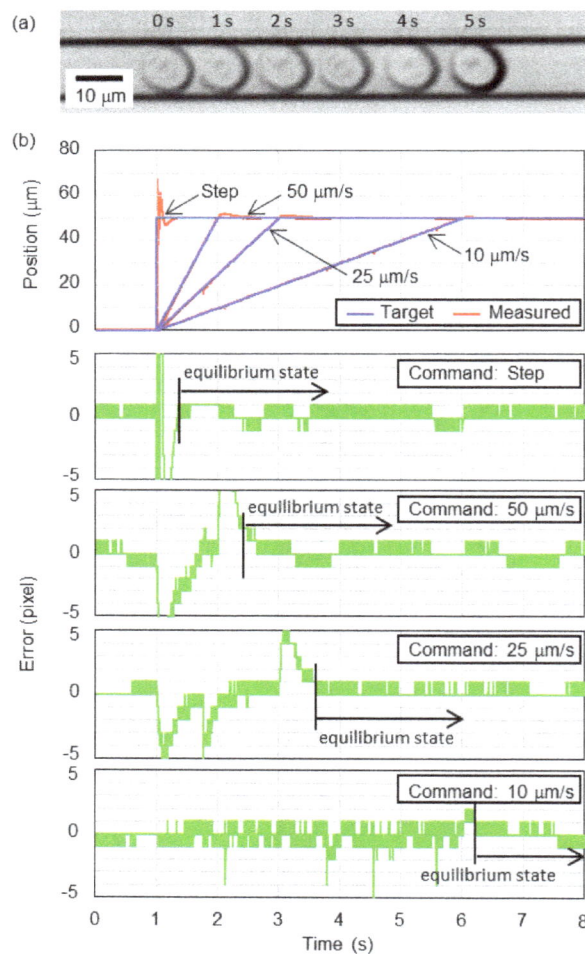

4. Application Example

4.1. Cell Extensibility

In this section, we show an application example of the developed actuation system. By changing the position command continuously, we can control the flowing velocity of the target cell. In this section, we evaluate the RBC's extensibility, which is an index of the deformability of the cell [9,20–22]. Figure 6a shows an example of the extensibility evaluation test of the RBC. The microchannel has a narrow throat, whose width and length are 3 μm and 25 μm, respectively, and the target RBC is compressed by the throat. We evaluate the extensibility index (*EI*) by Equation (17),

$$EI = \frac{L_v}{D_0} \tag{17}$$

where D_0 and L_v show the original diameter of the RBC and the length of the RBC in the narrow throat, respectively, and the subscript v shows the command velocity. The command velocities are given as 1, 5, 10, 50 and 100 μm/s, respectively. Figure 6b shows the relationship between *EI* and the cell velocity. The cell length increases with respect to the control velocity. The plots show the average value of the fifty measurement data, and the error bars show the standard deviation of that. This result indicates that the extensibility is strongly dependent on the cell velocity. While the cell velocity changes according to the cell deformability under the constant pressure difference between the entrance and the exit of the channel, we note that the velocity by using the developed system can be accurately controlled. By considering that *EI* depends on the cell velocity, we set the cell velocity to 100 μm/s in the following experiments.

Figure 6. Extensibility evaluation. (**a**) An example of the measurement of cell length and (**b**) *EI* with respect to the cell velocity.

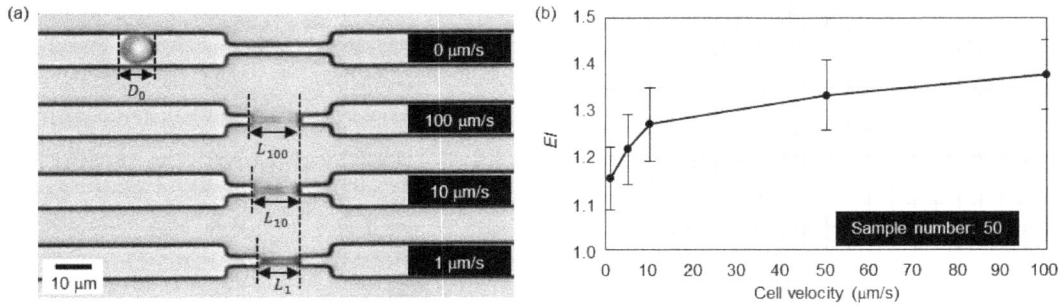

4.2. Extensibility Map

We can evaluate the effect of positioning resolution by using an extensibility map. Figures 7 and 8 show the heat map where the color means the frequency of the data, as shown in Figure 1.

Figure 7 shows the extensibility map for the same blood where Figure 7a,b correspond to the positioning resolution of ±240 nm and ±1200 nm, respectively. To tune the positioning resolution, we purposely changed the resolution of the displacement of the pump δ_1 as 1×10^{-3} μm and 4 μm for the experiments of Figure 7a,b, respectively. We evaluated the standard deviation as a dispersion index. In Figure 7a, the standard deviations of the original diameter of RBCs and *EI* are 0.37 μm and 0.06, respectively, while those standard deviations are 0.39 μm and 0.11, respectively, in Figure 7b. Since we use the same blood sample for both experiments, the standard deviations of the original diameter in Figures 7a,b should be close to each other. However, the standard deviations of *EI* show a different value in each experiment, while the standard deviations of the diameters of RBCs are almost the same. This means that the system with high positioning resolution can suppress the dispersion of the map, as expected. We believe that this is one of the biggest advantages for utilizing such a high resolution positioning system.

Figure 8 shows another example of an extensibility map, where the sample blood is taken from a subject who has a blood transfusion within 24 h. An interesting observation is that the shape of the map is completely different form those shown in Figure 7. We can see that the effect of blood transfusion still remains for the shape of the map in Figure 8, while this tendency can also be observed by the distribution of EI; we can see it at a glance through the extensibility map, where there are multiple groups of EI. These behaviors perhaps come from the remaining effect of the blood transfusion. An important note is that such a true characteristic can be enhanced only by a high resolution cell positioning system.

Figure 7. Extensibility map for the same blood. (**a**) Positioning resolution: ±240 nm; (**b**) positioning resolution: ±1200 nm.

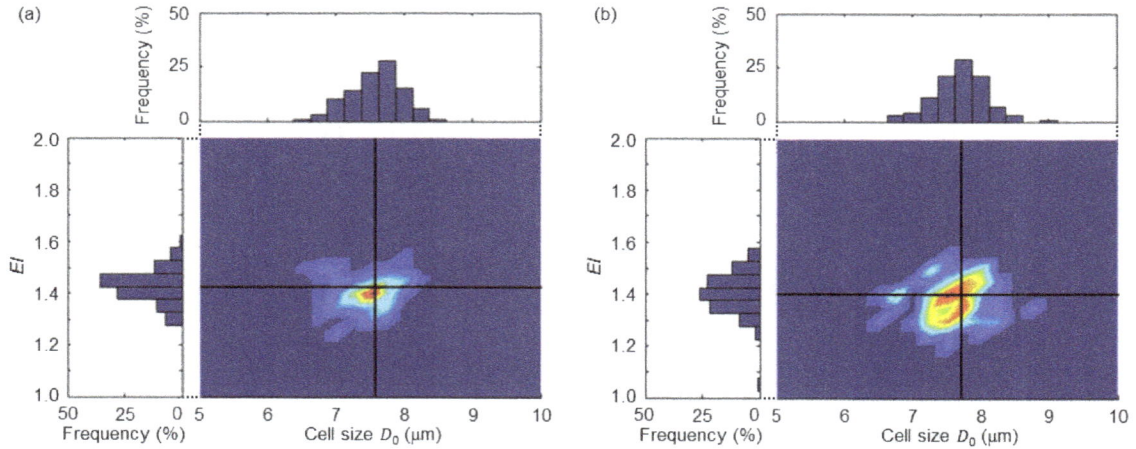

Figure 8. Extensibility map of a subject who received a blood transfusion within 24 h.

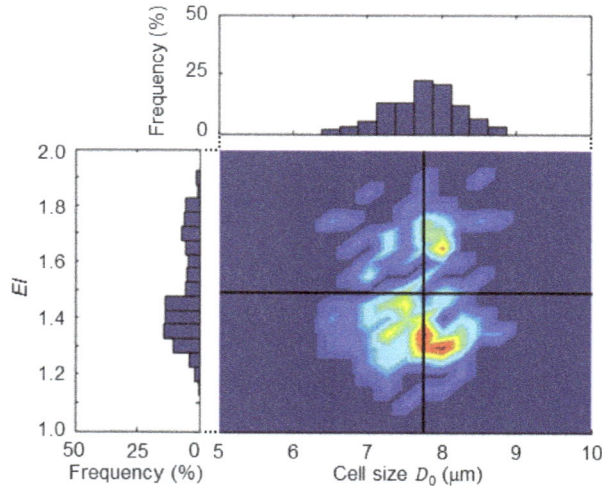

5. Discussions

Since the difference between R_{Com} and R_{Incom} is $\Delta t / \tau$, we can regard it as an additional parameter when the elasticity of the microfluidic chip is considered. Now, let us define $\hat{R}$ by the ratio between the reduction ratio under compressibility and incompressibility.

$$\hat{R} = \frac{R_{Com}}{R_{Incom}} = \frac{\Delta t}{\tau} \tag{18}$$

$\hat{R} = 1$ means that there is no effect by including the elasticity in the microfluidic chip. Our intention is to achieve $\hat{R} \ll 1$. From Figure 4, under $\Delta t = 1$ ms, for example, $\Delta t / \tau \simeq 10^{-7}$, and therefore, the position resolution δ_2 is in the order of $10^{-1}\delta_1$. This means that the resolution of cell positioning can be achieved with even higher resolution than that of the piezoelectric actuator. This is because the resolution of the sensing system is not high enough to support the target resolution.

6. Conclusions

In this paper, we proposed a novel concept, the flow reduction mechanism, for high resolution cell positioning in a microfluidic chip. The mechanism is based on the elasticity embedded in the fluid circuit of a microfluidic system and plays an essential role for the high resolution of cell positioning. There is great advantage in that we can easily construct a high resolution cell manipulation system by combining a macro-level actuator and a macro-level position sensor, even though the resolution of the actuator is larger than the desired resolution for cell manipulation. Focusing on this reduction mechanism, we successfully achieved cell position control with a resolution of 240 nm, corresponding to the limitation of the vision system. We also showed the importance of high resolution cell positioning through the extensibility map.

Acknowledgments

A part of this work was supported by the "Hyper Bio Assembler", the "Nanotechnology Platform Project (Nanotechnology Open Facilities in Osaka University)", Grant-in-Aid for Challenging Exploratory Research (26630098) of the Ministry of Education, Culture, Sports, Science and Technology", Creating Hybrid Organs of the Future" at Osaka University and the Japan Society for the Promotion of Science.

Author Contributions

Shinya Sakuma, Keisuke Kuroda, Fumihito Arai and Makoto Kaneko performed conception and design of the study, collection of data, analysis and interpretation of data, drafting of the manuscript and critical revision of the manuscript for important intellectual content. Tatsunori Taniguchi, Tomohito Ohtani and Yasushi Sakata performed design of the study and interpretation of data. All authors read and approved the final manuscript.

Conflicts of Interest

The authors declare no conflict of interest.

References

1. Sun, Y.; Wan, K.T.; Roberts, K.P.; Bischof, J.C.; Nelson, B.J. Mechanical property characterization of mouse zona pellucida. *IEEE Trans. NanoBiosci.* **2003**, *2*, 279–286.

2. Bremmell, K.E.; Evans, A.; Prestidge, C.A. Deformation and nano-rheology of red blood cells: An AFM investigation. *Colloids Surf. B Biointerfaces* **2006**, *50*, 43–48.

3. Sakuma, S.; Arai, F. Cellular force measurement using a nanometric-probe-integrated microfluidic chip with a displacement reduction mechanism. *J. Robot. Mechatron.* **2013**, *25*, 277–284.

4. Shelby, J.P.; White, J.; Ganesan, K.; Rathod, P.K.; Chiu, D.T. A microfluidic model for single-cell capillary obstruction by plasmodium-falciparuminfected erythrocytes. *Proc. Natl. Acad. Sci. USA* **2003**, *100*, 14618–14622.

5. Hirose, Y.; Tadakuma, K.; Higashimori, M.; Arai, T.; Kaneko, M.; Iitsuka, R.; Yamanishi, Y.; Arai, F. A new stiffness evaluation toward high speed cell sorter. In Proceedings of the IEEE Internatinal Conference on Robotics and Automation, Anchorage, AK, USA, 3–7 May 2010; pp. 4113–4118.

6. Adamo, A.; Sharei, A.; Adamo, L.; Lee, B.; Mao, S.; Jensen, K.F. Microfluidics-based assessment of cell deformability. *Anal. Chem.* **2012**, *84*, 6438–6443.

7. Zheng, Y.; Shojaei-Baghini, E.; Azad, A.; Wang, C.; Sun, Y. High-throughput biophysical measurement of human red blood cells. *Lab Chip* **2012**, *12*, 2560–2567.

8. Fukui, W.; Kaneko, M.; Sakuma, S.; Kawahara, T.; Arai, F. μ-Cell fatigue test. In Proceedings of the IEEE Internatinal Conference on Robotics and Automation, Saint Paul, MN, USA, 14–18 May 2012; pp. 4600–4605.

9. Sakuma, S.; Kuroda, K.; Tsai, C.H.D.; Fukui, W.; Arai, F.; Kaneko, M. Red blood cell fatigue evaluation based on the close-encountering point between extensibility and recoverability. *Lab Chip* **2014**, *14*, 1135–1141.

10. Pamme, N.; Wilhelm, C. Magnetism and microfluidics. *Lab Chip* **2006**, *6*, 24–38.

11. Maruo, S.; Inoue, H. Optically driven micropump produced by three-dimensional two-photon microfabrication. *Appl. Phys. Lett.* **2006**, *89*, 144101.

12. Dao, M.; Lim, C.T.; Suresh, S. Mechanics of the human red blood cell deformed by optical tweezers. *J. Mech. Phys. Solids* **2003**, *51*, 2259–2280.

13. Lintel, H.T.G.V.; Pol, F.C.M.V.D.; Bouwstra, S. A piezoelectric micropump based on micromachining of silicon. *Sens. Actuators* **1988**, *15*, 153–167.

14. Krüger, J.; Singh, K.; O'Neill, A.; Jackson, C.; Morrison, A.; O'Brien, P. Development of a microfluidic device for fluorescence activated cell sorting. *J. Micromech. Microeng.* **2002**, *12*, 486–494.

15. Chung, B.G.; Flanagan, L.A.; Rhee, S.W.; Schwartz, P.H.; Lee, A.P.; Monuki, E.S.; Jeon, N.L. Human neural stem cell growth and differentiation in a gradient-generating microfluidic device. *Lab Chip* **2005**, *5*, 401–406.

16. Mahalanabis, M.; Muayad, H.A.; Kulinski, M.D.; Altman, D.; Klapperich, C.M. Cell lysis and DNA extraction of gram-positive and gram-negative bacteria from whole blood in a disposable microfluidic chip. *Lab Chip* **2009**, *9*, 2811–2817.

17. Ichikawa, A.; Tanikawa, T.; Akagi, S.; Ohba, K. Automatic cell cutting by high-precision microfluidic control. *J. Robot. Mechatron.* **2011**, *23*, 13–18.

18. Gomez-Sjoberg, R.; Leyrat, A.A.; Pirone, D.M.; Chen, C.S.; Quake, S.R. Versatile, fully automated, microfluidic cell culture system. *Anal. Chem.* **2007**, *79*, 8557–8563.

19. Tai, C.H.; Hsiung, S.K.; Chen, C.Y.; Tsai, M.L.; Lee, G.B. *Biomed. Microdevices* **2007**, *9*, 533–543.

20. Tsukada, K.; Sekizuka, E.; Oshio, C.; Minamitani, H. Direct measurement of erythrocyte deformability in diabetes mellitus with a transparent microchannel capillary model and high-speed video camera system. *Microvasc. Res.* **2001**, *61*, 231–239.

21. Lee, S.S.; Yim, Y.; Ahn, K.H.; Lee, S.J. Extensional flow-based assessment of red blood cell deformability using hyperbolic converging microchannel. *Biomed. Microdevices* **2009**, *11*, 37–43.

22. Braunmuller, S.; Schmid, L.; Sackmann, E.; Franke, T. Hydrodynamic deformation reveals two coupled modes/time scales of red blood cell relaxation. *Soft Matter* **2012**, *8*, 11240–11248.

5

Classification of Cells with Membrane Staining and/or Fixation Based on Cellular Specific Membrane Capacitance and Cytoplasm Conductivity

Song-Bin Huang [1,†], Yang Zhao [2,†], Deyong Chen [2], Shing-Lun Liu [1], Yana Luo [2], Tzu-Keng Chiu [3], Junbo Wang [2,*], Jian Chen [2,*] and Min-Hsien Wu [1,*]

[1] Graduate Institute of Biochemical and Biomedical Engineering, Chang Gung University, Taoyuan 333, Taiwan; E-Mails: angel94901111@yahoo.com.tw (S.-B.H.); nsx10241030@hotmail.com (S.-L.L.)

[2] State Key Laboratory of Transducer Technology, Institute of Electronics, Chinese Academy of Sciences, Beijing 100190, China; E-Mails: zhaoyang110@mails.ucas.ac.cn (Y.Z.); dychen@mail.ie.ac.cn (D.C.); luoyana88@126.com (Y.L.)

[3] Department of Chemical and Materials Engineering, Chang Gung University, Taoyuan 333, Taiwan; E-Mail: b74225@hotmail.com

[†] These authors contributed equally to this work.

[*] Authors to whom correspondence should be addressed; E-Mails: jbwang@mail.ie.ac.cn (J.W.); chenjian@mail.ie.ac.cn (J.C.); mhwu@mail.cgu.edu.tw (M.-H.W.)

Academic Editor: Phillipe Renaud

Abstract: Single-cell electrical properties (e.g., specific membrane capacitance ($C_{\text{specific membrane}}$) and cytoplasm conductivity ($\sigma_{\text{cytoplasm}}$)) have been regarded as potential label-free biophysical markers for the evaluation of cellular status. However, whether there exist correlations between these biophysical markers and cellular status (e.g., membrane-associate protein expression) is still unknown. To further validate the utility of single-cell electrical properties in cell type classification, $C_{\text{specific membrane}}$ and $\sigma_{\text{cytoplasm}}$ of single PC-3 cells with membrane staining and/or fixation were analyzed and compared in this study. Four subtypes of PC-3 cells were prepared: untreated PC-3 cells, PC-3 cells with anti-EpCAM staining, PC-3 cells with fixation, and fixed PC-3 cells with anti-EpCAM staining. In experiments, suspended single cells were aspirated through microfluidic constriction channels with raw impedance data quantified and translated to $C_{\text{specific membrane}}$ and $\sigma_{\text{cytoplasm}}$. As to experimental results, significant differences in $C_{\text{specific membrane}}$ were observed for both live and fixed PC-3 cells with and without membrane staining, indicating that membrane staining proteins can contribute to electrical properties of cellular membranes. In addition, a significant decrease in $\sigma_{\text{cytoplasm}}$ was located for PC-3 cells with and without fixation, suggesting that cytoplasm protein crosslinking during the fixation process can alter the cytoplasm conductivity. Overall, we have demonstrated how to classify single cells based on cellular electrical properties.

Keywords: single-cell analysis; cellular electrical properties; specific membrane capacitance; cytoplasm conductivity; microfluidics

---

1. Introduction

Single-cell electrical properties (e.g., specific membrane capacitance ($C_{\text{specific membrane}}$) and cytoplasm conductivity ($\sigma_{\text{cytoplasm}}$)) are promising biophysical markers for understanding cellular functions and status [1–4], enabling cell type classification (e.g., tumor cells [5–9], stem cells [10], red blood cells [11,12], and white blood cells [13,14]).

Techniques capable of characterizing cellular $C_{\text{specific membrane}}$ and $\sigma_{\text{cytoplasm}}$ include patch clamping, electrorotation, and micro electrical impedance spectroscopy (µEIS) [4]. Although powerful, these techniques cannot obtain statistically meaningful data (data points from hundreds and even thousands of cells for each cell type) due to limited assay throughput (e.g., 20 cells per cell type based on electrorotation [15] and 23 cells per cell type based on µEIS [16]).

By combining µEIS with flow cytometry, Renaud *et al.* [17–20] and Morgan *et al.* [21–23] demonstrated the high-throughput characterization of single-cell electrical properties. Due to the lack of corresponding electrical models, raw size-dependent electrical data cannot be translated to size-independent intrinsic electrical parameters, limiting their functionalities in cell status evaluation and cell type classification.

Recently, we proposed microfluidic platforms to aspirate single cells through the constriction channel (cross sectional area smaller than cells) continuously with two-frequency impedance data sampled and translated to $C_{\text{specific membrane}}$ and $\sigma_{\text{cytoplasm}}$ based on equivalent lumped [24] or distributed [25] electrical models. Leveraging these platforms, intrinsic cellular electrical properties of hundreds of single cells were obtained, enabling the classification of wild-type tumor cells and their counterparts with single oncogenes under regulation [6]. However, whether there exist correlations between label-free biophysical markers (e.g., $C_{\text{specific membrane}}$ and $\sigma_{\text{cytoplasm}}$) and cellular biochemical properties (e.g., membrane-associate protein expression) was still unknown.

To address this issue, in this study, the electrical properties of untreated and treated human prostate tumor cells (PC-3) (e.g., untreated PC-3 cells ($G_{\text{NS\&NF}}$), PC-3 cells with anti-EpCAM staining ($G_{\text{S\&NF}}$), PC-3 cells with fixation ($G_{\text{NS\&F}}$), and fixed PC-3 cells with anti-EpCAM staining ($G_{\text{S\&F}}$)) were characterized and compared (see Figure 1).

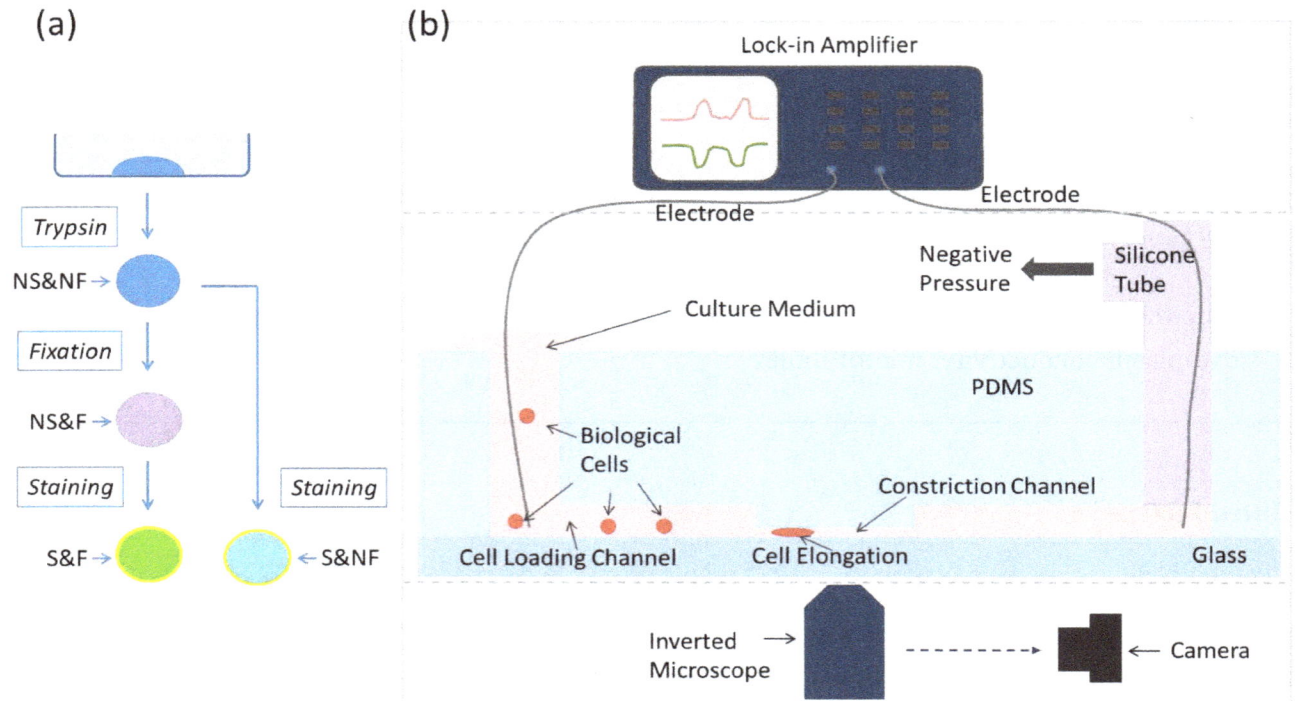

Figure 1. (**a**) The cell preparation flow chart where four sub-types of human prostate tumor cells (PC-3) were prepared: untreated PC-3 cells ($G_{NS\&NF}$), PC-3 cells with anti-EpCAM staining ($G_{S\&NF}$), PC-3 cells with fixation ($G_{NS\&F}$), and fixed PC-3 cells with anti-EpCAM/FITC staining ($G_{S\&F}$). (**b**) Schematic of the microfluidic system for single-cell electrical property characterization where single cells are aspirated continuously through the constriction channel with impedance data and cell elongation length measured and translated to $C_{specific\ membrane}$ and $\sigma_{cytoplasm}$.

2. Materials and Methods

2.1. Materials

Unless otherwise indicated, all cell-culture reagents were purchased from Life Technologies Corporation (Carlsbad, CA, USA). Materials required for device fabrication included a SU-8 photoresist (MicroChem Corporation, Newton, MA, USA) and a 184 silicone elastomer (Dow Corning Corporation, Midland, MI, USA).

2.2. Cell Culture, Membrane Staining, and Fixation

Human prostate tumor cells (PC-3) were cultured with RPMI-1640 medium supplemented with 10% fetal bovine serum and 1% penicillin and streptomycin. As shown in Figure 1a, untreated PC-3 cells were harvested by trypsinization to form suspended single cells, which were then fixed (mixture with 0.1% paraformaldehyde for 3 min) and/or stained with anti-EpCAM/FITC (mixture with 0.05 µg/µL anti-Human EpCAM Alexa Fluor 488 (ebioscience, cat. 11-5791-82) for 30 min).

2.3. Device Fabrication and Operation

The two-layer polydimethylsiloxane (PDMS) device (constriction channel cross-section area of 10 μm × 10 μm) was replicated from a double-layer SU-8 mold based on conventional lithography. Briefly, the first layer of SU-8 5 was to form the constriction channel (10 μm) and the second layer of SU-8 25 was to form the cell loading channel (25 μm). PDMS prepolymer and curing agent were mixed, degassed, poured on channel masters, and baked in an oven. PDMS channels were then peeled from the SU-8 masters with reservoir holes punched through and bonded to a glass slide.

The operation process was summarized as follows (see Figure 1b). The cell samples were pipetted to the entrance of the cell loading channel of the microfluidic device, where a negative pressure of 1 kPa was applied to aspirate cells continuously through the constriction channel with two-frequency impedance data (1 kHz + 100 kHz) and images recorded.

2.4. Data Analysis

The detailed procedures for data analysis were described previously [24,25], and are summarized as follows. Raw impedance data at 1 kHz were used to evaluate the sealing properties of deformed cells with constriction channel walls, and raw impedance data at 100 kHz were used to quantify equivalent cellular membrane capacitance and cytoplasm resistance, respectively. By combining cell elongation length during its traveling process within the constriction channel based on image processing, equivalent membrane capacitance and cytoplasm resistance were further translated to $C_{specific\ membrane}$ and $\sigma_{cytoplasm}$.

2.5. Statistical Analysis

All results were expressed as means ± standard deviations. In the statistical analysis, the Student's t-test was used for two group comparisons. $P < 0.01$ was considered statistically significant.

3. Results and Discussion

The experimental results of fixed PC-3 cells with and without anti-EpCAM staining ($G_{NS\&F}$ vs. $G_{S\&F}$) were first compared since the only difference between these two sub-cell types is the stained membrane protein (anti-Human EpCAM Alexa Fluor 488), which selectively binds EpCAM expressed in PC-3 cell membranes. There is no disturbance of endocytosis and other cytoplasm activities due to the fixation of PC-3 cells. Figure 2a shows the bright field and fluorescent images of fixed and stained PC-3 cells ($G_{S\&F}$), which indicated the existence of anti-Human EpCAM Alexa Fluor 488.

Figure 2b shows a scatterplot of $C_{specific\ membrane}$ vs. $\sigma_{cytoplasm}$ for $G_{NS\&F}$ ($n_{cell} = 208$) and $G_{S\&F}$ ($n_{cell} = 252$) cells, which were quantified as 2.16 ± 0.72 vs. 1.66 ± 0.46 μF/cm^2 of $C_{specific\ membrane}$ and 0.59 ± 0.10 vs. 0.59 ± 0.10 S/m of $\sigma_{cytoplasm}$ (Figure 2c). No significant difference in $\sigma_{cytoplasm}$ was observed, indicating that for fixed PC-3 cells, surface antigen staining cannot alter cytoplasm properties. In addition, a significant difference in $C_{specific\ membrane}$ for $G_{NS\&F}$ vs. $G_{S\&F}$ cells was observed, indicating the contribution of anti-EpCAM on cellular membrane electrical properties as a correlation between a specific membrane protein and $C_{specific\ membrane}$.

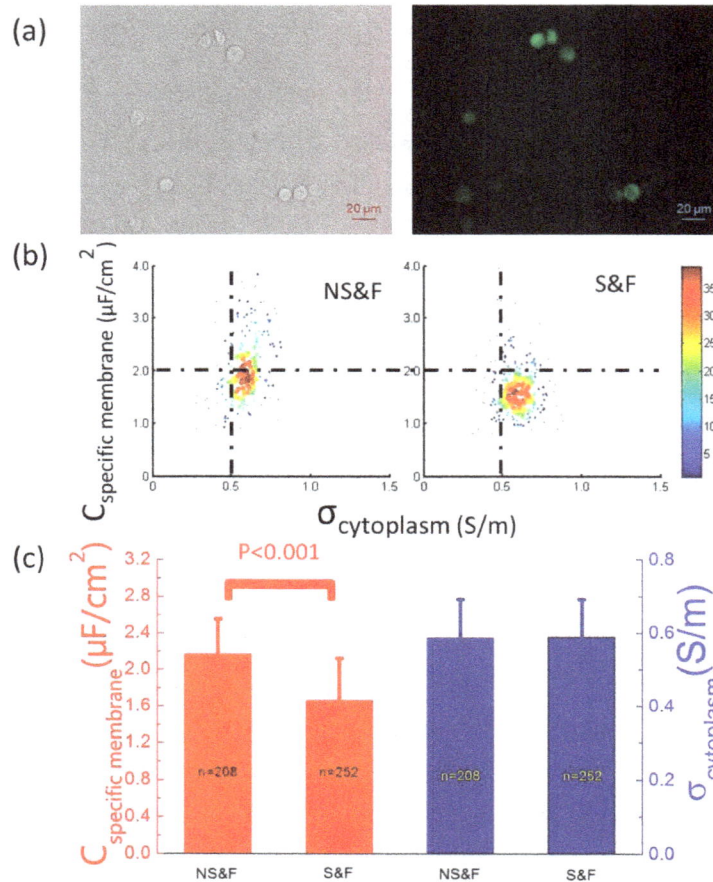

Figure 2. (a) The bright field and fluorescent images of fixed and stained PC-3 cells ($G_{S\&F}$), which confirmed the existence of anti-Human EpCAM Alexa Fluor 488. **(b)** A scatterplot of $C_{specific\ membrane}$ *vs.* $\sigma_{cytoplasm}$ for $G_{NS\&F}$ ($n_{cell} = 208$) and $G_{S\&F}$ ($n_{cell} = 252$) cells, which were quantified as 2.16 ± 0.72 *vs.* 1.66 ± 0.46 $\mu F/cm^2$ of $C_{specific\ membrane}$ and 0.59 ± 0.10 *vs.* 0.59 ± 0.10 S/m of $\sigma_{cytoplasm}$, respectively. **(c)** No significant difference of $\sigma_{cytoplasm}$ between $G_{NS\&F}$ and $G_{S\&F}$ was observed, suggesting that for fixed PC-3 cells, surface antigen staining cannot alter cytoplasm properties. In addition, a significant difference in $C_{specific\ membrane}$ was observed, indicating the contribution of anti-Human EpCAM Alexa Fluor 488 on cellular membrane electrical properties as a correlation between a specific type of membrane proteins and $C_{specific\ membrane}$.

The experimental results of live PC-3 cells with and without anti-EpCAM staining ($G_{NS\&NF}$ *vs.* $G_{S\&NF}$) were then compared to investigate the effect of surface antigen staining on electrical properties of live cells. Figure 3a shows the bright field and fluorescent images of stained PC-3 cells without fixation ($G_{S\&NF}$), which suggested the existence of anti-Human EpCAM Alexa Fluor 488.

Figure 3b shows a scatterplot of $C_{specific\ membrane}$ *vs.* $\sigma_{cytoplasm}$ for $G_{NS\&NF}$ ($n_{cell} = 415$) and $G_{S\&NF}$ ($n_{cell} = 417$) cells, which were quantified as 2.21 ± 0.49 *vs.* 1.97 ± 0.39 $\mu F/cm^2$ of $C_{specific\ membrane}$ and 0.77 ± 0.15 *vs.* 0.90 ± 0.13 S/m of $\sigma_{cytoplasm}$, respectively (Figure 3c). A significant difference in $C_{specific\ membrane}$ for $G_{NS\&NF}$ *vs.* $G_{S\&NF}$ cells was observed, indicating the contribution of anti-EpCAM on $C_{specific\ membrane}$ values. Compared to the difference of $C_{specific\ membrane}$ between $G_{NS\&F}$ and $G_{S\&F}$ (2.16 ± 0.72 *vs.* 1.66 ± 0.46 $\mu F/cm^2$), a smaller difference of $C_{specific\ membrane}$ was observed between $G_{NS\&NF}$ and $G_{S\&NF}$ (2.21 ± 0.49 *vs.* 1.97 ± 0.39 $\mu F/cm^2$), which may result from the endocytosis of anti-EpCAM after binding with surface antigen EpCAM. In addition, a significant difference in $\sigma_{cytoplasm}$ for $G_{NS\&NF}$ $vs.$ $G_{S\&NF}$ cells was observed, further suggesting the possibility of surface antigen endocytosis and the triggering of the downstream signal pathways [26].

As to the effects of fixation on cellular electrical properties, a significant difference in $\sigma_{cytoplasm}$ (0.77 ± 0.15 $vs.$ 0.59 ± 0.10 S/m) rather than $C_{specific\ membrane}$ (2.21 ± 0.49 $vs.$ 2.16 ± 0.72 µF/cm^2) was observed for $G_{NS\&NF}$ ($n_{cell} = 415$) $vs.$ $G_{NS\&F}$ ($n_{cell} = 208$) (see Figure 4). This significant decrease in $\sigma_{cytoplasm}$ of fixed cells compared to live counterparts may result from the crosslinking of cytoplasm proteins in the fixing process, which was consistent with previous results [19].

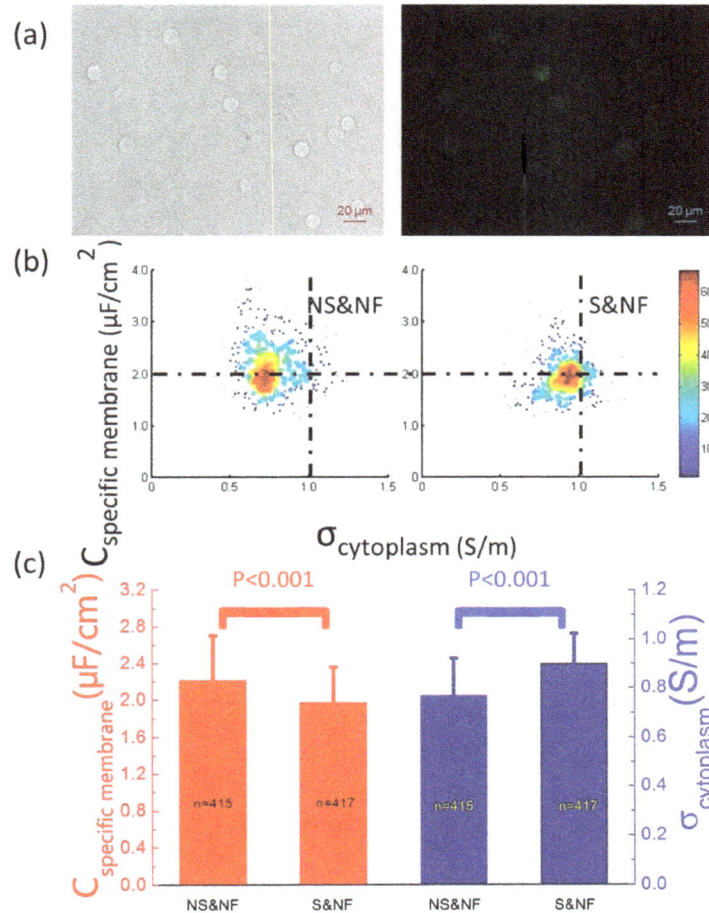

Figure 3. (**a**) The bright field and fluorescent images of stained PC-3 cells without fixation ($G_{S\&NF}$), confirming the existence of anti-Human EpCAM Alexa Fluor 488. (**b**) A scatterplot of $C_{specific\ membrane}$ $vs.$ $\sigma_{cytoplasm}$ for $G_{NS\&NF}$ ($n_{cell} = 415$) and $G_{S\&NF}$ ($n_{cell} = 417$) cells, which were quantified as 2.21 ± 0.49 $vs.$ 1.97 ± 0.39 µF/cm^2 of $C_{specific\ membrane}$ and 0.77 ± 0.15 $vs.$ 0.90 ± 0.13 S/m of $\sigma_{cytoplasm}$, respectively. (**c**) Significant differences in both $C_{specific\ membrane}$ and $\sigma_{cytoplasm}$ for $G_{NS\&NF}$ and $G_{S\&NF}$ cells were observed, suggesting the potential contribution of anti-Human EpCAM Alexa Fluor on $C_{specific\ membrane}$ and the possible existence of surface antigen endocytosis of anti-Human EpCAM Alexa Fluor, which may trigger downstream signal pathways in cytoplasm.

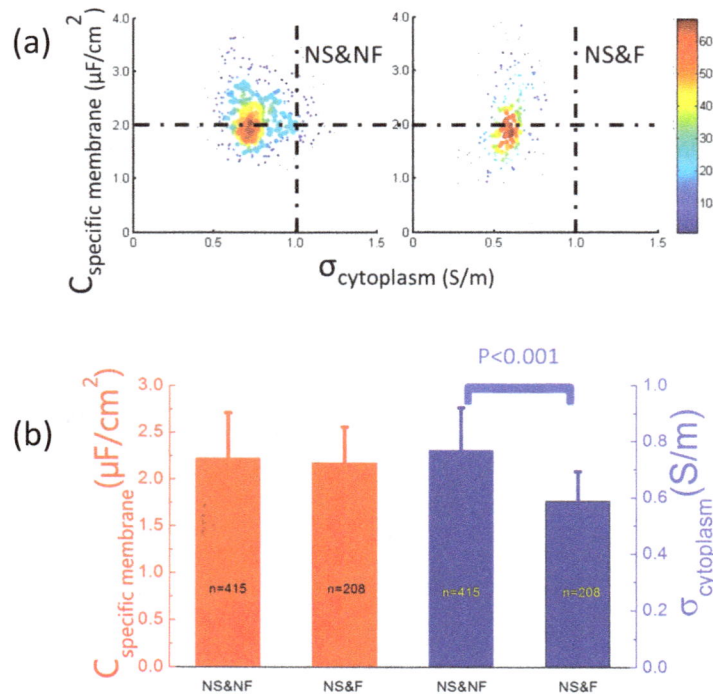

Figure 4. (**a**) A scatterplot of $C_{\text{specific membrane}}$ vs. $\sigma_{\text{cytoplasm}}$ for $G_{\text{NS\&NF}}$ ($n_{\text{cell}} = 415$) and $G_{\text{NS\&F}}$ ($n_{\text{cell}} = 208$) cells, which were quantified as 2.21 ± 0.49 vs. 2.16 ± 0.72 $\mu F/cm^2$ of $C_{\text{specific membrane}}$ and 0.77 ± 0.15 vs. 0.59 ± 0.10 S/m of $\sigma_{\text{cytoplasm}}$, respectively. (**b**) A significant difference in $\sigma_{\text{cytoplasm}}$ rather than $C_{\text{specific membrane}}$ for $G_{\text{NS\&NF}}$ vs. $G_{\text{NS\&F}}$ cells was observed, which may result from the crosslinking of cytoplasm proteins in the fixing process.

4. Conclusions

In summary, this paper reported the electrical properties of four sub-types of PC-3 cells: $C_{\text{specific membrane}}$ of 2.16 ± 0.72 $\mu F/cm^2$ and $\sigma_{\text{cytoplasm}}$ of 0.59 ± 0.10 S/m for $G_{\text{NS\&F}}$ cells ($n_{\text{cell}} = 208$), $C_{\text{specific membrane}}$ of 1.66 ± 0.46 $\mu F/cm^2$ and $\sigma_{\text{cytoplasm}}$ of 0.59 ± 0.10 S/m for $G_{\text{S\&F}}$ cells ($n_{\text{cell}} = 252$), $C_{\text{specific membrane}}$ of 2.21 ± 0.49 $\mu F/cm^2$ and $\sigma_{\text{cytoplasm}}$ of 0.77 ± 0.15 S/m for $G_{\text{NS\&NF}}$ cells ($n_{\text{cell}} = 415$), and $C_{\text{specific membrane}}$ of 1.97 ± 0.39 $\mu F/cm^2$ and $\sigma_{\text{cytoplasm}}$ of 0.90 ± 0.13 S/m for $G_{\text{S\&NF}}$ cells ($n_{\text{cell}} = 417$). For $G_{\text{NS\&F}}$ vs. $G_{\text{S\&F}}$, a significant difference was observed only in $C_{\text{specific membrane}}$ rather than $\sigma_{\text{cytoplasm}}$, as a correlation between anti-Human EpCAM Alexa Fluor and $C_{\text{specific membrane}}$. Furthermore, for $G_{\text{NS\&NF}}$ vs. $G_{\text{S\&NF}}$, significant differences in $C_{\text{specific membrane}}$ and $\sigma_{\text{cytoplasm}}$ were observed, indicating the contribution of the anti-Human EpCAM Alexa Fluor 488 on cellular membrane electrical properties and the possible trigger of downstream signal pathways in the process of live cell staining. In addition, for $G_{\text{NS\&NF}}$ vs. $G_{\text{NS\&F}}$, a significant decrease in $\sigma_{\text{cytoplasm}}$ was observed, suggesting the potential effect of fixation in cytoplasm protein crosslinking, leading to decreased ion transportation capabilities. On the whole, we have demonstrated how to classify single cells based on cellular electrical properties.

Acknowledgments

The authors would like to acknowledge financial support from the National Basic Research Program of China (973 Program, Grant No. 2014CB744602), the Natural Science Foundation of China (Grant No. 61201077, 61431019 and 81261120561), the National High Technology Research and Development

Program of China (863 Program, Grant No. 2014AA093408), the Beijing NOVA Program, and Chang Gung Memorial Hospital in Taiwan (CMRPD2D0041).

Author Contributions

Song-Bin Huang, Yang Zhao, and Deyong Chen designed and conducted experiments; Shing-Lun Liu and Yana Luo conducted cell culture; Tzu-Keng Chiu and Junbo Wang were responsible for data processing; and Jian Chen and Min-Hsien Wu organized the experiments and drafted the manuscript.

Conflicts of Interest

The authors declare no conflict of interest.

References

1. Morgan, H.; Sun, T.; Holmes, D.; Gawad, S.; Green, N.G. Single cell dielectric spectroscopy. *J. Phys. D Appl. Phys.* **2007**, *40*, 61–70.

2. Sun, T.; Morgan, H. Single-cell microfluidic impedance cytometry: A review. *Microfluid. Nanofluid.* **2010**, *8*, 423–443.

3. Valero, A.; Braschler, T.; Renaud, P. A unified approach to dielectric single cell analysis: Impedance and dielectrophoretic force spectroscopy. *Lab Chip* **2010**, *10*, 2216–2225.

4. Zheng, Y.; Nguyen, J.; Wei, Y.; Sun, Y. Recent advances in microfluidic techniques for single-cell biophysical characterization. *Lab Chip* **2013**, *13*, 2464–2483.

5. Liang, X.; Graham, K.A.; Johannessen, A.C.; Costea, D.E.; Labeed, F.H. Human oral cancer cells with increasing tumorigenic abilities exhibit higher effective membrane capacitance. *Integr. Biol.* **2014**, *6*, 545–554.

6. Zhao, Y.; Zhao, X.T.; Chen, D.Y.; Luo, Y.N.; Jiang, M.; Wei, C.; Long, R.; Yue, W.T.; Wang, J.B.; Chen, J. Tumor cell characterization and classification based on cellular specific membrane capacitance and cytoplasm conductivity. *Biosens. Bioelectron.* **2014**, *57*, 245–253.

7. Cho, Y.; Kim, H.S.; Frazier, A.B.; Chen, Z.G.; Shin, D.M.; Han, A. Whole-cell impedance analysis for highly and poorly metastatic cancer cells. *J. Microelectromechan. Syst.* **2009**, *18*, 808–817.

8. Han, K.H.; Han, A.; Frazier, A.B. Microsystems for isolation and electrophysiological analysis of breast cancer cells from blood. *Biosens. Bioelectron.* **2006**, *21*, 1907–1914.

9. Kang, G.; Kim, Y.-J.; Moon, H.-S.; Lee, J.-W.; Yoo, T.-K.; Park, K.; Lee, J.-H. Discrimination between the human prostate normal cell and cancer cell by using a novel electrical impedance spectroscopy controlling the cross-sectional area of a microfluidic channel. *Biomicrofluidics* **2013**, *7*, 044126.

10. Song, H.; Wang, Y.; Rosano, J.M.; Prabhakarpandian, B.; Garson, C.; Pant, K.; Lai, E. A microfluidic impedance flow cytometer for identification of differentiation state of stem cells. *Lab Chip* **2013**, *13*, 2300–2310.

11. Du, E.; Ha, S.; Diez-Silva, M.; Dao, M.; Suresh, S.; Chandrakasan, A.P. Electric impedance microflow cytometry for characterization of cell disease states. *Lab Chip* **2013**, *13*, 3903–3909.

12. Zheng, Y.; Shojaei-Baghini, E.; Azad, A.; Wang, C.; Sun, Y. High-throughput biophysical measurement of human red blood cells. *Lab Chip* **2012**, *12*, 2560–2567.

13. Watkins, N.N.; Hassan, U.; Damhorst, G.; Ni, H.; Vaid, A.; Rodriguez, W.; Bashir, R. Microfluidic CD4$^+$ and CD8$^+$ T lymphocyte counters for point-of-care HIV diagnostics using whole blood. *Sci. Transl. Med.* **2013**, *5*, 214ra170.

14. Han, X.; van Berkel, C.; Gwyer, J.; Capretto, L.; Morgan, H. Microfluidic lysis of human blood for leukocyte analysis using single cell impedance cytometry. *Anal. Chem.* **2012**, *84*, 1070–1075.

15. Becker, F.F.; Wang, X.B.; Huang, Y.; Pethig, R.; Vykoukal, J.; Gascoyne, P.R.C. Separation of human breast-cancer cells from blood by differential dielectric affinity. *Proc. Natl. Acad. Sci. USA* **1995**, *92*, 860–864.

16. Tan, Q.; Ferrier, G.A.; Chen, B.K.; Wang, C.; Sun, Y. Quantification of the specific membrane capacitance of single cells using a microfluidic device and impedance spectroscopy measurement. *Biomicrofluidics* **2012**, *6*, 034112.

17. Gawad, S.; Schild, L.; Renaud, P. Micromachined impedance spectroscopy flow cytometer for cell analysis and particle sizing. *Lab Chip* **2001**, *1*, 76–82.

18. Gawad, S.; Cheung, K.; Seger, U.; Bertsch, A.; Renaud, P. Dielectric spectroscopy in a micromachined flow cytometer: Theoretical and practical considerations. *Lab Chip* **2004**, *4*, 241–251.

19. Cheung, K.; Gawad, S.; Renaud, P. Impedance spectroscopy flow cytometry: On-chip label-free cell differentiation. *Cytom. Part A* **2005**, *65A*, 124–132.

20. Shaker, M.; Colella, L.; Caselli, F.; Bisegna, P.; Renaud, P. An impedance-based flow microcytometer for single cell morphology discrimination. *Lab Chip* **2014**, *14*, 2548–2555.

21. Holmes, D.; Pettigrew, D.; Reccius, C.H.; Gwyer, J.D.; van Berkel, C.; Holloway, J.; Davies, D.E.; Morgan, H. Leukocyte analysis and differentiation using high speed microfluidic single cell impedance cytometry. *Lab Chip* **2009**, *9*, 2881–2889.

22. Holmes, D.; Morgan, H. Single cell impedance cytometry for identification and counting of CD4 T-cells in human blood using impedance labels. *Anal. Chem.* **2010**, *82*, 1455–1461.

23. Barat, D.; Spencer, D.; Benazzi, G.; Mowlem, M.C.; Morgan, H. Simultaneous high speed optical and impedance analysis of single particles with a microfluidic cytometer. *Lab Chip* **2012**, *12*, 118–126.

24. Zhao, Y.; Chen, D.; Li, H.; Luo, Y.; Deng, B.; Huang, S.B.; Chiu, T.K.; Wu, M.H.; Long, R.; Hu, H.; *et al.* A microfluidic system enabling continuous characterization of specific membrane capacitance and cytoplasm conductivity of single cells in suspension. *Biosens. Bioelectron.* **2013**, *43C*, 304–307.

25. Zhao, Y.; Chen, D.; Luo, Y.; Li, H.; Deng, B.; Huang, S.-B.; Chiu, T.-K.; Wu, M.-H.; Long, R.; Hu, H.; *et al.* A microfluidic system for cell type classification based on cellular size-independent electrical properties. *Lab Chip* **2013**, *13*, 2272–2277.

26. Schnell, U.; Cirulli, V.; Giepmans, B.N. Epcam: Structure and function in health and disease. *Biochim. Biophys. Acta Biomembr.* **2013**, *1828*, 1989–2001.

6

A Rapid and Low-Cost Nonlithographic Method to Fabricate Biomedical Microdevices for Blood Flow Analysis

Elmano Pinto [1,2], Vera Faustino [1], Raquel O. Rodrigues [1,3], Diana Pinho [1,2], Valdemar Garcia [1], João M. Miranda [2] and Rui Lima [1,2,4,*]

[1] School of Technology and Management (ESTiG), Polytechnic Institute of Bragança (IPB), Campus de Santa Apolónia, 5300-253 Bragança, Portugal; E-Mails: elmanopinto@hotmail.com (E.P.); verafaustino@ipb.pt (V.F.); raquel.rodrigues@ipb.pt (R.O.R.); diana@ipb.pt (D.P.); valdemar@ipb.pt (V.G.)

[2] Transport Phenomena Research Center, Department of Chemical Engineering, Engineering Faculty, University of Porto, Rua Dr. Roberto Frias, 4200-465 Porto, Portugal; E-Mail: jmiranda@fe.up.pt

[3] Laboratory of Catalysis and Materials—Associate Laboratory LSRE/LCM, Engineering Faculty, University of Porto, Rua Dr. Roberto Frias, 4200-465 Porto, Portugal

[4] Department of Mechanical Engineering, Minho University, Campus de Azurém, 4800-058 Guimarães, Portugal

* Author to whom correspondence should be addressed; E-Mail: ruimec@ipb.pt

Academic Editor: Cheng Luo

Abstract: Microfluidic devices are electrical/mechanical systems that offer the ability to work with minimal sample volumes, short reactions times, and have the possibility to perform massive parallel operations. An important application of microfluidics is blood rheology in microdevices, which has played a key role in recent developments of lab-on-chip devices for blood sampling and analysis. The most popular and traditional method to fabricate these types of devices is the polydimethylsiloxane (PDMS) soft lithography technique, which requires molds, usually produced by photolithography. Although the research results are extremely encouraging, the high costs and time involved in the production of molds by photolithography is currently slowing down the development cycle of these types of devices. Here we present a simple, rapid, and low-cost nonlithographic technique to create microfluidic systems for biomedical applications. The results demonstrate the ability of the proposed method to perform cell free layer (CFL) measurements and the formation of microbubbles in continuous blood flow.

Keywords: low cost biochips; nonlithographic technique; xurography; blood flow; bifurcations; microbubbles; biomicrofluidics

---

1. Introduction

Microfluidic systems have become an increasingly popular interdisciplinary technology with applications in many fields, such as engineering, biology, and medicine [1–4]. Generally, these systems have characteristic dimensions of one to 1000 μm, and combine electrical and mechanical components at the micro-scale level. Microfluidic systems work with small sample volumes, shorter reaction times, and can perform massive amount of parallel operations. By integrating several steps into a single system, microfluidic devices allow performing rapid measurements in a single step and, consequently, they offer great potential to develop portable and point-of-care devices [3,5].

One of most popular and traditional methods to fabricate microfluidic devices is the soft lithography technique, with polydimethylsiloxane (PDMS) structures made from SU-8 molds [6–8]. The SU-8 molds are produced by photolithography. The main attraction of this technology is due, mainly, to its high-resolution capabilities, low material cost, gas permeability, and optical transparency [6]. However, the photolithographic production of molds usually requires a clean-room environment and specialized equipment that can be quite costly, and the process is time consuming. These drawbacks are currently slowing down development, especially in research institutions without specialized facilities to produce molds by photolithography. Access to rapid microfabrication methods is essential to improve development cycles and to test new chip designs and research ideas. Hence, it is crucial to develop simple, rapid, and low-cost nonlithographic techniques to fabricate microfluidic systems.

Nonlithographic techniques, also known as print-and-peel (PAP) [9], were first reported at the beginning of this century, using a xerographic process to fabricate molds for PDMS microfluidic microdevices [10]. Later on, Branham *et al.* [11] demonstrated the application of a laser jet printer to fabricate PDMS microdevices. A few years later, Bao *et al.* [12] used this PAP technique to perform capillary electrophoresis. Chen *et al.* [13] proposed the use of pre-stressed thermoplastic sheets, which shrink when heated, as a base material for low cost microfluidics. Bonyár *et al.* [14] demonstrated the ability of 3D rapid prototyping technology to produce functional microfluidic devices and molds for PDMS microchannels. Other low cost non-lithographic technologies were reviewed by Waldbaur *et al.* [15], including laser direct machining [16] and Inkjet 3D printing [17]. A PAP method, known as xurography, has shown to be an effective, novel, and rapid prototyping technique to fabricate microfluidic channels [18]. Xurography uniquely uses a cutting plotter machine and adhesive vinyl films to generate the master molds, or to directly fabricate the microchannels, and, consequently, does not involve any photolithographic processes. Recently, Sundberg *et al.* [19] demonstrated the application of this PAP approach to create microfluidic channels to perform DNA melting analysis. However, no work has been performed yet to test biomimetic separations.

An important application of microfluidics is blood rheology in microchannels, which has played a key role in recent developments of lab-on-chip devices for blood sampling and analysis [20–23]. Blood is a non-Newtonian fluid, containing an extremely rich amount of information about the physiological and pathological state of the human body. Hence, there is an evident interest to develop microfluidic devices for the diagnosis of major diseases, such as cancer, diabetes mellitus, and cardiovascular disorders. The traditional method for creating the microfluidic devices is soft lithography, using molds obtained by photolithography. By using this fabrication technique Shevkoplyas *et al.* [24] developed a microdevice to isolate white blood cells (WBCs) from a blood sample by using the margination effect, whereas Hou *et al.* [25] have very recently proposed a biomimetic separation device to separate normal and malaria infected red blood cells (RBCs). Other researchers, such as Faivre *et al.* [26], Sollier *et al.* [27], and Pinho *et al.* [22], have demonstrated that the cell free layer (CFL) could be enhanced by using a microchannel containing a constriction followed by a sudden expansion to separate plasma from the whole *in vitro* blood.

Although the results are extremely encouraging, the high costs and time involved in the traditional photolithography process are currently slowing down the development and diffusion of biomimetic separation biochips among a large research and industrial community. Hence, there is a demand for low-cost and short fabrication time methods to produce microfluidic systems. Here, we introduce a low-cost nonlithographic technique, known as bioxurography, due to its fast fabrication time, independence from clean room facilities and low material and equipment cost, as an ideal solution to manufacture affordable microfluidic devices to work on the scales required to perform blood cell analysis in continuous flow.

2. Materials and Methods

2.1. Geometries of the Microchannels

The molds for the microfluidic devices tested in this study were fabricated using a xurography technique and consist of two main geometries: a microchannel with a diverging and converging bifurcation and a flow-focusing microfluidic device. Figures 1 and 2 show the geometries and main dimensions of the microfluidic devices tested in this study. For the first case (see Figure 1), the geometries were selected taking into account a previous study performed on the blood flow behavior through microchannels with bifurcations and confluences fabricated by a soft lithography technique [7]. Here, due to the limitation of the xurography technique, the minimum dimension of the microchannel width was around 150 µm. Hence, the parent microchannel had a minimum width of 300 µm and maximum width of 1000 µm whereas the two branches of the bifurcation and confluence correspond to 50% of parent channel width. Chips with three different channel dimensions were studied; with parent channel dimensions of 300, 500 and 1000 µm.

The geometry used for the second biomedical application was based on a flow-focusing device often used to generate droplets [28]. For this particular example, we tested this device to generate bubbles of air flowing through *in vitro* blood.

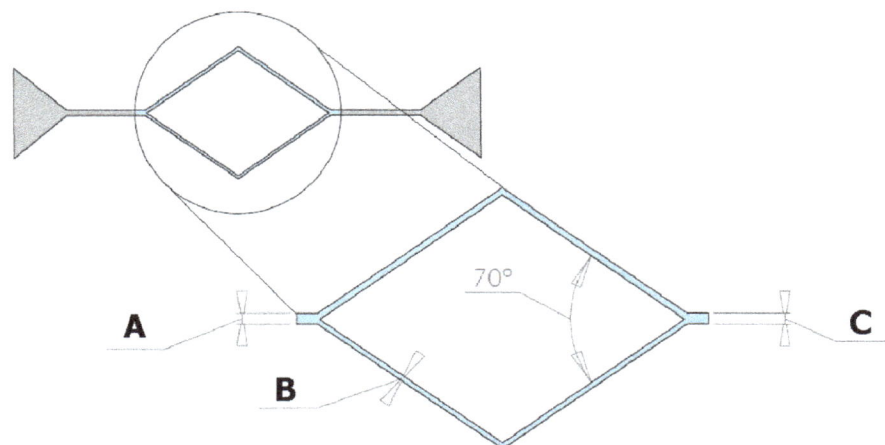

Figure 1. Schematic diagram of the microchannel geometry with a diverging and converging bifurcation where the width of channels A and C are 300, 500, or 1000 μm, and the widths of respective ramifications B are 150, 250, or 500 μm.

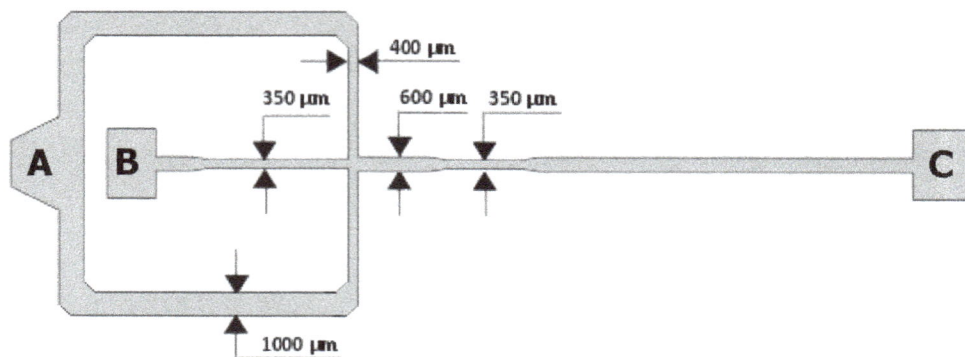

Figure 2. Schematic diagram of the microfluidic flow-focusing device geometry used to generate bubbles where continuous phase (*in vitro* blood) and dispersed phase (air) are introduced in A and B, respectively, and C is the outlet of mixture of both phases.

2.2. Fabrication of the Microfluidic Devices

The tested microchannels were fabricated by a xurography technique, which can be performed at low cost in a straightforward manner. Specifically, the fabrication process, which is outlined in Figure 3, was as follows:

(a) The microchannel geometries were drawn using a CAD software, and then the molds were cut by using a cutting plotter Jaguar II (GCC *Innovation*, Capelle aan den Ijssel, The Netherlands).
(b) Adhesive paper was used to transfer and place the mold master inside the petri dish.
(c) The PDMS prepolymer was prepared by mixing a commercial prepolymer and a curing agent (10:1 ratio) and poured onto master mold in the petri dish and cured in an oven at 80 °C for 20 min.
(d) The PDMS (20:1 ratio) was spin coated over a glass slide and cured in an oven at 80 °C for 20 min.
(e) By using a blade, the microchannels were cut off and the inlet/outlet holes of the fluid were done by using a fluid-dispensing tip.
(f) Finally, the channels were sealed by the PDMS covered glass slides. To have a strong adhesion of the materials, the device was placed in the oven at 80 °C for 24 h.

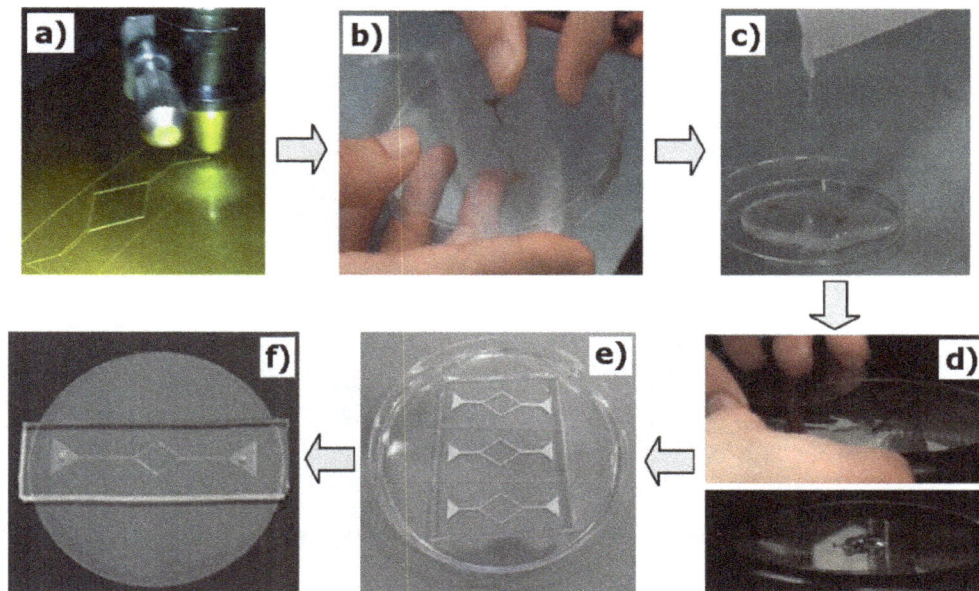

Figure 3. Main steps of the fabrication procedure: (**a**) Cutting the CAD geometries in the vinyl paper by means of a cutting plotter; (**b**) transfer the mold to the petri dish using an adhesive; (**c**) pouring the polymer PDMS over the mold to obtain the microchannel; (**d**) spin coating a slide glass with PDMS; (**e**) remove the cured PDMS microchannel from the vinyl mold; (**f**) three-dimensional PDMS microchannel with the input/output ports sealed with the spin-coated glass slide.

2.3. Working Fluids

The working fluid used in the first tested device (see Figure 1) was ovine red blood cells suspended with dextran 40 (Dx40). In this study we have used a hematocrit (Hct) of 5% and 15%. For the case of the flow-focusing device, the working fluids were air and Dx40 containing about 10% Hct of ovine RBCs. Briefly, blood was collected from a healthy ovine; heparin was added in order to prevent coagulation. The RBCs were separated from bulk blood by centrifugation (2000 rpm for 15 min at 4 °C) and then the plasma and buffy coat were removed by aspiration. The RBCs were then washed twice with a physiological saline solution and diluted with Dx40 to make up the required RBC concentration. All blood samples were stored hermetically at 4 °C until the experiments were performed at a room temperature of approximately 20 °C.

2.4. Experimental Set-Up

The high-speed video microscopy system used in both studies is shown in Figure 4. The main part of this system consists of an inverted microscope (IX71, Olympus, Tokyo, Japan) and a high-speed camera (i-SPEED LT, Olympus). All the microfluidic devices were placed on the stage of the inverted microscope. A syringe pump (Harvard Apparatus PHD ULTRA, Holliston, MA, USA) was used to produce a constant flow rate. Note that, for the flow-focusing device, an additional pressure pump (Elveflow AF1 PG1113, Paris, France) was used to inject air at a constant pressure in the microchannel. Figure 4 shows the main experimental equipment used to control and visualize the fluids flowing through the microchannels.

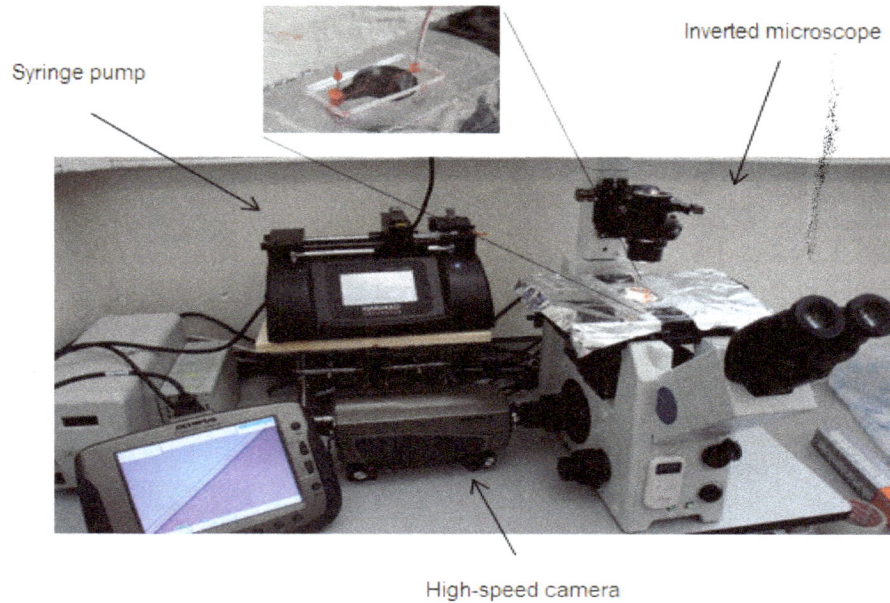

Figure 4. Main experimental equipment used to control and visualize the flow in microchannels produced by xurography.

2.5. Image Analysis

A manual tracking plugin (MTrackJ) of the image analysis software ImageJ [29] was used to track individual RBC flowing around the boundary of the RBCs core (Point 2 in Figure 5). By using the MTrackJ plugin, the centroid of the selected RBC was automatically computed [30]. Besides, the nearest wall position was measured (Point 1 in Figure 5) so that the distance between the tracked RBCs and the wall can be calculated, which consequently draws the CFL thickness (Figure 5).

To visualize the CFL formation in the microchannels, the captured videos were converted to a sequence of static images (stack) and then, by using "Z project" function in ImageJ, it was possible to obtain a single image having a sum of all static images. In this image, the darker region indicates RBC core and brighter region indicates CFL.

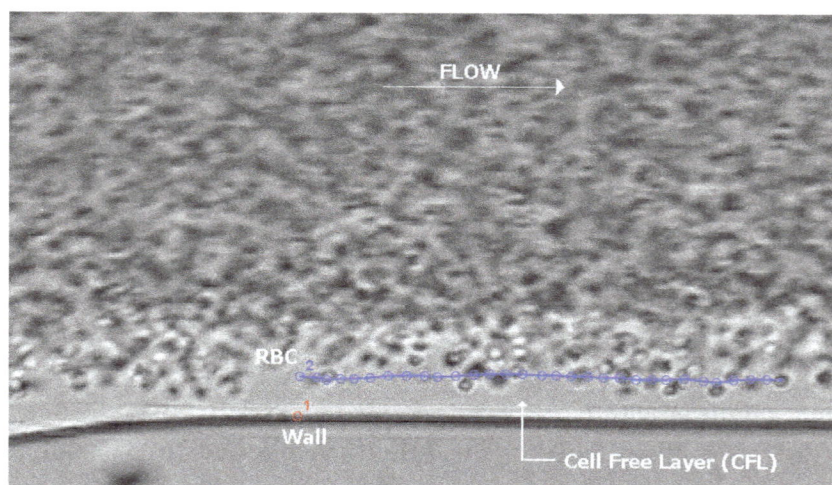

Figure 5. Tracking an individual red blood cell (RBC) flowing around the boundary of the RBCs core.

3. Results and Discussion

3.1. Characteristics of the Microchannels

To evaluate the geometrical quality of the mold masters and correspondent microchannels, several microscopic images were obtained along a device with a simple geometry. Light micrographs of the PDMS microchannels, fabricated using the proposed technique are shown in Figure 6. It also shows a schematic representation of the sections where the microscopic images were taken.

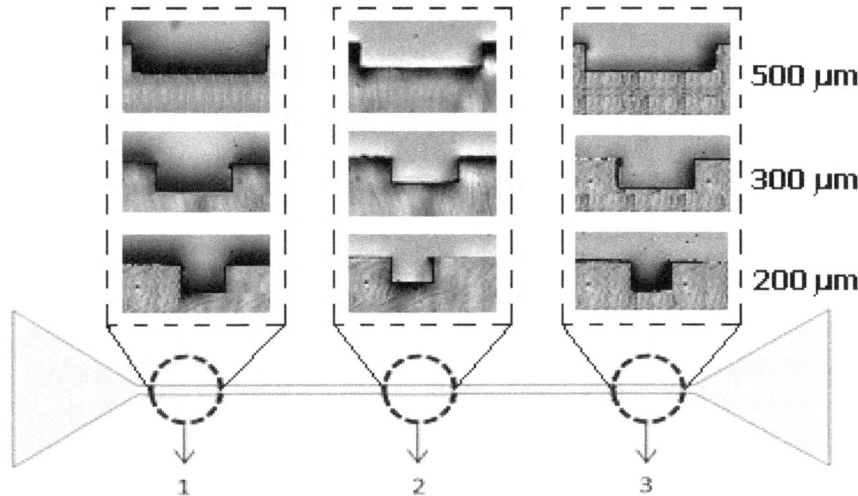

Figure 6. Schematic representation of the microchannel geometry and location of the sections where the microscopic images were collected to evaluate the geometrical quality. From left to right we have: (**1**) entrance, (**2**) middle, and (**3**) exit.

The geometrical qualities of the PDMS microchannels were evaluated using ImageJ and the ranges of values measured for all sections along the microchannels are shown in Figures 7 and 8. Figure 7 shows the comparison between the AutoCAD projected values and actual width measurements of the PDMS microchannels and consequently from the vinyl master molds. Figure 8 shows the depth measurements of the PDMS microchannels at three different sections, *i.e.*, entrance (1), middle (2), and exit (3).

Detailed microscopic visualizations have shown that the difference between the microchannel and the actual dimensions of the PDMS microchannel tends to increase by decreasing the size of the geometry. For instance, in Figure 7, it is possible to observe that the highest difference occurs for the smallest microchannels. This is mainly related to the limitation of our cutting plotter to cut precisely geometries with dimensions smaller than 500 μm, *i.e.*, microchannels with 300 and 200 μm widths. In contrast, no significant difference was observed for the depth of microchannels (99.4 ± 1.4 μm).

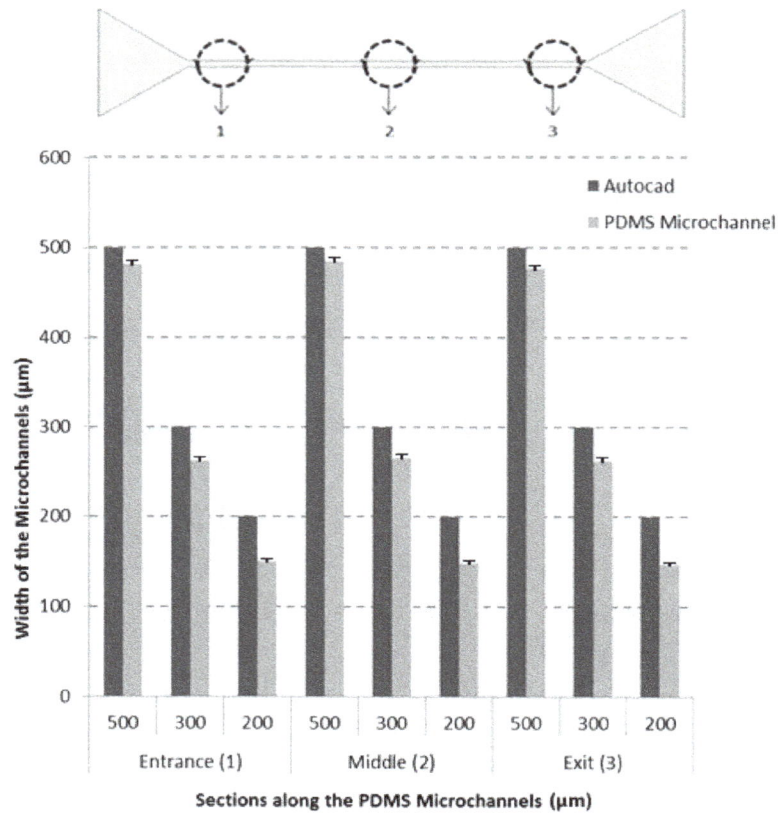

Figure 7. Comparison between the values projected from AutoCAD and width measurements of the PDMS microchannels at three different sections: entrance (**1**), middle (**2**), and exit (**3**). The measurements are expressed as the mean ± standard deviation according to a *t*-test analysis at 95% confidence interval.

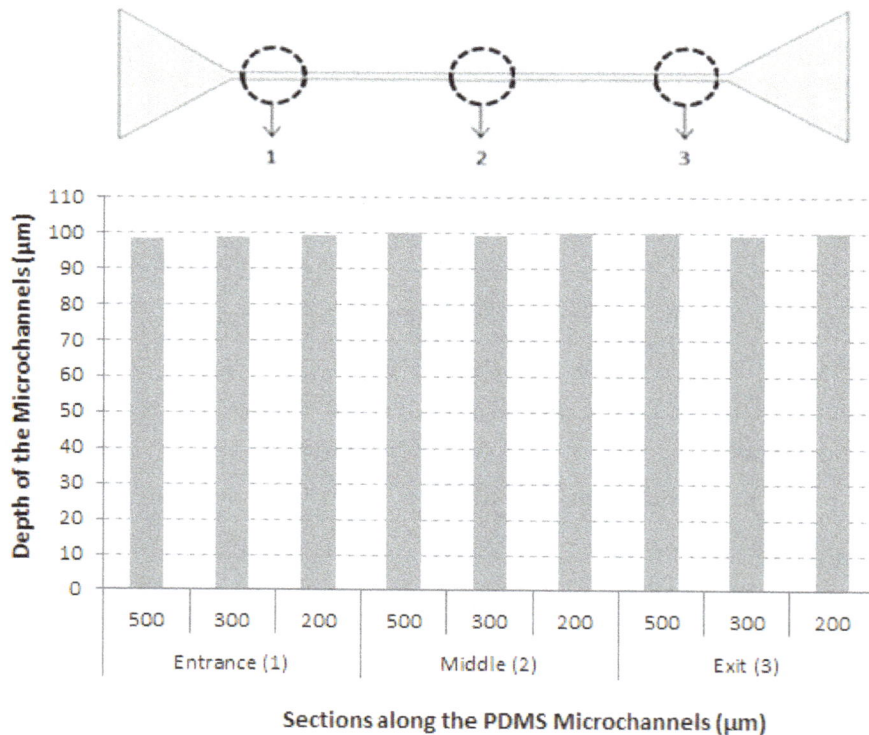

Figure 8. Depth measurements in three different sections: entrance (**1**), middle (**2**), and exit (**3**).

3.2. Cell-Free Layer (CFL) Measurements and Visualization

In addition to a tendency of the RBCs to undergo axial migration induced by the cells' tank treading motion, high wall shear stress forces the RBCs to move towards the center of the vessel. As a result, there is a formation of CFL with extremely low concentration of cells along the walls of the microchannels. In straight microchannels, the CFL can be defined as the distance between the microchannel wall and the boundary region of the RBCs core. Although the formation of the CFL *in vivo* [31–33] has been of great interest over many years, little information is available about this phenomenon mainly due to the limitation of the measurement techniques. Recently, several researchers [7,22,34] have attempted to replicate this phenomenon *in vitro* using biomedical microdevices in order to better understand it and explore its potential as a new diagnostic tool. Here, we show the effect of the Hct and geometry of a microcahnnel fabricated by xurography on CFL formation (Figure 9).

Figure 10 shows the results for CFL thickness for a constant flow rate of 10 µL/min, of 5% Hct and 15% Hct in a microchannel with a diverging and converging bifurcation for a width of the parent channel (Wp) of 300, 500, and 1000 µm. Note that all the CFL measurements were executed where the CFL had a steady formation, *i.e.*, at the regions where the CFL is parallel to the wall. The results show that the CFL thickness is not strongly affected by the bifurcation, whereas it decreases with increasing Hct. This latter result corroborates with the past results performed in both *in vivo* [31,33] and *in vitro* [34] experiments.

Figure 11 shows the visualization of *in vitro* blood with 15% Hct flowing in a microchannel with a bifurcation followed by a confluence. All original images were converted by means of the "Z project" function (ImageJ) to a single image having a sum of all static images. The qualitative results show clearly the formation of a CFL not only around the walls but also in the region immediately after the confluence.

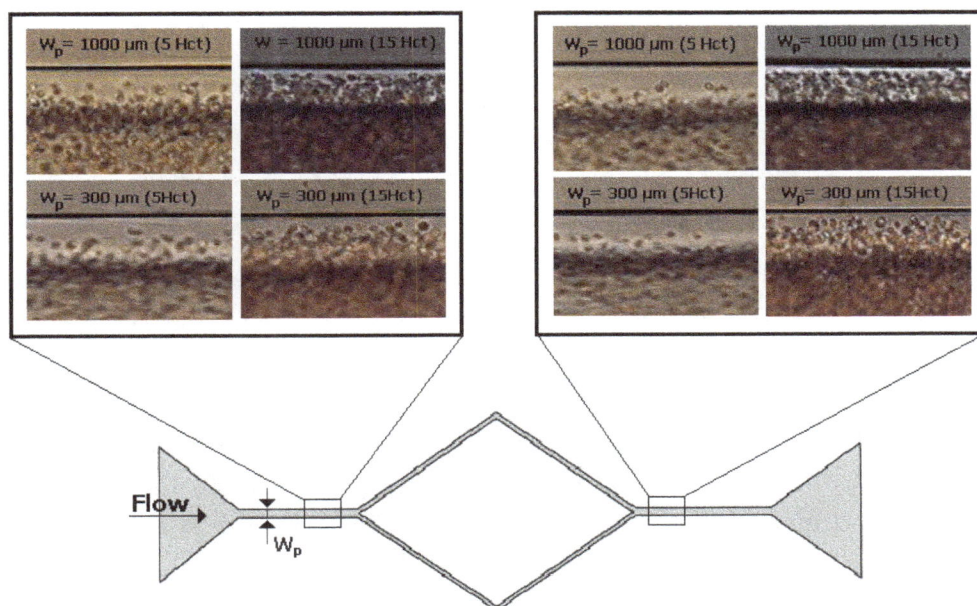

Figure 9. Visualization of *in vitro* blood with 5% Hct and 15% Hct flowing in a microchannel geometry with a diverging and converging bifurcation for a width of the parent channel (Wp) of 300 and 1000 µm. The flow rate was 10 µL/min.

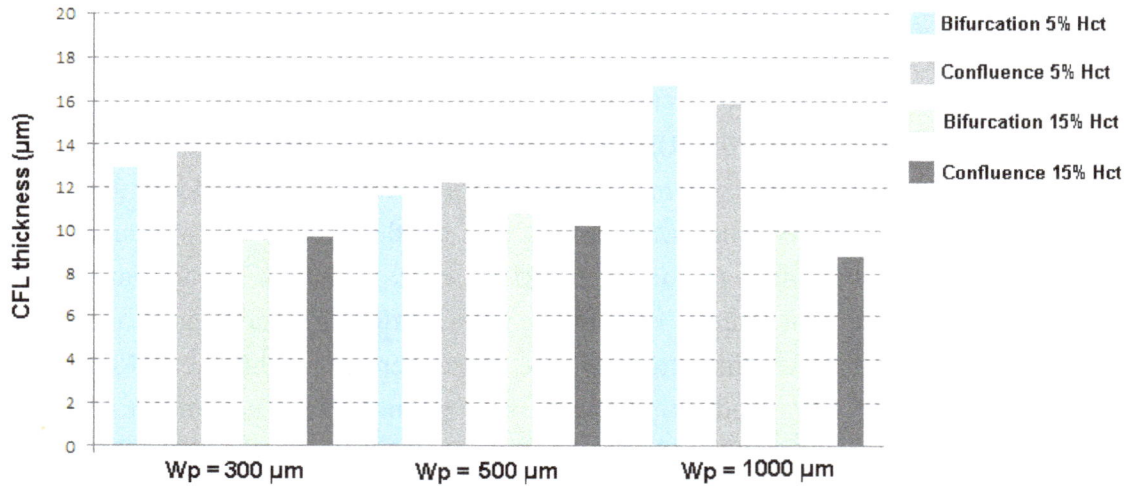

Figure 10. CFL thickness measured in blood flows of 5% Hct and 15% Hct in a microchannel with a diverging and converging bifurcation for a width of the parent channel (Wp) of 300, 500, and 1000 μm. The flow rate was 10 μL/min.

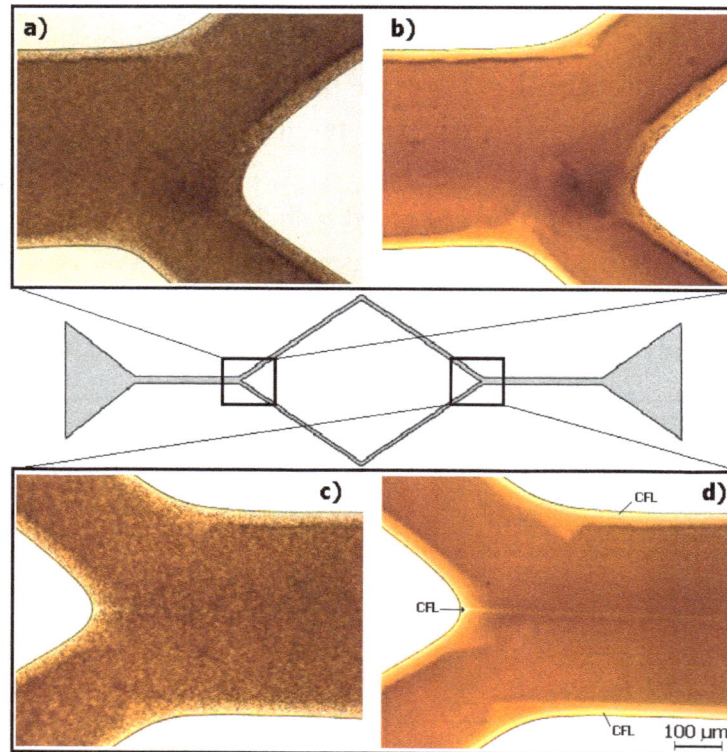

Figure 11. Visualization of *in vitro* blood with 15% Hct flowing in a microchannel geometry with (**a**) diverging bifurcation (original image); (**b**) diverging bifurcation (the sum of all images); (**c**) converging bifurcation or confluence (original image), and (**d**) converging bifurcation or confluence (the sum of all images). The flow rate was 10 μL/min.

These results corroborate the works of Ishikawa *et al.* [35] and Leble *et al.* [7] that found a formation of a CFL in the region of the confluence apex. This phenomenon was observed in a microchannel three times smaller than the one used in the current study as they used a photolithographic technique to fabricate the molds for the microchannels. The current work has shown evidence that at width dimensions up to 500 μm there still is a formation of a CFL around the apex of the confluence. Hence, by using xurography it is possible to create an artificial CFL under appropriate hemodynamic and geometrical conditions, and as a result, to develop low cost microfluidic systems for biomedical applications, such as blood cell separation and analysis.

3.3. Bubbles Generation and Visualization

An air embolism is a pathological condition caused by bubbles trapped in a blood vessel [36,37]. During surgeries and other medical procedures, small amounts of air may get into the blood stream accidentally. Air bubbles may also be formed in the blood stream due to cavitation in prosthetic heart valves or during decompression after scuba diving activities [37]. When an air bubble flows through a blood vessel, it may block small arteries and stop blood supply to a certain area of the body. Although it could be fatal if a large amount of bubbles flows through blood arteries, there is not enough information about the amount of air bubbles, which might cause death. It is known that the rate of reabsorption of microbubbles may be accelerated by the injection in the blood stream of surfactants [38]. Hence, it is important to improve our understanding on the flow of bubbles in microchannels with dimensions similar to *in vivo* blood vessels. There is also a demand for biochips for the development of new strategies and new surfactant based drugs to treat air embolisms. The xurographic technique was, therefore, used to developed microfluidic systems capable of generating air bubbles moving through the microchannel (Figure 12).

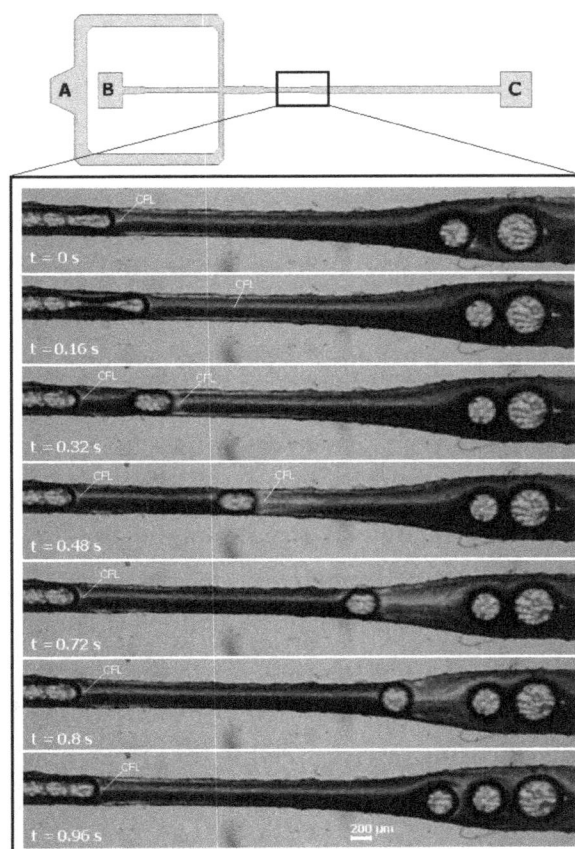

Figure 12. Visualization of bubbles generation and *in vitro* blood flow containing bubbles through a flow-focusing device fabricated by xurography.

Figure 12 shows bubbles generation and *in vitro* blood flow containing bubbles in a flow-focusing device fabricated by xurography. The liquid phase (*in vitro* blood) enters at inlet A with a flow rate of 3 μL/min, whereas the gas phase (air) enters at inlet B with a pressure of 25 mbar. The gas and liquid streams flow along the main channel and two side channels, respectively, meeting at the cross-junction. Two-phase flow occurs in the main channel and exits at the outlet C. The size of the cross-section of the main channel is 600 μm wide and 100 μm high, but at the section near the junction point the width is 350 μm (Figure 12). The bubbles' generation is as follows: the dispersed phase (air) is squeezed by two counter-streaming blood flows of the carrier phase, forcing the gas to breakup and to generate bubbles. Under these operating conditions, the bubbles were formed within the 350 μm wide microchannel, as shown in Figure 12. Two kinds of bubbles can be observed: slug bubbles, also known as Taylor bubbles, and spherical bubbles. First, bubbles with a slug shape are formed and keep their shape until they reach the 600 μm smooth expansion, where they acquired a circular shape. These slug bubbles flow through the microchannel, separated from each other by liquid slugs and from the wall by a thin liquid film. As expected, due to a lower cross section area of the 350 μm microchannel, the velocity of slug bubbles is higher than the spherical bubbles located in the expansion region of the microchannel. It is worth mentioning the formation of large bubbles in the expansion region of the microchannel. These bubbles are formed mainly due to the collision between them, which leads to coalescence. Another point of interest is the formation of a cell free layer among bubbles at the center of the microchannel. This phenomenon is presented and discussed in more detail elsewhere [39].

The qualitative results demonstrate that the proposed nonlithographic technique can generate air bubbles in a flow-focusing device and consequently enables the study of air embolism at a microscale level. In the near future, by using this technique, we plan to study in detail the flow behavior of air bubbles flowing in blood suspensions through geometries similar to *in vivo* microvessels.

4. Conclusions

Xurography was used to produce molds for the fabrication of microchannels by PDMS soft lithography. The accuracy of the method was evaluated by comparing the dimensions of the PDMS channels obtained with the nominal dimension of the AutoCAD design. The microchannels produced were applied to the study of two problems related to the blood flow in microchannels, the formation a CFL in bifurcations and confluences and the flow of microbubbles mimicking air embolisms.

Detailed analysis of the geometries has shown that the quality of the microchannel tends to decrease as its size decreases. This is mainly related to the limitation of our cutting plotter to cut precise geometries with dimensions smaller than 500 μm.

The molds and correspondent PDMS microchannels had good enough quality to study blood flow phenomena at the microscale level. It was possible to perform experiments to measure the CFL thickness with results consistent with the ones reported elsewhere in the literature. It was also possible to generate air bubbles by a flow-focusing device suitable for research of air embolisms.

Acknowledgments

The authors acknowledge the financial support provided by PTDC/SAU-BEB/105650/2008, PTDC/SAU-ENB/116929/2010, EXPL/EMS-SIS/2215/2013 and scholarship SFRH/BD/89077/2012 and SFRH/BD/97658/2013 from FCT (Science and Technology Foundation), COMPETE, QREN and European Union (FEDER).

Author Contributions

Elmano Pinto developed the fabrication process and performed the main experimental work. All authors have analyzed the data. Elmano Pinto has written the main manuscript and all authors reviewed the manuscript.

Conflicts of Interest

The authors declare no conflict of interest.

References

1. Garcia, V.; Dias, R.; Lima, R. *In vitro* blood flow behaviour in microchannels with simple and complex geometries. In *Applied Biological Engineering—Principles and Practice*; Naik, G.R., Ed. InTech: Rijeka, Croatia, 2012; pp. 393–416.
2. Lima, R.; Nakamura, M.; Omori, T.; Ishikawa, T.; Wada, S.; Yamaguchi, T. *Advances in Computational Vision and Medical Image Processing*; Springer: Dordrecht, The Netherlands, 2009; pp. 203–220.
3. Nguyen, N.T.; Wereley, S.T. *Fundamentals and Applications of Microfluidics*, 2nd ed.; Artech House: Boston, MA, USA, 2006.
4. Stone, H.A.; Kim, S. Microfluidics: Basic issues, applications, and challenges. *AIChE J.* **2001**, *47*, 1250–1254.
5. Beebe, D.J.; Mensing, G.A.; Walker, G.M. Physics and applications of microfluidics in biology. *Ann. Rev. Biomed. Eng.* **2002**, *4*, 261–286.
6. Duffy, D.C.; McDonald, J.C.; Schueller, O.J.A.; Whitesides, G.M. Rapid prototyping of microfluidic systems in poly(dimethylsiloxane). *Anal. Chem.* **1998**, *70*, 4974–4984.
7. Leble, V.; Lima, R.; Dias, R.; Fernandes, C.; Ishikawa, T.; Imai, Y.; Yamaguchi, T. Asymmetry of red blood cell motions in a microchannel with a diverging and converging bifurcation. *Biomicrofluidics* **2011**, *5*, doi:10.1063/1.3672689.
8. Lima, R.; Wada, S.; Tanaka, S.; Takeda, M.; Ishikawa, T.; Tsubota, K.; Imai, Y.; Yamaguchi, T. *In vitro* blood flow in a rectangular pdms microchannel: Experimental observations using a confocal micro-piv system. *Biomed. Microdevices* **2008**, *10*, 153–167.
9. Thomas, M.S.; Millare, B.; Clift, J.M.; Bao, D.; Hong, C.; Vullev, V.I. Print-and-peel fabrication for microfluidics: What's in it for biomedical applications? *Ann. Biomed. Eng.* **2010**, *38*, 21–32.
10. Tan, A.; Rodgers, K.; Murrihy, J.; O'Mathuna, C.; Glennon, J.D. Rapid fabrication of microfluidic devices in poly(dimethylsiloxane) by photocopying. *Lab Chip* **2001**, *1*, 7–9.
11. Branham, M.L.; Tran-Son-Tay, R.; Schoonover, C.; Davis, P.S.; Allen, S.D.; Shyy, W. Rapid prototyping of micropatterned substrates using conventional laser printers. *J. Mater. Res.* **2002**, *17*, 1559–1562.

12. Bao, N.; Zhang, Q.; Xu, J.J.; Chen, H.Y. Fabrication of poly(dimethylsiloxane) microfluidic system based on masters directly printed with an office laser printer. *J. Chromatogr. A* **2005**, *1089*, 270–275.

13. Chen, C.-S.; Breslauer, D.N.; Luna, J.I.; Grimes, A.; Chin, W.C.; Lee, L.P.; Khine, M. Shrinky-dink microfluidics: 3D polystyrene chips. *Lab Chip* **2008**, *8*, 622–624.

14. Bonyár, A.; Sántha, H.; Ring, B.; Varga, M.; Gábor Kovács, J.; Harsányi, G. 3D rapid prototyping technology (rpt) as a powerful tool in microfluidic development. *Procedia Eng.* **2010**, *5*, 291–294.

15. Waldbaur, A.; Rapp, H.; Länge, K.; Rapp, B.E. Let there be chip—Towards rapid prototyping of microfluidic devices: One-step manufacturing processes. *Anal. Methods* **2011**, *3*, 2681–2716.

16. Klank, H.; Kutter, J.P.; Geschke, O. Co 2-laser micromachining and back-end processing for rapid production of pmma-based microfluidic systems. *Lab Chip* **2002**, *2*, 242–246.

17. Kaigala, G.V.; Ho, S.; Penterman, R.; Backhouse, C.J. Rapid prototyping of microfluidic devices with a wax printer. *Lab Chip* **2007**, *7*, 384–387.

18. Bartholomeusz, D.A.; Boutte, R.W.; Andrade, J.D. Xurography: Rapid prototyping of microstructures using a cutting plotter. *J. Microelectromech. Syst.* **2005**, *14*, 1364–1374.

19. Sundberg, S.O.; Wittwer, C.T.; Greer, J.; Pryor, R.J.; Elenitoba-Johnson, O.; Gale, B.K. Solution-phase DNA mutation scanning and snp genotyping by nanoliter melting analysis. *Biomed. Microdevices* **2007**, *9*, 159–166.

20. Abkarian, M.; Faivre, M.; Horton, R.; Smistrup, K.; Best-Popescu, C.A.; Stone, H.A. Cellular-scale hydrodynamics. *Biomed. Mater.* **2008**, *3*, doi:10.1088/1748-6041/3/3/034011.

21. Gossett, D.R.; Weaver, W.M.; Mach, A.J.; Hur, S.C.; Tse, H.T.; Lee, W.; Amini, H.; Di Carlo, D. Label-free cell separation and sorting in microfluidic systems. *Anal. Bioanal. Chem.* **2010**, *397*, 3249–3267.

22. Pinho, D.; Yaginuma, T.; Lima, R. A microfluidic device for partial cell separation and deformability assessment. *Biochip J.* **2013**, *7*, 367–374.

23. Yaginuma, T.; Oliveira, M.S.; Lima, R.; Ishikawa, T.; Yamaguchi, T. Human red blood cell behavior under homogeneous extensional flow in a hyperbolic-shaped microchannel. *Biomicrofluidics* **2013**, *7*, doi:10.1063/1.4820414.

24. Shevkoplyas, S.S.; Yoshida, T.; Munn, L.L.; Bitensky, M.W. Biomimetic autoseparation of leukocytes from whole blood in a microfluidic device. *Anal. Chem.* **2005**, *77*, 933–937.

25. Hou, H.W.; Bhagat, A.A.; Chong, A.G.; Mao, P.; Tan, K.S.; Han, J.; Lim, C.T. Deformability based cell margination—A simple microfluidic design for malaria-infected erythrocyte separation. *Lab Chip* **2010**, *10*, 2605–2613.

26. Faivre, M.; Abkarian, M.; Bickraj, K.; Stone, H.A. Geometrical focusing of cells in a microfluidic device: An approach to separate blood plasma. *Biorheology* **2006**, *43*, 147–159.

27. Sollier, E.; Cubizolles, M.; Fouillet, Y.; Achard, J.L. Fast and continuous plasma extraction from whole human blood based on expanding cell-free layer devices. *Biomed. Microdevices* **2010**, *12*, 485–497.

28. Baroud, C.N.; Gallaire, F.; Dangla, R. Dynamics of microfluidic droplets. *Lab Chip* **2010**, *10*, 2032–2045.

29. Abramoff, M.D.; Magalhaes, P.J.; Ram, S.J. Image processing with ImageJ. *Biophoton. Int.* **2004**, *11*, 36–42.

30. Meijering, E.; Smal, I.; Danuser, G. Tracking in molecular bioimaging. *IEEE Signal Proc. Mag.* **2006**, *23*, 46–53.

31. Kim, S.; Ong, P.K.; Yalcin, O.; Intaglietta, M.; Johnson, P.C. The cell-free layer in microvascular blood flow. *Biorheology* **2009**, *46*, 181–189.

32. Maeda, N.; Suzuki, Y.; Tanaka, J.; Tateishi, N. Erythrocyte flow and elasticity of microvessels evaluated by marginal cell-free layer and flow resistance. *Am. J. Physiol.* **1996**, *271*, 2454–2461.

33. Tateishi, N.; Suzuki, Y.; Soutani, M.; Maeda, N. Flow dynamics of erythrocytes in microvessels of isolated rabbit mesentery: Cell-free layer and flow resistance. *J. Biomech.* **1994**, *27*, 1119–1125.

34. Lima, R.; Oliveira, M.S.; Ishikawa, T.; Kaji, H.; Tanaka, S.; Nishizawa, M.; Yamaguchi, T. Axisymmetric polydimethysiloxane microchannels for *in vitro* hemodynamic studies. *Biofabrication* **2009**, *1*, doi:10.1088/1758-5082/1/3/035005.

35. Ishikawa, T.; Fujiwara, H.; Matsuki, N.; Yoshimoto, T.; Imai, Y.; Ueno, H.; Yamaguchi, T. Asymmetry of blood flow and cancer cell adhesion in a microchannel with symmetric bifurcation and confluence. *Biomed. Microdevices* **2011**, *13*, 159–167.

36. Barak, M.; Katz, Y. Microbubbles: Pathophysiology and clinical implications. *Chest. J.* **2005**, *128*, 2918–2932.

37. Papadopoulou, V.; Tang, M.X.; Balestra, C.; Eckersley, R.J.; Karapantsios, T.D. Circulatory bubble dynamics: From physical to biological aspects. *Adv. Colloid Interface Sci.* **2014**, *206*, 239–249.

38. Branger, A.B.; Eckmann, D.M. Accelerated arteriolar gas embolism reabsorption by an exogenous surfactant. *Anesthesiology* **2002**, *96*, 971–979.

39. Sousa, L. Estudo de Embolias Gasosas em Microcanais. Master's Thesis, Instituto Politécnico de Bragança, Bragança, Portugal, 2013. (In Portuguese)

7

Sub-Micrometer Size Structure Fabrication Using a Conductive Polymer

Junji Sone [1,†,*], **Katsumi Yamada** [1,†], **Akihisa Asami** [2] **and Jun Chen** [1]

[1] Faculty of Engineering, Tokyo Polytechnic University, 1583 Iiyama Atsugi, Kanagawa 243-0297, Japan; E-Mails: kyamada@chem.t-kougei.ac.jp (K.Y.); chen@mega.t-kougei.ac.jp (J.C.)

[2] Graduation School of Engineering, Tokyo Polytechnic University, 1583 Iiyama Atsugi, Kanagawa 243-0297, Japan; E-Mail: siegs-welch@docomo.ne.jp

† These authors contributed equally to this work.

* Author to whom correspondence should be addressed; E-Mail: sone@cs.t-kougei.ac.jp

Academic Editor: Cheng Luo

Abstract: Stereolithography that uses a femtosecond laser was employed as a method for multiphoton-sensitized polymerization. We studied the stereolithography method, which produces duplicate solid shapes corresponding to the trajectory of the laser focus point and can be used to build a three-dimensional (3D) structure using a conductive polymer. To achieve this, we first considered a suitable polymerization condition for line stereolithography. However, this introduced a problem of irregular polymerization. To overcome this, we constructed a support in the polymerized part using a protein material. This method can stabilize polymerization, but it is not suited for building 3D shapes. Therefore, we considered whether heat accumulation causes the irregular polymerization; consequently, the reduction method of the repetition rate of the femtosecond laser was used to reduce the heating process. This method enabled stabilization and building of a 3D shape using photo-polymerization of a conductive polymer.

Keywords: stereolithography; conductive polymer; multi-photon; femto second laser; protein; repetition rate

1. Introduction

Conductive polymers, such as polypyrrole, polythiophene, polyaniline, and their derivatives, are very useful materials for opto-electronics and nanotechnology. Their applications include molecular wires, semiconductors, display devices, biosensors, and molecular actuators [1]. Many conducting polymers have been prepared mainly by chemical polymerization or electrochemical polymerization. On the other hand, their polymerization and patterning were simultaneously realized by photochemical polymerization. Using photo-polymerization, a two dimensional pattern of conductive materials can be obtained on the substrate [2–4]. Their process resolution was of micrometer order.

In a recent study, multi-photon stereolithography (SL) using a femtosecond laser has been investigated [5–7]. Focused illumination of such a laser limits the multi-photon absorption process to a space narrower than the laser wavelength. In a previous study conducted by our group, we reported two-dimensional patterning [8] and three-dimensional patterning [9] on a substrate as well as three-dimensional photo-polymerization of a conductive polymer (CP) in a transparent Nafion polymer sheet [10]. However, it is difficult to dissolve a Nafion sheet. Further, polymerization is very slow in an aqueous solution, and a long polymerization time leads to irregular and unstable solidification, which leads to difficulties in building complex three-dimensional (3D) shapes with high accuracy [9]. Improvements have been made in the spatial resolution of SL through temperature control of the resin [11]. Moreover, SL in a solid material that is soluble has many advantages for removing non-polymerized area and spatial resolution [12]. Another factor of unstable fabrication is repetition rate of the femtosecond laser. A high repetition rate resulted in the accumulation of heat [13].

CP is a new material for femtosecond SL. Therefore, we examined the SL condition using basic shapes. First, we searched for the optimal line SL condition using an 8 MHz repetition rate of the laser. In this case, we used high speed scanning to avoid irregular polymerization, but stabilization of the polymerization was difficult in achieving the designed thickness. We tried to achieve high accuracy of the 3D shape by considering a support material. Consequently, we used a soluble protein film (an aqueous solid) that includes the CP, a catalyst and gelatine. In this case, we could stabilize the solidification by repeatedly drawing each trajectory. However, we could not improve the height accuracy of the SL parts. Finally, we deemed that heat accumulation was the reason for the irregular polymerization. Thus, we searched for a more suitable condition of 3D SL by varying the scan speed, number of trajectory repeats and the repetition rate of the laser. Next, we realized the SL condition that produces similar height SL parts as designed shapes. In addition, we considered a shrinkage compensation method for the top layer. Furthermore, we considered the use of our SL parts for metamaterial [14] and so on.

2. Experimental Methods

2.1. Aqueous Solution

An aqueous solution was used as the polymerization solution that contains 0.1 mM of lithium tetrafluoroborate, 0.2 M of pyrrole, 1 mM of methylviologen, and 1 mM of $Ru(bpy)_3^{2+}$. The Glass with Au electrode was immersed in the polymerization solution.

2.2. Stereolithography System

Figure 1 shows construction of SL system. Excitation was provided by a mode-locked Ti/sapphire laser. The repetition rate was 8 MHz and wavelength is 850 nm. The laser pulses have a pulse width of 150 fs with a repetition rate of 80 MHz. The laser beam was tightly focused by a water immersion objective lens (NA = 0.95, Nikon Plan Apo 100×, Nikon, Tokyo, Japan). The illuminated areas were transferred under computer-control by shuttering the beam and driving the substrate using XYZ stage.

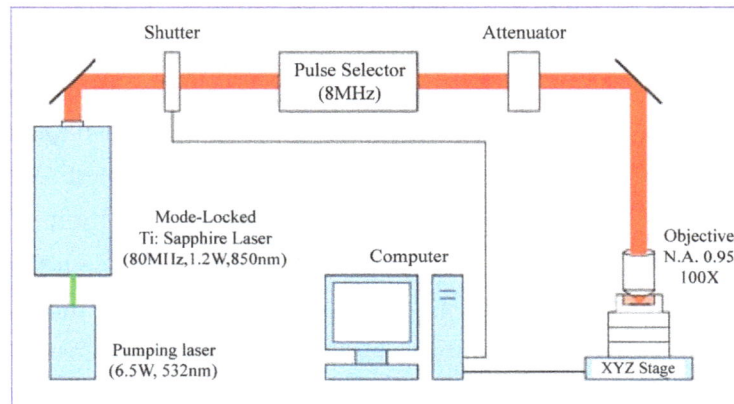

Figure 1. Stereolithography system.

2.3. Fabrication Method

2.3.1. Line SL

We searched line SL optimal condition before make a three dimensional structure. Repetition rate of femotosecond laser is 8 MHz. Figure 2 shows the trajectory of the line SL. The first layer was three lines with a horizontal interval pitch Xp and a second layer consisting of two lines with a horizontal interval pitch Xp and height interval pitch Zp. We used a high scanning speed of 2–8 μm/s to avoid irregular polymerization. Moreover, we tried drawing five times on each trajectory to produce a thick polymerization shape.

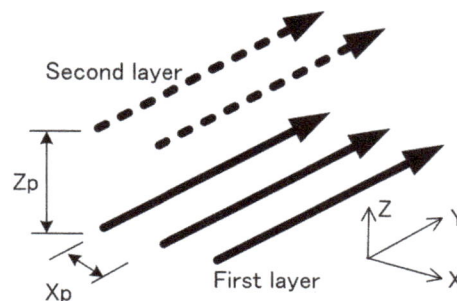

Figure 2. Trajectory of Line stereolithography (SL).

2.3.2. Support Method

To stabilize the irregular polymerization, we used the dissolve support, which can clear away after SL experiment. Note that the SL method necessitates supporting structures under overhanging parts.

Although constructing such supports is feasible using micro SL, it is very difficult to eliminate these parts afterwards. We thus considered that using a dissolvable support would mean we could break up the supports easily after the SL experiment concluded. Then, we considered two methods. In the first method, we used a 20-μm collagen film that included an aqueous solution. The setup is shown in Figure 3a.

In the second method, we added 2% gelatine by weight to the aqueous solution and cooled the mixture to 10 °C for 30 min to achieve solidification. After the SL experiment, the SL part was washed using 40 °C deionized water to dissolve the exception of the SL part. The setup for the SL system is shown in Figure 3b. In this system, the temperature of the aqueous solid was controlled at 10 °C using a thermoelectric module. In the following contents, we described this method as support case.

Figure 2 shows the trajectory of the line SL. The first layer was three lines with a horizontal interval pitch Xp and a second layer consisting of two lines with a horizontal interval pitch Xp and height interval pitch Zp.

We confirmed three dimensional SL using simple trajectory. Figure 4 shows the 3D SL trajectory. The first layer has three lines with a horizontal interval pitch, Xp. The second layer has two lines with a horizontal interval pitch Yp and height interval pitch of Zp. Each path was scanned 15 times, with a scanning speed of 8 μm/s and a laser power of 15 mW.

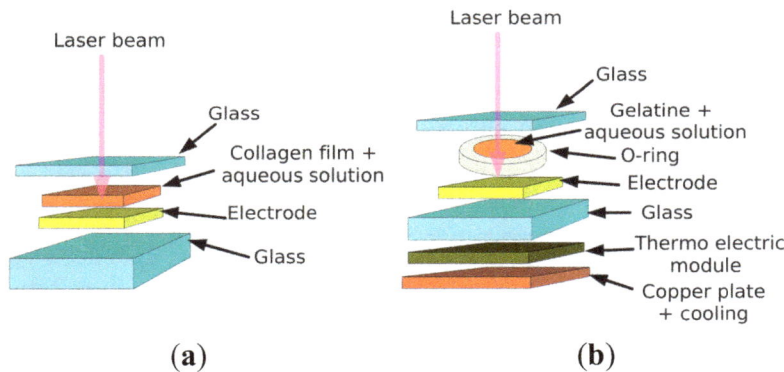

Figure 3. Setting protein film method in SL: (**a**) Collagen film and (**b**) gelatine.

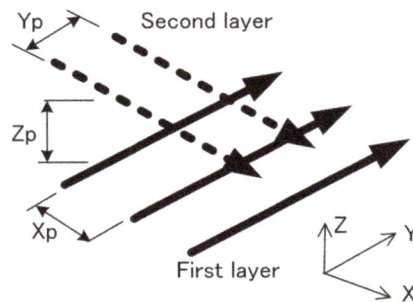

Figure 4. 3D stereolithography (SL) trajectory.

2.3.3. Reduction of the Repetition Rate of a Laser

To produce a higher accuracy construction with SL, another method was employed. When the laser repetition rate of a femtosecond laser is high, heat build-up is added to the two-photon polymerization. As a result, the polymerization process was slow (compared with [15]), and the accumulation effect caused by the heating process was high. Thus, we concluded that the heating process produced the irregular polymerization. Hence, the repetition rate of the laser was reduced.

A Pulse picker was used; the pulse energy was lower due to the reduction in the repetition rate. To confirm this assumption, an emission spectrum with the peak near the wavelength of 600 nm (orange color) was visually observed by focusing the femtosecond laser (ex. 850 nm) into the Nafion sheet containing $Ru(bpy)_3^{2+}$. The emission indicated the radiative relaxation of the metal complex from the excited state to the ground state. Figure 5 shows the relationship between the emission intensity at the wavelength of 600 nm and the repetition rate of the laser. The incident laser power was a constant value of 0.28 mW for each repetition rate using a polarizer attenuator. The emission intensities at 100 and 200 kHz were clearly higher than those at 400 kHz to 8 MHz. These results suggested that the radiative relaxation process was depressed at the higher repetition rates by the nonradiative thermal process due to the over-excitation.

Figure 6 shows the trajectory of the 3D SL, constructed using three layers. Each layer has four lines; the first and second lines as well as the second and third lines are orthogonal. Vertical lines were set at intersections. The horizontal interval pitch Xp was 1.75 μm and the vertical interval pitch, Zp was 0.4 μm. The height of the third layer was 0.8 μm.

Abnormal small whiskers are generated by slow scanning speed. Then scan speed is a key factor of SL. Two methods were used to find the optimal scanning method. The first method involved drawing one time on each trajectory and the second method involved drawing six times, repeatedly on each trajectory. Both methods took almost the same laser power in one trajectory. The draw speed of the second method was six times higher than in the first method.

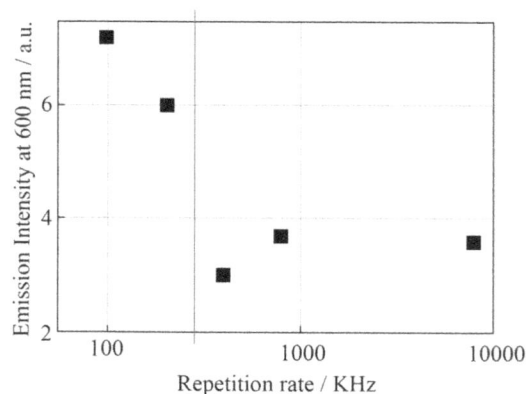

Figure 5. Relationships between repetition rate and emission intensity of $Ru(bpy)_3^{2+}$ in the Nafion sheet [16].

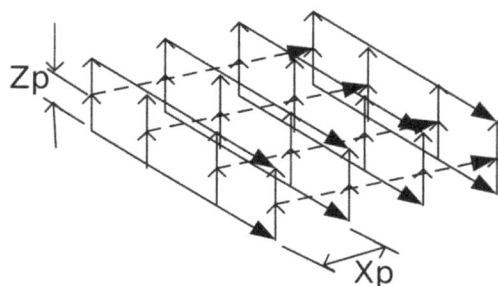

Figure 6. Trajectory of fabrication.

3. Results

3.1. Line SL

Figure 7 shows the SEM photograph of the line SL result. We used the trajectory of fabrication in Figure 2. Each path was scanned one time and the laser power was 15 mW. We set Xp at 0.3, 0.5, 0.7, 0.9 μm. Zp was 0.3, 0.6 μm and the scan speed was 2, 4, 8 μm/s. (a) In the case of a one-time repeat, small abnormal whiskers were visible at a low speed of 2–4 μm/s. (b) Each path was repeated five times with a laser power of 15 mW. The width of the SL part obviously changed with the variation in the horizontal interval pitch, Xp. We observed a small abnormal whisker in all cases. From these results, it is difficult to build thick polymerized parts without avoiding irregular polymerization.

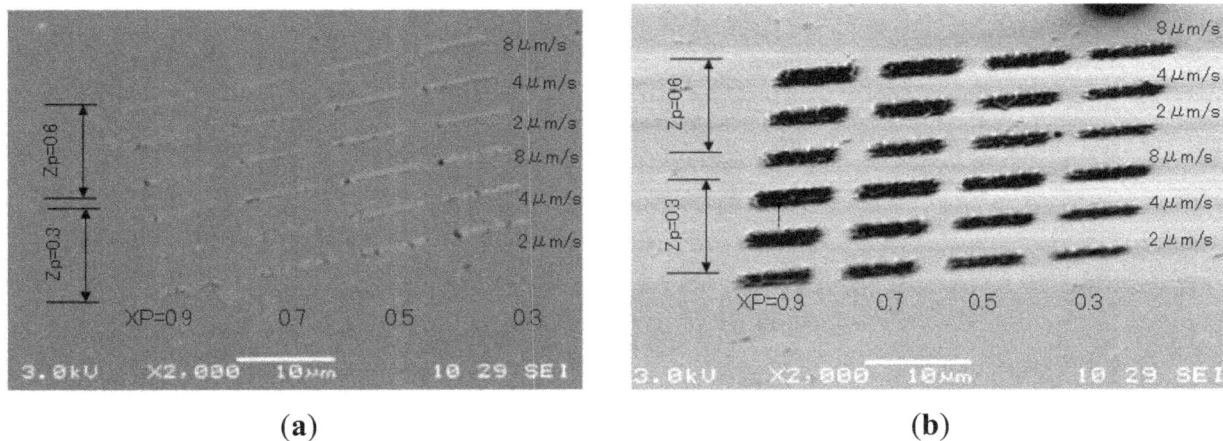

(a)

(b)

Figure 7. Line stereolithography (SL) result: (**a**) Case of one time repeat and (**b**) case of fine times repeat.

3.2. Support Method

3.2.1. Case of Collagen Film

Figure 8 shows the results of the SL lines using collagen film that included an aqueous solution, with a laser power of 30 mW and scan speeds as shown in figure. The edge-of-line was indistinct, the velocity of the SL process was still slow and it was very difficult to handle 20-μm film without wrinkles.

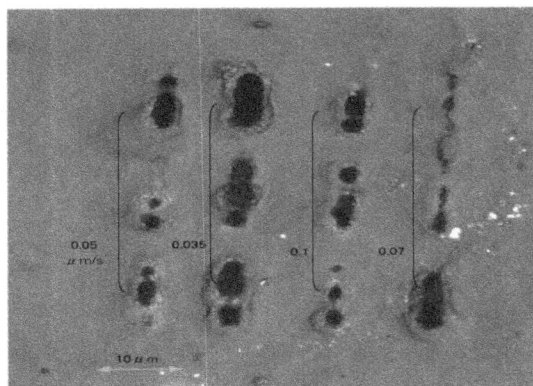

Figure 8. Line stereolithography (SL) result in collagen film.

3.2.2. Case of Gelatine

(1) Case of line SL. Figure 9 shows the SEM photograph of the SL result. We used the trajectory of fabrication in Figure 2. Each path was scanned one time and the laser power was 15 mW. We set Xp at 0.3, 0.5, 0.7, and 0.9 µm. Zp was 0.3, 0.6 µm and the scan speed was 2, 4, and 8 µm/s. (a) In the case of a one-time repeat, although the fabricated part was stable with no whisker, the thickness of the polymerized part was lower than the trajectory. (b) Each path was repeated five times with a laser power of 15 mW. The SL part was stable with no whisker. The width of the SL part did not show an obvious change with variations in the horizontal interval pitch, Xp. In this method, irregular polymerization was inhibited, and we inferred that the laser power was absorbed in the gelatine and the outer part was slightly dissolved by the warm deionized water.

(2) Case of 3D SL. Figure 10 shows the SEM photograph of the SL result with a support case. We used the trajectory of fabrication in Figure 4. We set Xp at 3 µm and Yp at 6 µm. Scanning 15 times is the limit of stable SL. The height was changed following a change in Zp from 1.0 to 1.6 and 2.0 µm. The Zp case was 3.2 µm, the height of the polymerized material was not changed from that of the case with a Zp of 2.0 µm.

Using a gelatine support case, solidification was stabilized; however, multiple repeats of each trajectory to build the designed shape and the precision of the SL shape were not enough to produce a high accuracy construction using sub-micrometre width lines.

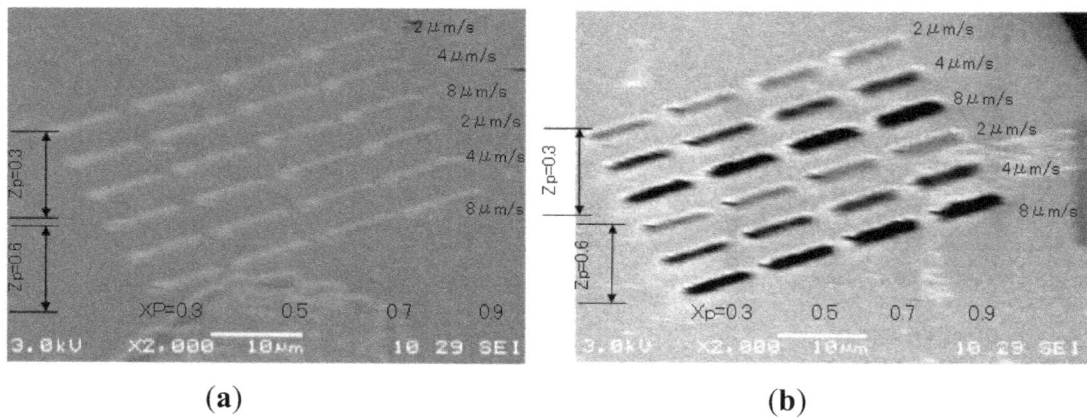

(a)   (b)

Figure 9. Line stereolithography (SL) result for support case: (**a**) Case of one time repeat and (**b**) case of fine times repeat.

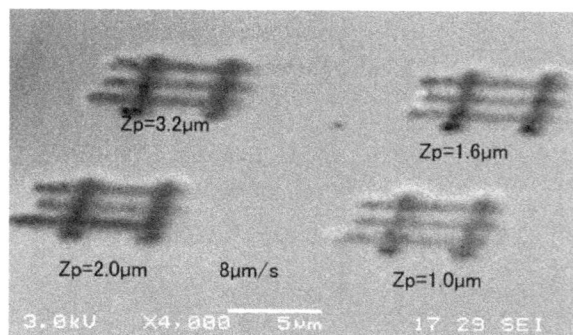

Figure 10. 3D stereolithography (SL) result with support case (repeated 15 times).

3.3. Reduction of the Repetition Rate of a Laser

3.3.1. First Method (Single Repeat)

To confirm heat accumulation, we tried SL experiments using a slow scan speed and a high repetition rate. Figure 11 shows the SL result when the repetition rate of the femtosecond laser was fixed at 8 MHz and the laser power was varied from 12 to 3 mW. We used the trajectory of fabrication in Figure 6 and repeated once on each trajectory. The scan speed of the left part was 0.33 μm/s and that of the right part was 0.29 μm/s. The solidification part using 12 to 9 mW of laser power was not stable. The laser power was higher for solidification. Although 3 mW was stable, the height of solidification was less than 0.5 μm. The three dimensional fabrication was difficult using this repetition rate.

Figure 12 shows the results of fabrication by reducing the repetition rate from 800 kHz to 20 kHz. The trajectory used is shown in Figure 6 and repeated once on each trajectory. The left scan speed was 0.29 μm/s and the right speed was 0.33 μm/s. The laser powers were 800 kHz: 0.3 mW, 400 kHz: 0.15 mW, 200 kHz: 0.075 mW, 100 kHz: 0.0375 mW, 50 kHz: 0.01875 mW and 20 kHz: 0.00075 mW. The solidification using 800 to 200 kHz was not stable; however, using 50 to 20 kHz was stable. The solidification height was maintained at 1 to 1.5um.

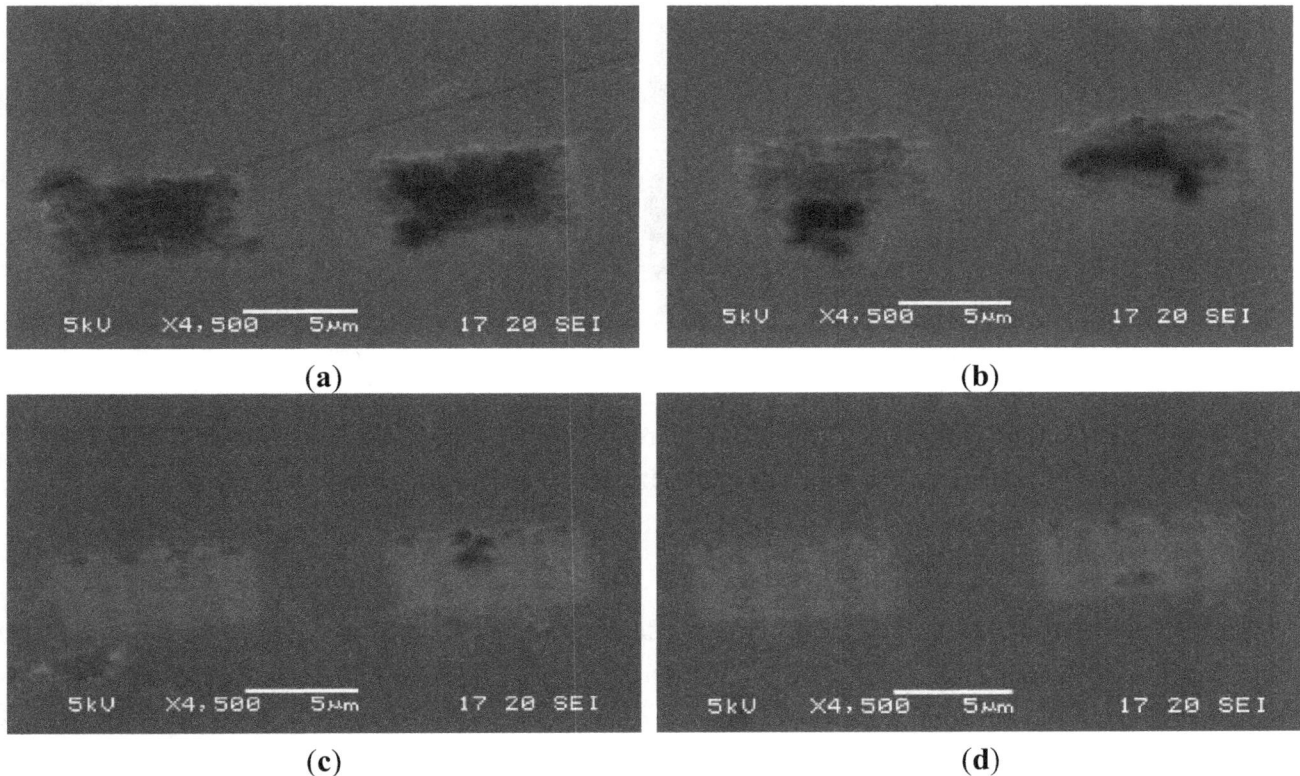

Figure 11. Stereolithography (SL) result when the repetition rate was 8 MHz, with variation in laser power. **(a)** Laser power is 12 mW. **(b)** Laser power is 9mW. **(c)** Laser power is 6 mW. **(d)** Laser power is 3 mW.

(a)

(b)

(c)

(d)

(e)

(f)

Figure 12. Stereolithography (SL) result of reducing the repetition rate from 800 kHz to 20 kHz: **(a)** 800 kHz, **(b)** 400 kHz, **(c)** 200 kHz, **(d)** 100 kHz, **(e)** 50 kHz and **(f)** 20 kHz.

3.3.2. Second Method (Six Repeats)

Figure 13 shows the SL result of the repetition rate of a femtosecond laser, fixed to 8 MHz. The laser power was varied from 12 to 3 mW. We used the trajectory of fabrication shown in Figure 6 and this was repeated six times on each trajectory. The left scan speed was 1.54 μm/s and the right speed was 2.0 μm/s. This scan speed was adjusted to the equivalent power, repeating each trajectory once. Solidification using 12 to 9 mW laser power was not stable, which indicates that higher laser power was higher for solidification. Although 3 mW was stable in (a) to (d), the height of solidification is less than 0.5 μm. Three dimensional fabrication is difficult using this repetition rate.

Figure 14 shows the fabrication result by reducing the repetition rate from 800 to 20 kHz. The trajectory used is shown in Figure 6 and repeated six times for each trajectory. The left scan speed was 1.54 μm/s and the right scan speed was 2.0 μm/s. Solidification using 800 kHz to 200 kHz was not stable; however, solidification using 100 to 50 kHz was stable and the solidification height was maintained at 1 to 1.5 μm.

Therefore, we confirmed that the reduction of the repetition rate of the femtosecond laser can stabilize the SL and this method can use three dimensional SL. At 100 kHz, scanning method with drawing six times is stable than drawing one time.

Figure 15 shows the relation between shape magnification and repetition rate of the laser. We used a base height of 1 µm because the highest trajectory is 0.8 µm and the height of one line is 0.4 µm. Shape magnification is calculated by dividing the highest dimension by the base height for each SL result. Abnormal solidification is reduced by decreasing the repetition rate to below 100 kHz. The scanning method of repeating each trajectory six times was a little less stable than a one-time pass. We infer that SL includes sputtering phenomena at 8 MHz.

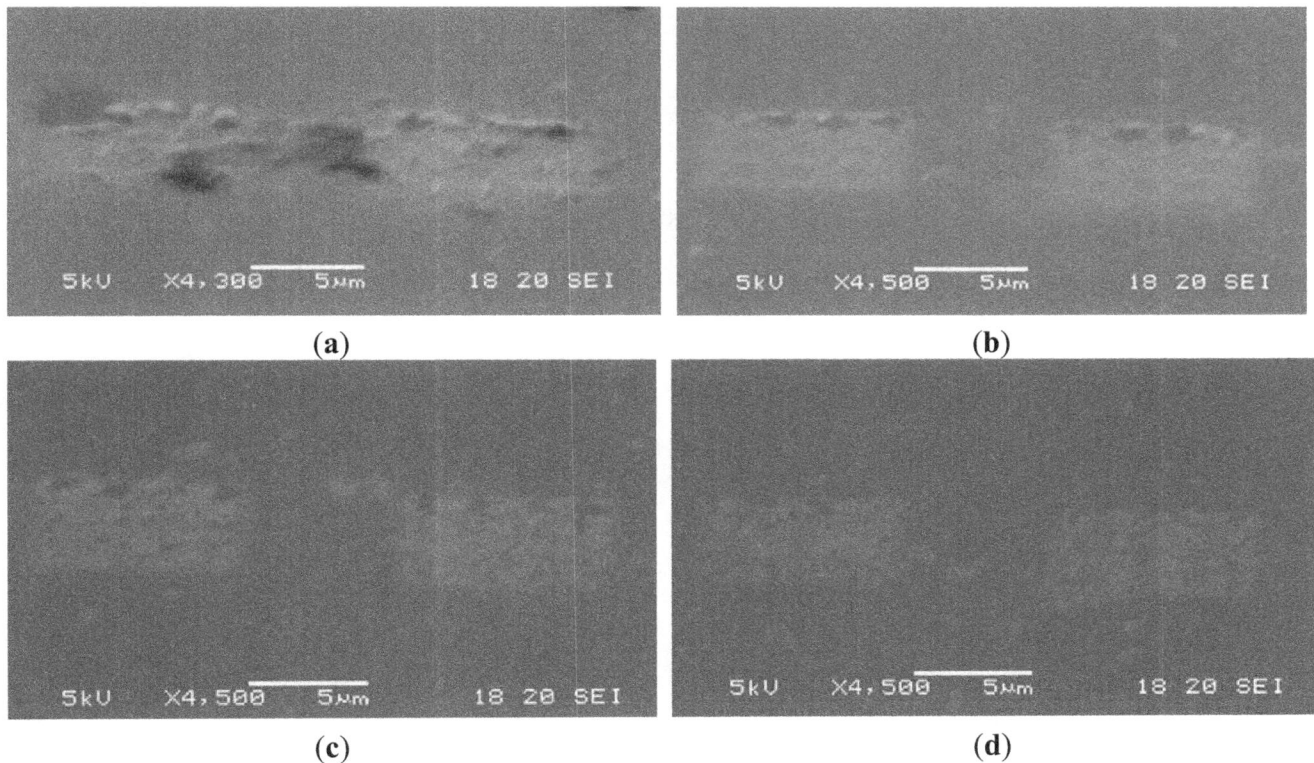

(a)   **(b)**

(c)   **(d)**

Figure 13. Stereolithography (SL) result when the repetition rate was 8 MHz, with variation in laser power: **(a)** Laser power is 12 mW, **(b)** laser power is 9 mW, **(c)** laser power is 6 mW and **(d)** laser power is 3 mW.

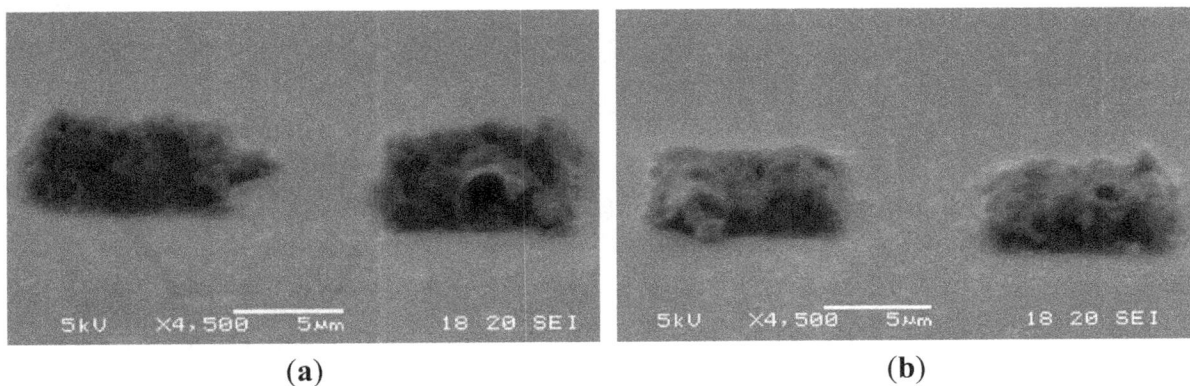

(a)   **(b)**

Figure 14. *Cont.*

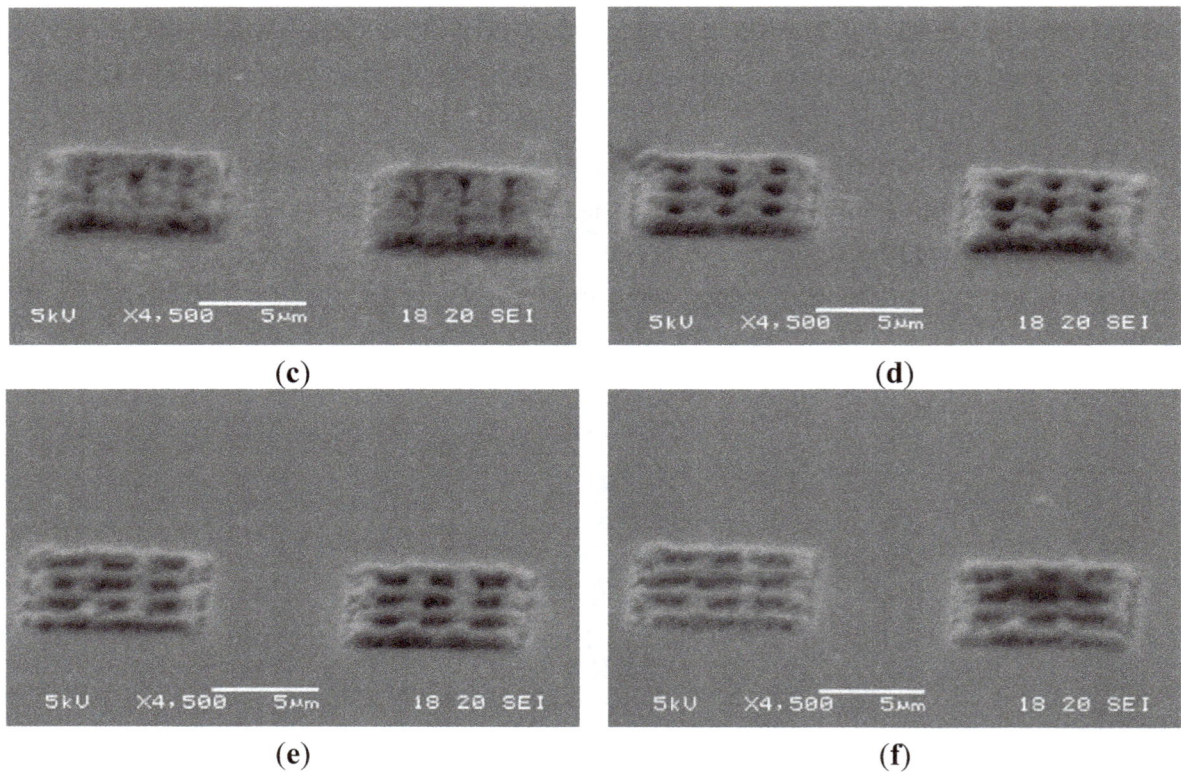

Figure 14. Stereolithography (SL) result of reducing the repetition rate from 800 kHz to 20 kHz: (**a**) 800 kHz, (**b**) 400 kHz, (**c**) 200 kHz, (**d**) 100 kHz, (**e**) 50 kHz and (**f**) 20 kHz.

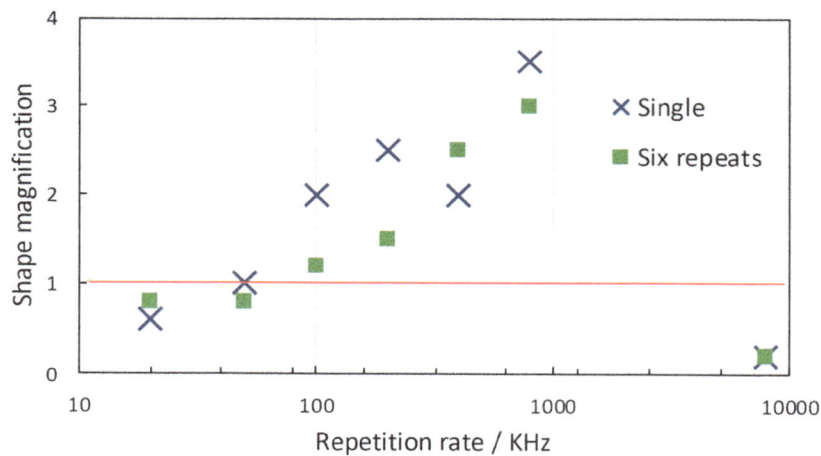

Figure 15. Shape accuracy result with changing repetition rate.

3.3.3. Compensation Method of Shrinkage

Figure 16 shows the SL result of nine layers: Figure 16a is one repeat of each trajectory, with a laser repetition rate of 50 kHz and other SL conditions are the same as in Section 3.2.1. Figure 16b is six repetitions of each trajectory, with a laser repetition rate of 50 kHz and other SL conditions are the same as in Section 3.2.2. The top trajectory height was 3.2 μm. In both cases, the SL part has shrinkage around the top layer. In this study, we considered the shrinkage compensation method. The second to nine layer trajectory is expanded 20% at the center from the first trajectory. Figure 17 shows the results of the shrinkage compensation. We reduced the shrinkage at the top layer. These results show that the photo fabrication can be stabilized by reducing the repetition rate and constructing a three dimensional shape almost similar to this trajectory. This indicates that the resolution for each trajectory needs be to improved in order to build functional NEMS. In the previous study, the polymerization area was ellipsoidal with a depth that was almost five times greater than the horizontal length [17]. We need to solve these problems to achive high accuracy.

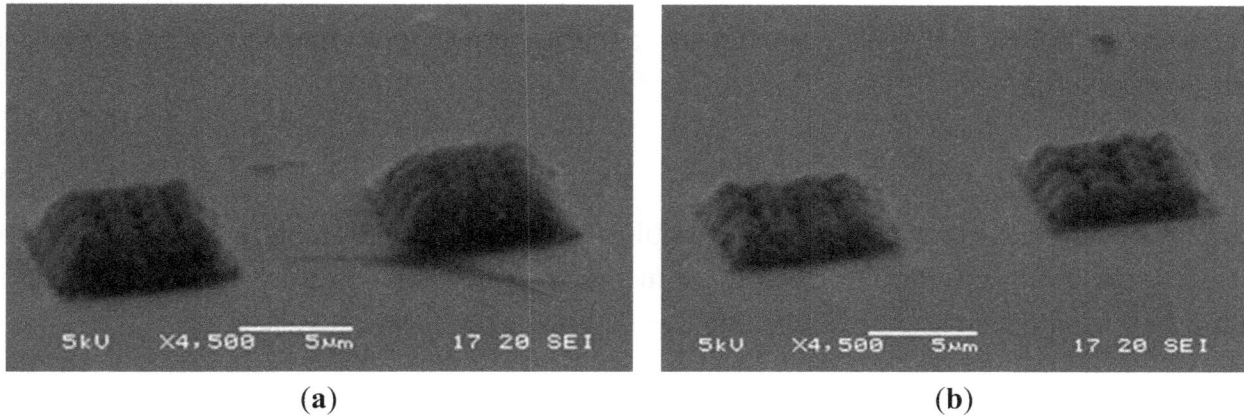

(a)   **(b)**

Figure 16. Nine layer stereolithography (SL) result. (**a**) One time for each trajectory, 50 kHz. (**b**) Six time for each trajectory, 50 kHz.

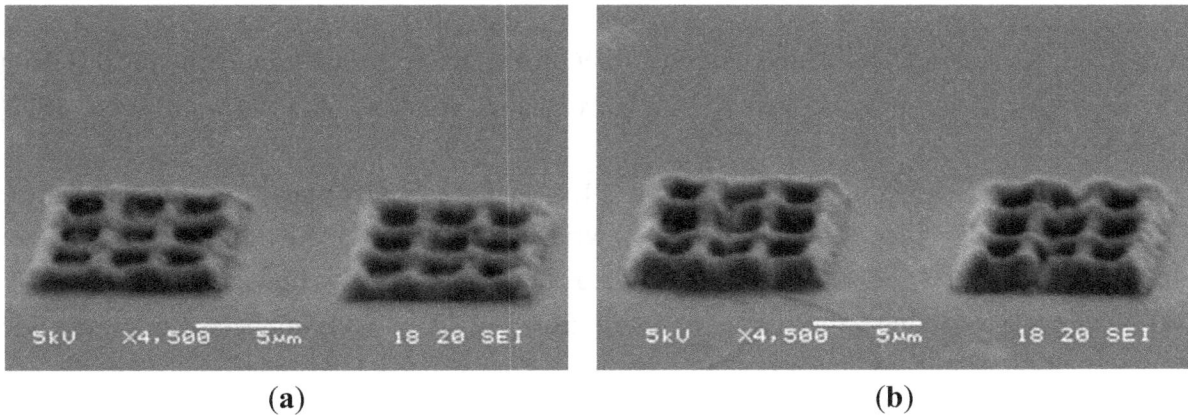

(a)   **(b)**

Figure 17. Shrinkage compensation result. (**a**) One time for each trajectory, 50 kHz. (**b**) Six time for each trajectory, 50 kHz.

4. Conclusions

In this study, we determined optimal conditions and a suitable method of multiphoton stereolithography using a conductive polymer. First, we tried a high scan speed and multiple repeats on each trajectory using the original aqueous solution and an 8 MHz repetition rate for the laser. However, irregular polymerization was produced when trying to build the designed thickness of the polymerized part. To avoid the irregular polymerization, we used a soluble support. Two percent gelatin by weight was added to the aqueous solution to compose the support for the solidified material. Although this method can reduce irregular polymerization, it required many repeats of each trajectory to build the designed shape and the accuracy of the detail was reduced. Finally, we tried to solve irregular polymerization caused by heat accumulation. We reduced the repetition rate of the femtosecond laser. A repetition rate of 50–100 kHz was stable for our system and the CP. In addition, we were able to produce a 3D structure with almost the same height as the designed trajectory by utilizing a compensation method for shrinkage during polymerization.

Photo fabrication is stabilized by the reduction of the repetition rate. This SL method can be used to build several micrometer sized three dimensional actuators with micrometer resolution. This method is not archived sub-micrometer resolution, then it is difficult to build high efficiency metamaterials. Further research should be done to find more applications for micrometer/sub-micrometer size three-dimensional conductive polymers.

Author Contributions

Junji Sone designed the experimental methodologies and made the SL trajectory. Katsumi Yamada developed chemical reaction of electro conductive polymer and measured emission intensity. Akihisa Asami executed the experiments. Jun Chen supported laser optical system.

Conflicts of Interest

The authors declare no conflict of interest.

References

1.   Kobayashi, N.; Yamada, K.; Hirohashi, R. Effect of Anion Species on Electrochemical Behavior of Poly(aniline)s Electropolymerized in Dichloroethane Solution. *Electrochim. Acta* **1992**, *37*, 2101–2102.

2.   Okano, M.; Itoh, K.; Fujishima, A.; Honda, K. Generation of organic conducting patterns on semiconductors by photoelectrochemical polymerization of pyrrole. *Chem. Lett.* **1986**, *15*, 469–472.

3.   Iyoda, T.; Toyoda, H.; Fujitsuka, M.; Nakahara, R.; Tsuchiya, H.; Honda, K.; Shimidzu, T. The 100-.ANG.-order depth profile control of polypyrrole-poly(3-methylthiophene) composite thin film by potential-programmed electropolymerization. *J. Phys. Chem.* **1991**, *95*, 5215–5220.

4.   Teshima, K.; Yamada, K.; Kobayashi, N.; Hirohashi, R. Photopolymerization of aniline with a tris(2,2'-bipyridyl)ruthenium complex—Methylviologen polymer bilayer electrode system. *Chem. Commun.* **1996**, *1996*, 829–830.

5.   Cumpston, B.H.; Ananthavel, S.P.; Barlow, S.; Dyer, D.L.; Ehrlich, J.E.; Erskine, L.L.; Heikal, A.A.; Kuebler, S.M.; Sandy Lee, I.-Y.; McCord-Maughon, D.; Qin, J.; Röckel, H.; Rumi, M.; Wu, X.L.; Marder, S.R.; Perry, J.W. Two-photon polymerization initiators for three-dimensional optical data storage and microfabrication. *Nature* **1999**, *398*, 51–54.

6.   Kawata, S.; Sun, H.-B.; Tanaka, T.; Takada, K. Finer Features for Functional Microdevices. *Nature* **2001**, *412*, 697–698.

7.   Li, L.; Fourkas, J. Multiphoton Polymerization. *Mater. Today* **2007**, *10*, 30–37.

8.   Yamada, K.; Kimura, Y.; Suzuki, S.; Chen, J.; Sone, J. Multiphoton-sensitized polymerization of pyrrole. *Chem. Lett.* **2006**, *35*, 908–909.

9.   Sone, J.; Asami, A.; Kimura, G.; Yamada, K. Feasibility study of micro-actuator using electro conducting polymers. *IEEJ Trans. Sens. Micromach.* **2009**, *129*, 81–82.

10. Yamada, K.; Sone, J.; Chen, J. Three-Dimensional Photochemical Microfabrication of Conductive Polymers in Transparent Polymer Sheet. *Opt. Rev.* **2009**, *16*, 208–212.

11. Takada, K.; Kaneko, K.; Li, Y.; Kawata, S.; Chen, Q.; Sun, H. Temperature effects on pinpoint photopolymerization and polymerized micronanostructures. *Appl. Phys. Lett.* **2008**, *92*, 041902.

12. Juodkazis, S.; Mizeikis, V.; Seet, K.; Miwa, M.; Misawa, H. Two-photon lithography of nanorods in SU-8 photoresist. *Nanotechnology* **2005**, *16*, 846–849.

13. Eaton, S.; Zhang, H.; Herman, P.; Yoshino, F.; Shah, L.; Bovatsek, J.; Arai, A. Heat accumulation effects in femtosecond laserwritten waveguides with variable repetition rate. *Opt. Express* **2005**, *13*, 4708–4716.

14. Tanaka, T. Plasmonic metamaterials produced by two-photon-induced photoreduction technique. *J. Laser Micro/Nanoeng.* **2008**, *3*, 152–156.

15. Sun, H.; Kawata, S. Two-Photon Photopolymerization and 3D Lithographic Microfabrication. *NMR 3D Anal. Photopolym.* **2004**, *170*, 169–273.

16. Yamada, K.; Watanabe, M.; Sone, J. Three-Dimensional Printing of Conducting Polymer Microstructures into Transparent Polymer Sheet: Relationship between Process Resolution and Illumination Conditions. *Opt. Rev.* **2014**, *21*, 679–682.

17. Yamada, K.; Kyoya, A.; Sone, J.; Chen, J. Evaluations of Vertical Resolution of Conductive Polymer Three-Dimensional Microstructures Photofabricated in Transparent Polymer Sheet. *Opt. Rev.* **2011**, *18*, 162–165.

8

A New Concept of a Drug Delivery System with Improved Precision and Patient Safety Features

Florian Thoma *, Frank Goldschmidtböing and Peter Woias

Design of Microsystems, Department of Microsystems Engineering (IMTEK), University of Freiburg, Georges-Koehler-Allee 103, 79110 Freiburg, Germany; E-Mails: fgoldsch@imtek.de (F.G.); woias@imtek.de (P.W.)

* Author to whom correspondence should be addressed; E-Mail: florian.thoma@imtek.de

Academic Editors: Joost Lötters and Miko Elwenspoek

Abstract: This paper presents a novel dosing concept for drug delivery based on a peristaltic piezo-electrically actuated micro membrane pump. The design of the silicon micropump itself is straight-forward, using two piezoelectrically actuated membrane valves as inlet and outlet, and a pump chamber with a piezoelectrically actuated pump membrane in-between. To achieve a precise dosing, this micropump is used to fill a metering unit placed at its outlet. In the final design this metering unit will be made from a piezoelectrically actuated inlet valve, a storage chamber with an elastic cover membrane and a piezoelectrically actuated outlet valve, which are connected in series. During a dosing cycle the metering unit is used to adjust the drug volume to be dispensed before delivery and to control the actually dispensed volume. To simulate the new drug delivery concept, a lumped parameter model has been developed to find the decisive design parameters. With the knowledge taken from the model a drug delivery system is designed that includes a silicon micro pump and, in a first step, a silicon chip with the storage chamber and two commercial microvalves as a metering unit. The lumped parameter model is capable to simulate the maximum flow, the frequency response created by the micropump, and also the delivered volume of the drug delivery system.

Keywords: drug delivery system; lumped parameter model; micropump

1. Introduction

Modern medical treatments should ensure a defined level of a drug in the blood stream. This defined level may vary in a narrow therapeutic window [1–5]. Therefore conventional injections start with a higher concentration c_{max} to keep the drug in the therapeutic window over a certain time [3]. This high concentration may lead to adverse side effects. Implantable drug delivery systems open new possibilities, by frequently dosing a defined volume to enhance the healing process and to minimize undesirable side effects, by keeping the drug level permanently inside the therapeutic window. Besides frequent and defined dosing, patient safety is an essential requirement. Pump based drug delivery systems lack in accuracy of flow rate, showing dosing errors of at least 10%, see Table 1. In order to satisfy the requirements of modern therapy, the delivered drug should be measured before release and a possibility to verify the correct measurement should be implemented.

In this paper a drug delivery system is shown, which extends a silicon micropump by a storage chamber, delimited by a membrane, and an outlet valve. While the outlet valve is closed, the pump can pressurize the storage chamber, see Figure 1. This pressure deflects the membrane on top of the storage chamber. The deflection can be measured and thereby the stored volume can be calculated very precisely. These two parts permit a measurement of the drug volume and a frequent delivery of the drug, by opening and closing the outlet valve. This concept of a dosing unit add-on, see Figure 1, can be used for every pump based drug delivery system, which is established at the market.

Table 1. Commercial drug infusion pumps for clinical applications [4].

Manufactor's Data	Medtronic	Medtronic	Codman
Device name	Synchromed II	IsoMed	Medstream
Flow rate	48–1000 µL/h	20.8–62.5 µL/h	4–167 µL/h
Flow rate accuracy	14.5% error	15% error	10% error

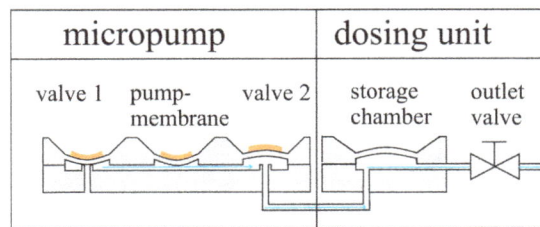

Figure 1. Working principle of the drug delivery system. The storage chamber is filled by the micro-pump. During the filling procedure the outlet valve is closed. The membrane, which caps the storage chamber, bends due to the increased pressure. The outlet valve can be opened on demand, so that the bended membrane relaxes.

2. Experimental Section

In this section, the concept of the storage chamber will be explained along with the working principle of the whole system. The improvement in the dosing accuracy compared to commonly used micropumps will be discussed. After that a lumped parameter model of the drug delivery system is introduced and the measurement setup is shown.

2.1. Working Principle

The drug delivery system consists of a silicon micropump a storage chamber and an outlet valve (see Figure 1). The piezoelectrically actuated micropump is a displacement pump [6,7]. A detailed description of the fabrication process for the pump can be found in [6]. The working principle of the micropump is shown in Figure 2. The piezoelectric actuators are mounted on top of the membranes and ensure a maximum positive and negative displacement of the membranes. The volume change under the pump-membrane defines the volume flow per cycle [8–10].

If the piezoelectric actuators follow a certain sequence, a positive flow, from the inlet to the outlet of the micropump, can be generated.

To prevent a strong variation of the flow rate generated by the micropump, the storage chamber and the outlet valve were added. At the beginning of a dosing cycle, the pump pressurizes the storage chamber, while the outlet valve is closed, see Figure 1. The thin membrane which delimits the storage chamber is deflected by the pressure generated by the micropump. This deflection can be used to measure the additional volume stored in the chamber. The time-dependent deflection of the membrane during filling and delivery process can be used to double check the correct measurement of the deflection and to detect parasitic capacitances like air bubbles, see Figure 3. In this state the deformed membrane exerts a certain pressure onto the stored drug. After opening of the outlet valve, a defined volume of the drug is squeezed out by the preloaded membrane of the storage chamber. A second measurement of the membrane deflection at the end of the dosing cycle will allow calculating the amount of the dispensed drug. With the additional storage chamber, variations in the flow rate generated by the micropump are not an issue. Due to the time-dependent displacement during filling and release of the drug, these unwanted effects can be detected and compensated.

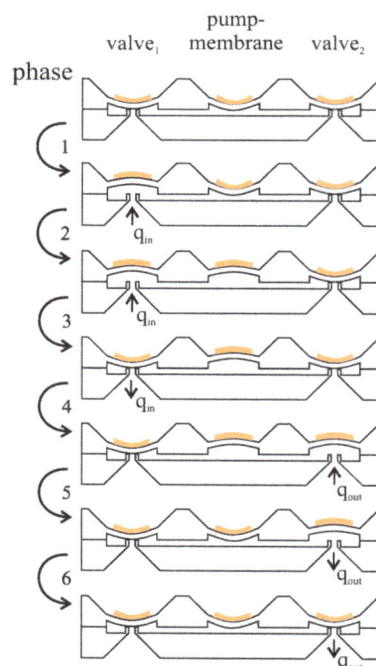

Figure 2. Working principle of the micropump: To create a fluidic flow from the left to the right side the membranes have to be driven in the depicted order. The manufacturing process of the pump can be found in [11].

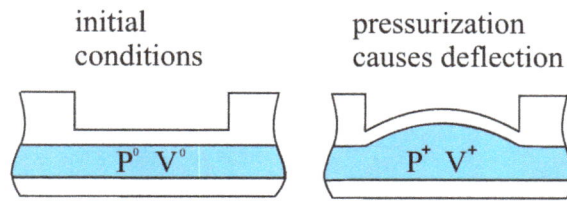

Figure 3. Pressurized storage chamber: The deflection of the thin membrane can be used to measure the stored volume. To release the drug the preloaded membrane can be used to push the drug into the body.

2.2. Lumped Parameter Model

A lumped parameter model can be used to model a fluidic system [6,7,12–15] . Based on the equation of a mechanically oscillating system, Equation (1) describes an incompressible laminar flow for a Newtonian fluid in steady state [16]:

$$p = \frac{m}{A^2}\frac{d\dot{V}}{dt} + \frac{\chi}{A^2}\dot{V} + \frac{k}{A^2}\int \dot{V}dt \tag{1}$$

In Equation (1), p is the pressure drop, V is the volume, m the mass, and A the cross sectional area. χ is the friction coefficient and k the stiffness factor of the membrane. If we now compare the equation of an electric resistor, inductor and capacitor (RCL)-series-oscillator with Equation (1), we can define constants of the volume flow as fluidic capacitance C, resistance R and inductance L, see Equation (2) [14]:

$$\frac{k}{A^2} = C; \ \frac{\chi}{A^2} = R; \ \frac{m}{A^2} = L \tag{2}$$

The lumped parameter model of a fluidic system is build up similar to an electric circuit and consists of parts like fluidic resistors, capacitances, and inductances, see Equation (2). By subdividing the fluidic system into defined parts with known behavior Equation (1) can be used to build a network based on the mesh equations of an electric circuit. The lumped parameters can be calculated with the mathematics of pipe hydraulics and mechanics. The schematic of the lumped parameter model is shown in Figure 4. For simplification the schematic is drawn with the equivalent electric counterparts.

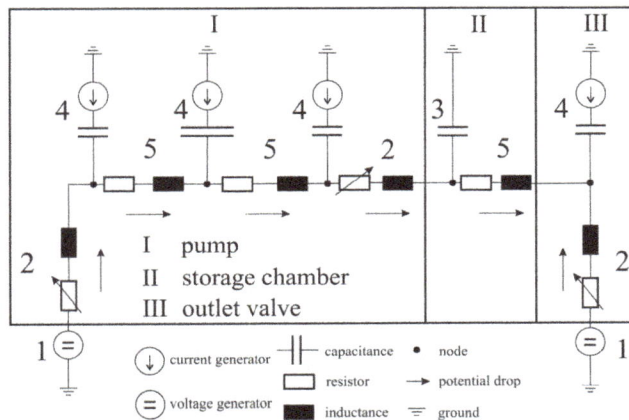

Figure 4. Schematic of the lumped parameter model. Element 1: source pressure applied at the inlet and outlet. Element 2 is the valve, and element 3 is the storage capacitance. Element 4 represents a piezo-electrically actuated membrane and element 5 is the fluidic resistance between two nodes.

The lumped parameter model, shown in Figure 4, is used to calculate the fluidic flow at a node, and the pressure drop between two nodes with the law of mass conservation, see Equation (3):

$$\frac{dV}{dt} = q_{in} - q_{out} \tag{3}$$

In Equation (3) q_{in} is the volume flow to a node and q_{out} the flow out of a node. The time-dependent change in volume $\dot{V}$ for a piezoelectrically actuated membrane can be written as a function of the applied pressure p and voltage U, see Equation (4):

$$\frac{dV}{dt} = \frac{\partial V}{\partial p}\Big|_U * \frac{\partial p}{\partial t} + \frac{\partial V}{\partial U}\Big|_p * \frac{\partial U}{\partial t} \tag{4}$$

By inserting Equation (3) in Equation (4), Equation (5) can be concluded:

$$\frac{\partial p}{\partial t} = \frac{q_{in} - q_{out} - \frac{\partial V}{\partial U}|p \times \frac{\partial U}{\partial t}}{\frac{\partial V}{\partial p}\Big|_U} \tag{5}$$

The piezoceramic actuator is driven with a square wave signal. To circumvent the mathematical problem of differentiation of this square wave function, Equation (5) is integrated, see Equation (6). Equation (6) is used in the lumped parameter model for the deflection of the piezoelectrically actuated membranes and the storage membrane, see elements 3 and 4 in Figure 4:

$$p = \int \frac{q_{in} - q_{out} - \frac{\partial V}{\partial U}|p \times \frac{\partial U}{\partial t}}{\frac{\partial V}{\partial p}\Big|_U} \, dt = \frac{V_{in} - V_{out} - \frac{\partial V}{\partial U}|p \times U}{C_p} \tag{6}$$

The denominator in Equation (6) is the capacitance of the membrane, and is hence written as C_p. After building up the drug delivery system with lumped parameters, the discrete parts have to be calculated, see Figure 4. To calculate the lumped parameters fluidic calculations and mechanical simulations have to be conducted. The inductance L can be calculated with the knowledge of the dimensions and the density of the fluid [8]. The constant resistance, see Figure 4 element 5, of the channels can be calculated by the knowledge of pipe hydraulics and a dimensionless factor for the cross sectional area. The resistance of a valve has a range between infinity and a minimum value, when the gap is fully opened. Due to this fact, a function has to be found which models the behavior of the valve dependent on the applied voltage and pressure. The capacitance C_p is defined as the change of volume due to an applied pressure while the applied voltage is zero. The current source, see Figure 4, defines the dependence of the displacement of the membrane of the applied voltage on the piezoelectric actuator. The capacitance and the current generator act additively and change the pressure under the membrane.

2.2.1. Fluidic Calculation

As described in the previous section, two types of resistors are used in the lumped parameter model. The resistances of the chamber or between the pump and storage chamber, depicted in Figure 4 as element 5, can be assumed to be constant. Assuming a laminar slit stream, the resistance of the pump chamber can be calculated by Equation (7) [17]:

$$R_k = \frac{8\eta l_s}{\pi \left(\frac{4b_m h_k}{2b_m + 2h_k}\right)^4} \tag{7}$$

In Equation (7) η is the dynamic viscosity, and l_s the length of the channel. The cross-sectional area is defined by the height h_k and the width b_m, see Figure 5.

Obviously the fluidic resistance of the valve cannot be constant. It is assumed that there is no leakage in the closed state. Therefore the resistance is infinite. In the open state the resistance is dependent on the thickness of the valve lips and the gap h_v between the membrane and the lips, depicted in Figure 6. In Equation (8), the relation between the geometrical dimensions and the resistance is shown [17]:

$$R_v = \begin{cases} \frac{6n}{\pi h_v^3} \ln\left(\frac{r_2}{r_1}\right), & h_v > 0 \\ \infty, & h_v = 0 \end{cases} \tag{8}$$

In order to solve the problem of a variable resistor a parametric function for the gap $h_v(p,U)$ has to be found. For that purpose a finite element method (FEM) simulation is established to find a function that links the displacement with the applied pressure p and voltage U, see Section 2.2.2.

Due to the acceleration of the liquid during pumping, mass inertia has to be taken into account. For frequencies below 40 Hz and length scales in the micro range Equation (9) can be used [18]:

$$p_1 - p_2 = \frac{3}{4}\rho l \frac{\partial q}{\partial t} = \frac{3}{4}\rho l \frac{\ddot{V}}{A} = L \times \ddot{V} \tag{9}$$

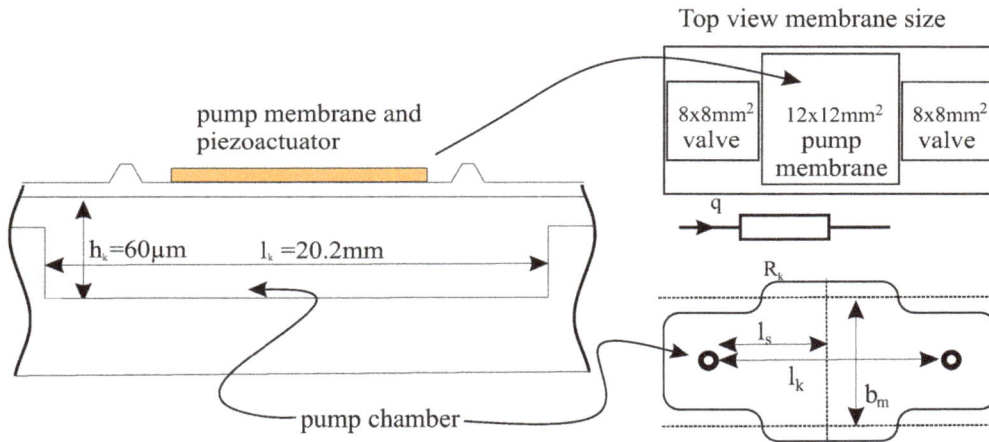

Figure 5. On the left side: side view of the pump chamber height and length from inlet to outlet. On the right side the top view of the pump chamber can be seen. The geometrical dimensions for the lumped parameter are l_k, h_k and b_m.

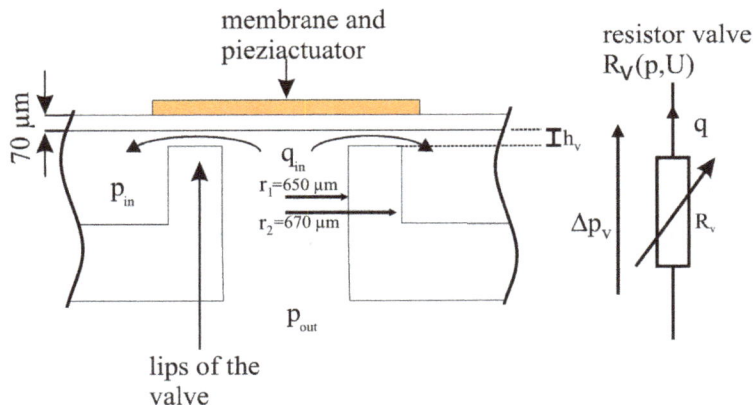

Figure 6. Dimensions of the valve. To calculate the fluidic resistance of the valve h_v, r_1 and r_2 have to be known.

2.2.2. Mechanical Simulation

In this section, a parametric fit function for the volume under the membrane and the distance between the valve lips and the membranes, caused by the applied pressure and voltage, has to be found. Therefore, an FEM-model was set up in ANSYS Multiphysics (ANSYS, Inc., Canonsburg, PA, USA). The assembly of the piezoelectrically driven membranes is modeled by four layers, see Figure 7. The membrane of the storage chamber has only two layers, due to the missing glue and piezoceramic layers. The edges of the membranes are built in. A linear model is assumed for the displaced volume, with respect to the applied pressure and voltage. The parametric equation for $V(p,U)$ is a superposition of the displacement due to pressure and voltage, and becomes $V(p,U) = C_P \times p + \frac{\partial V}{\partial U}|p \times U$, see Table 2. In a similar way we find a parametric equation for h_v, see Table 3.

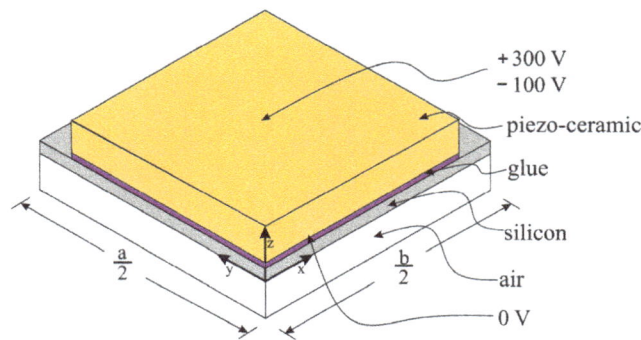

Figure 7. Assembly of the piezoelectrically driven valve- and pump-membrane.

Table 2. Factors of the parametric fit function of the simulation results for the generated volume under a membrane.

Factor	$dV/dp = C_p$	dV/dU
Generated volume: valve membrane	0.072 µL/bar	−0.0014 µL/V
Generated volume: pump membrane	1.027 µL/bar	−0.007 µL/V
Generated volume: storage membrane	18.8 µL/bar	0

Table 3. Factors of the parametric fit function of the simulation results for the gap of the valve.

Factor	dh/dp	dh/dU
Gap of the valve	2.97614 µm/bar	−0.05027 µm/V

2.3. Simulation Results

After the fluidic calculation and the mechanical simulation, the lumped parameter model is set up. The model of the drug delivery system is established in MATLAB Simulink. The initial value problem is solved with the fourth order Runge-Kutta-method, with a fixed step size of 1 µs. The pressurization of the storage chamber, generated by the pump, is simulated, see Figure 8. The rising pressure correlates with the voltage, applied to the pump-actuators. In the lumped parameter simulation, the pump was triggered to stop pumping, if a pressure of 100 mbar was reached in the storage chamber. Afterwards the pump stops and the outlet valve opens at 0.22 s (see Figures 9 and 10). Then, the pressure in the storage chamber drops with a capacitive discharge characteristic.

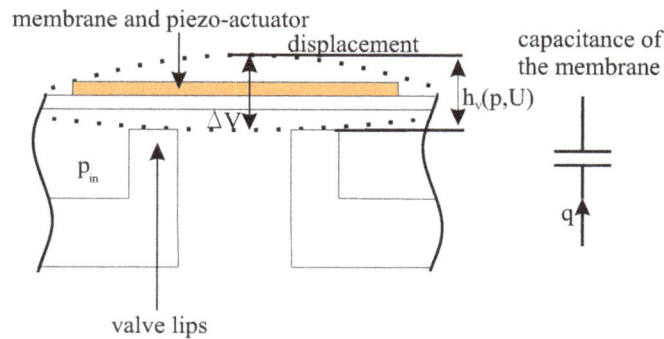

Figure 8. The values to be calculated, $h_v(p,U)$, C_p and the center displacement, are shown in the schematic drawing. The piezo-ceramic is a 200 μm thick Stelko PK21 actuator. For the valve a 7×7 mm^2 and for the pumping membrane it's a 10×10 mm^2 piezo-ceramic actuator is used.

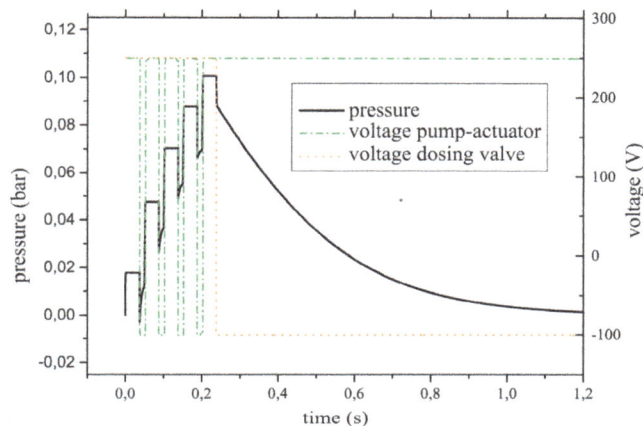

Figure 9. The voltage of the pump actuator can be seen in the green dash-dotted line and the voltage of the valve in the dotted line. The pressure in the storage chamber rises while the pump actuator is operated. After reaching a pressure of 0.1 bar the actuation of the pumps stops and the valve opens (see corresponding lines). The pressure drops with a capacitive discharge characteristic.

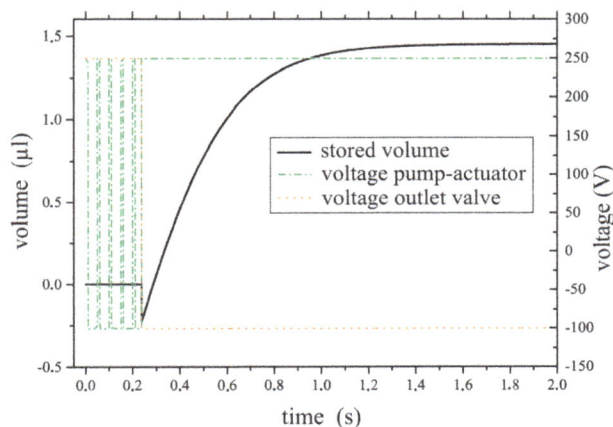

Figure 10. The voltage of the pump actuator can be seen in the green dash-dotted line and the voltage of the valve in the dotted line. After a certain time the valve is opened and the volume is released, (see corresponding line). The dispensed volume rises with a capacitive charge characteristic.

The dispensed volume, see Figure 10, corresponds to the pressure drop in the storage chamber. When the outlet valve opens, a negative flow in the pump is observed, and a negative pressure is generated. Due to this fact the fluid is forced to flow to this node. After reaching a state of stable equilibrium the excess pressure of the storage chamber forces the fluid to flow out of the drug delivery system, see Figure 10 at the time 0.23 s.

3. Results and Discussion

3.1. Measurement Setup

The measurements are separated into two parts. First a characterization of the micropump is performed to verify the simulation of the pump itself. After that the dispensed volume of the drug delivery system is gravimetrically measured.

3.1.1. Measurement Setup for Flow Rate Characterization

To drive the drug delivery system, a LabVIEW program (National Instruments Germany GmbH, Munich, Germany)generates a square wave for every phase of the pump and a signal for the valves, see Figure 2, and sends it to an amplifier (SVR 350-3 bip by Piezomechanik Dr. Lutz Pickelmann GmbH, Munich, Germany). To characterize the micropump a flow sensor (Piezomechanik Dr. Lutz Pickelmann GmbH, Munich, Germany) measures the generated flow while the LabVIEW software ramps the frequency of the driving signal in 5 Hz steps.

3.1.2. Measurement Setup for the Drug Delivery System

The dosing is measured by a gravimetrical method (Sartorius Cubis® micro scale, accuracy of reading 0.1 µg, Sartorius AG, Göttingen, Germany) while the displacement of the membrane of the storage chamber is measured by a laser triangulation sensor (AWL7 by Welotec GmbH, Laer, Germany) with a resolution of 0.4 µm, see Figure 3. The pump is operated with the setup introduced in Section 3.3.1 with a frequency of 20 Hz. The outlet valve is a "LHLA0521111H" magnetic valve manufactured by "The LEE Company (Westbrook, ME, USA)". The switching is triggered by the LabVIEW program at a certain time. The amplified switching signal is used to change the state of the valve, see Figure 11.

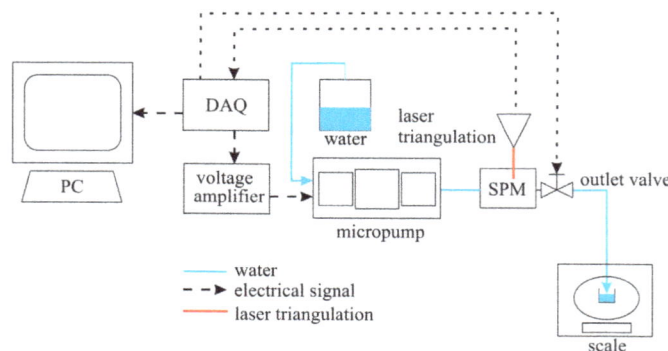

Figure 11. Measurement setup for the drug delivery system. The outlet valve can be opened on demand. The pump can stop at any time depending on the displacement, measured by the laser triangulation.

3.2. Measurement Results

First, a simulation of the pump itself is set up, to verify the lumped parameter model of the pump. Therefore, the frequency dependence of the pump itself, see Figure 2, is measured and compared with the simulation results, see Figure 12. Two flow measurements where performed. First the flow rate was measured three times for 10 s, and secondly the measurement was performed three times for 60 s, see Figure 12. The theoretical curve, simulated by the lumped parameter model, is given as the red curve. Compared with the 60 s measurement, the simulations show a satisfactory correlation below a frequency of 35 Hz. Higher actuation frequencies cannot be simulated with this model due to inertial effects that are not covered by the simple model of a fluidic inductance.

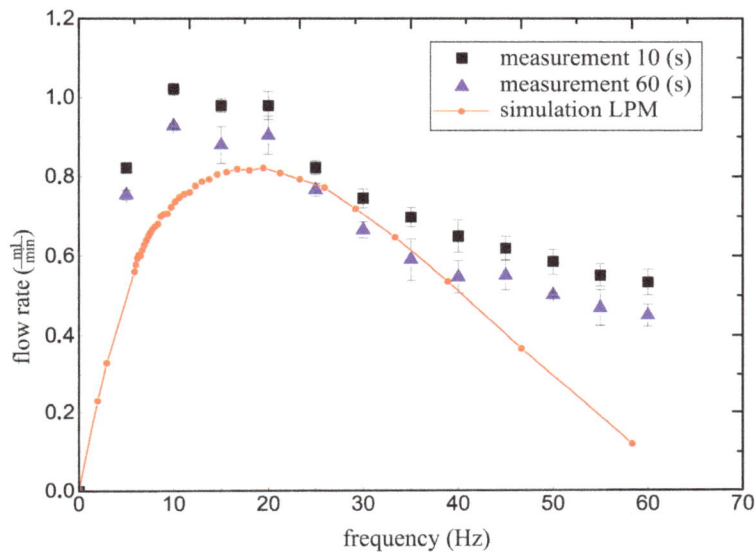

Figure 12. Comparison of the frequency dependence of the simulated and measured pump. The measuring is repeated three times for 10 and 60 s.

After the verification of the lumped parameter model of the pump, the dosing part is added. The simulated delivered volume is compared with the measurement of the drug delivery system. To measure the dispensed volume using water, evaporation effects of the small delivered droplets have to be taken into account. Therefore a single measurement is shown in Figure 13, to explain the approach for the calculation of the delivered volume. After finishing the filling process of the storage chamber, the center displacement and the weight of the dispensed fluid is calculated, see Figure 13. As can be seen in Figure 13 the height is constant, while the weight is decreasing. After a certain time the outlet valve is opened. Due to the preload of the membrane, the stored water volume is forced to flow out. The membrane moves to the original state, while the measured weight increases. The evaporation has a significant influence on the weight. Therefore we perform a linear fit to the measurement points and evaluate them at the time of the dispensing. The offset of the two calculations is taken as the weight of the delivered fluid. The center displacement is calculated by the offset of the mean of the two measurements before and after delivery. To prove the accuracy of the new drug delivery method 67 dosages where performed. It can be seen that the correlation between the center displacement and the delivered volume is very close to the calibration line, see Figure 14.

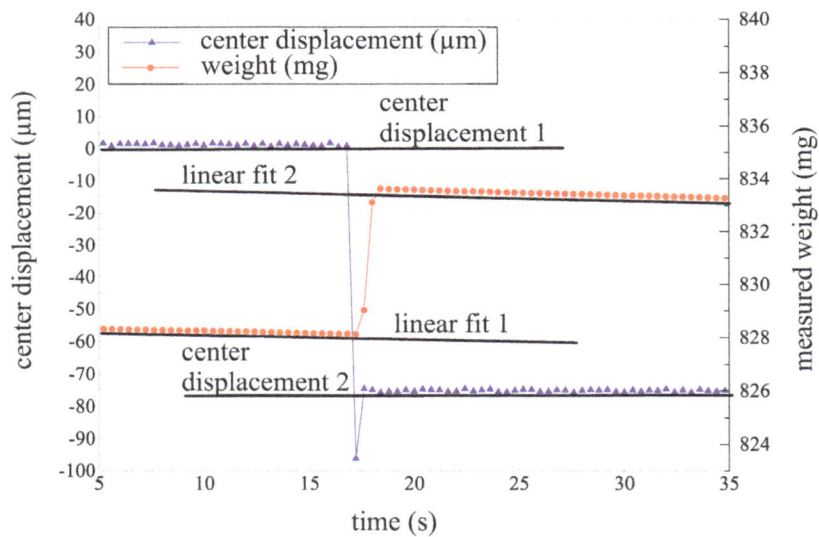

Figure 13. Measurement of the delivered volume and the center displacement at the same time. The center shows stable displacement result while the measured weight of the fluid changes with time. This is due to evaporation effects.

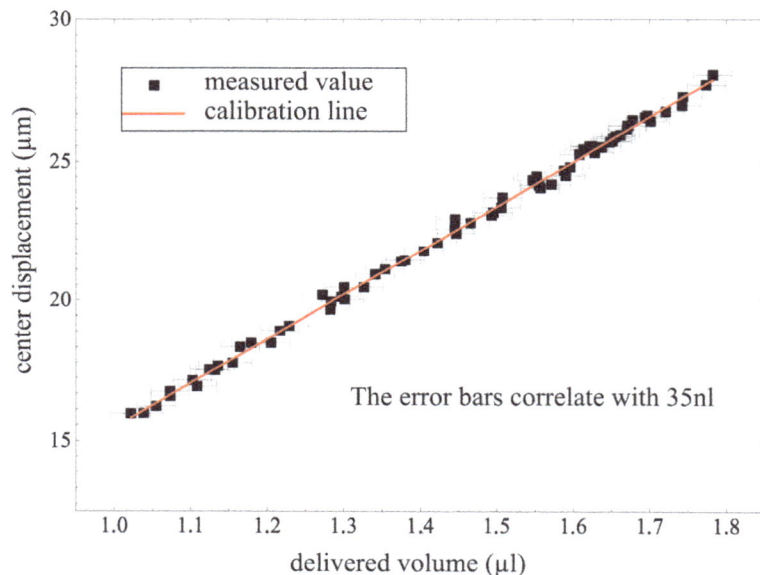

Figure 14. Delivered volumes in relation to the center displacement. The delivered volume deviates from the calibration line by a maximum of 35 nL.

The residual plot, see Figure 15, shows a normal distribution around the calculated value of the calibration line. The regular residual r_i is the observed value of the center displacement minus the predicted value of center displacement. The fluctuation around zero can be explained by the resolution of 0.4 µm of the laser triangulation measurement.

The gradient of the simulated dosing volume compared to the calibration line of the dispensed volumes deviates by 29%. As a result of the parasitic capacitances like tubes and connectors, calibration measurements of the whole system without the storage capacitance were performed. The results of this measurement where taken as the offset. The offset is subtracted from the measurements, depicted in Figure 16.

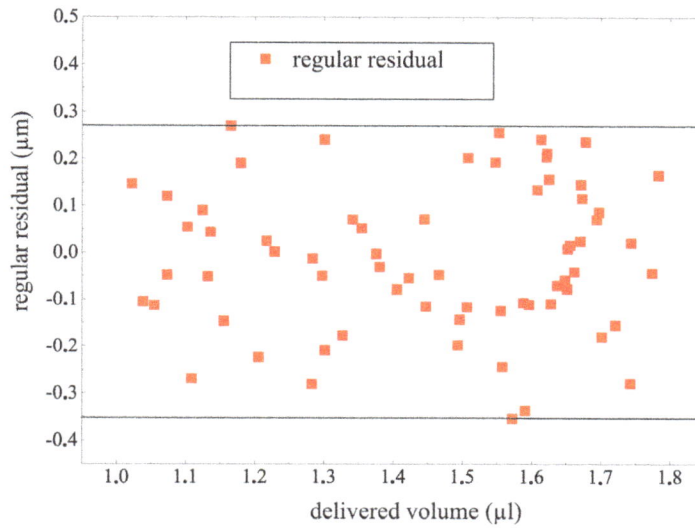

Figure 15. The regular residual of the dosing event between 1 and 1.8 µL with a constant variance.

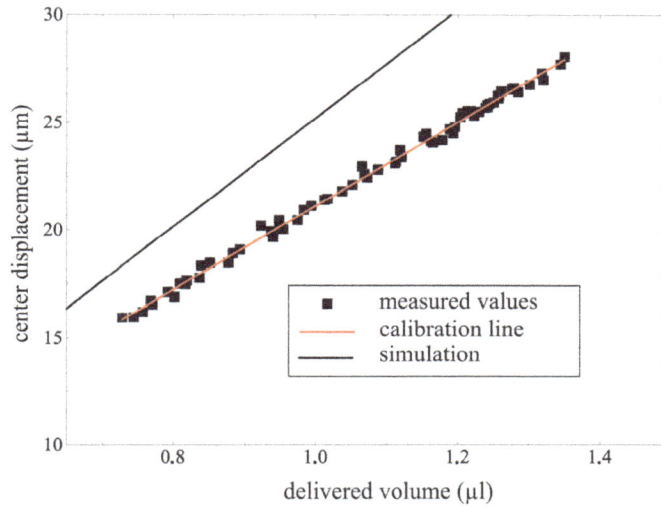

Figure 16. Simulated line and measured values fitted with a calibration regression.

In addition to the accurate delivery, the temporal profile of the center displacement during the filling and the dosing process of the fluid can be used as a safety feature. The center displacement is measured during a filling to the maximum achievable pressure, see Figure 17. The time dependent displacement is calculated by solving the differential Equation (10).

$$C_P \frac{\partial p}{\partial t} = q(p) = q_{max} \times \left(1 - \frac{q}{qmax}\right) \tag{10}$$

Solving differential Equation (10) results in Equation (11).

$$p = p_{max} * e^{-\frac{q_0}{R(C_{mem}+C_{para})}*t} \tag{11}$$

With the measurement of the flow at 20 Hz, see Figure 12, and the displacement of the membrane, which can be used to calculate the maximum pressure, C_{para} can be calculated. With p_{max} of 210 mbar and a flow rate at 20 Hz of 880 µL/min, $R \times (C_{mem} + C_{para})$ can be calculated. If we now assume that the resistance between the pump and the storage membrane is small and that the capacitance of the storage membrane is known, the parasitic capacitances can be derived from the delivered volume.

If we compare the exponential fits, of the center displacment in Figure 17 and the simulated pressure in Figure 18, it is possible to define a time constant for the exponential characteristic.

In the lumped parameter model C_{para} is assumed as zero. With the exponential fit curves of the simulated and measured filling process, we obtain the following values for the RC-modules for the storage chamber, see Equations (12) and (13) and section II in Figure 4.

$$\tau_{sim} = R * (C_{mem}) = 0.18 \text{ s} \tag{12}$$

$$\tau_{mes} = R * (C_{mem} + C_{para}) = 0.43 \text{ s} \tag{13}$$

If we now compare the simulated and the measured gradient of the delivered volume *vs.* center displacement, see Figure 16, we observe that a factor of 1.83 between them. $C_{mem} \times 0.83$ has to be substituted as C_{para} in Equation (11). With this assumption the time constant τ is then calculated to 0.46 s. Comparing the measured time constant 0.43 s, Equation (12), with the calculated τ of 0.46 s which is derived from the time dependent dosing event, the dispensed volume can be calculated with an error of 7%.

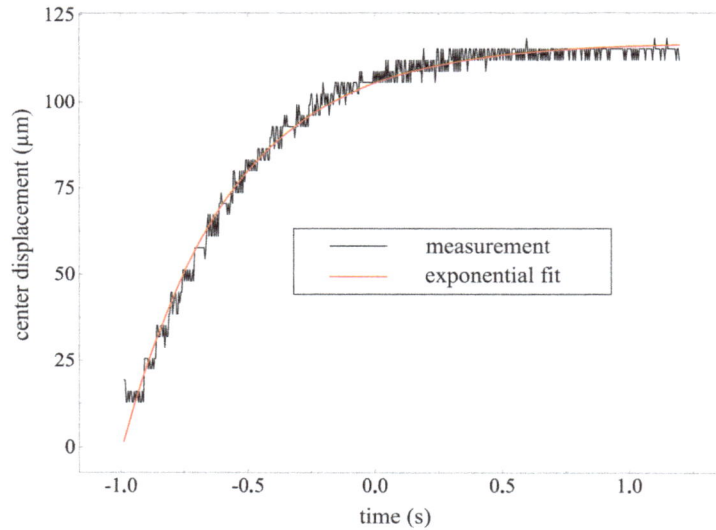

Figure 17. Time dependent center displacement of the storage chamber membrane.

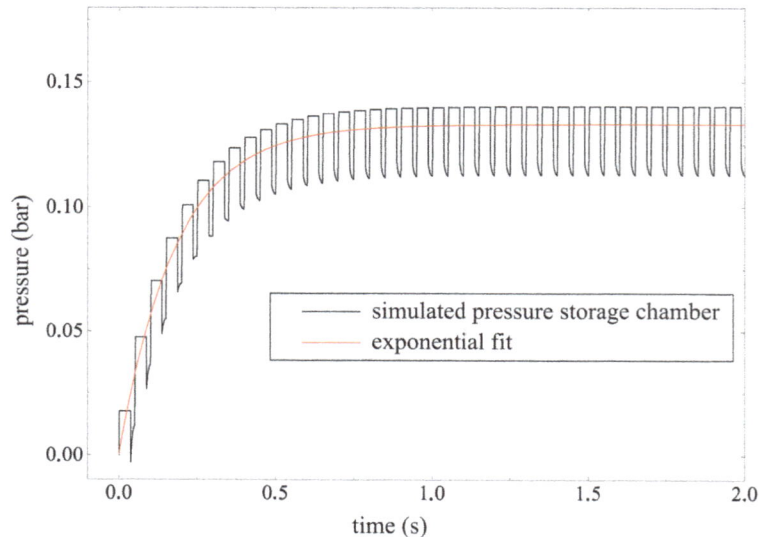

Figure 18. Simulated time dependent pressure under the storage chamber membrane.

3.3. Discussion

To summarize the results of our new design, we have to account for two things: First the validation of the lumped parameter model and second the accuracy of the developed drug delivery system. The lumped parameter model for the pump calculates the frequency dependent flow with an accuracy of 20%. The frequency, at which the maximum flow rate is generated, can be calculated very precisely. Taking into account the fact, that only the geometric parameters are used in the model, also the deviation of the electro-mechanical simulation and the fluidic calculations influence the results. The mechanical simulations of the piezoelectrically actuated membranes are not represented in this paper. The deviation between mechanical simulation and measurement are about 12%–20% for different membranes. Taking into account all these facts, the results of the lumped parameter model are satisfactory. Therefore this first principle modeling of the whole microfluidic system can be considered as a valuable tool for further optimizations.

The delivery of a defined volume is compared to a calibration line. The error has a maximum deviation of 35 nL in a dosing range of 1 to 1.8 μL. This error can be explained by the measurement uncertainty of the used equipment. A calibrated system has the potential to show a distinct improvement to the state of the art by reaching relative accurrancy of better than 3.5%. The currently used implanted drug delivery systems with an active micropump have accuracies between 10% and 15%, see Table 1 [5].

4. Conclusions

In a summary the new drug delivery system has the potential to be more accurate than the drug delivery systems on the market. Implementing a measurement technology before drug delivery furthermore increases the patient's safety. With the lumped parameter model, we have developed a tool that allows a fast calculation of the fluidic characteristics only with the knowledge of geometric and piezo-ceramic parameters.

5. Outlook

After the proof of concept, the next development step will be an integration of the measurement into a miniaturized system. First a reduction of the actuator voltage has to be established. This can be achieved with multilayer piezoactuators. The functionality of multilayer piezoelectrically driven micropumps is shown in [19,20]. With the reduction of the applied voltage almost the same performance can be expected. An integrated measurement of the additional volume in the storage chamber can be done with a pressure sensor [21], that can be integrated into the membrane of the storage chamber. Alternatively a capacitive measurement can be used [22]. With these development steps and the following examination of biocompatibility issues a miniaturized implantable drug delivery system with increased accuracy and safety seems feasible.

Author Contributions

Florian Thoma designed the dosing system, set up the lumped parameter model, carried out the experiments and wrote the paper. Peter Woias and Frank Goldschmidtböing advised the author concerning the simulation and the dosing technology.

Appendix A1. List of Symbols

Symbol	Definition	Unit
A	Area	m^2
C	Fluidic capacitance	m^5/N
h_v	Gap of the valve	m
k	Stiffness	N/m
L	Fluidic inductance	kg/m^4
m	Mass	kg
p	Pressure	bar
V	Volume	m^3
q	Fluidic flow	m^3/s
R	Fluidic resistance	$N \cdot s/m^5$
ρ	Density	kg/m^3
τ	Time constant	s
χ	Friction coefficient	$N \cdot s/m$

Conflicts of Interest

The authors declare no conflict of interest.

References

1. Wang, W.; Soper, S.A. *Bio-MEMS: Technologies and Applications*; CRC Press: Boca Raton, FL, USA, 2006.

2. Stevenson, C.L.; Santini, J.T., Jr.; Langer, R. Reservoir-based drug delivery systems utilizing microtechnology. *Adv. Drug Deliv. Rev.* **2012**, *64*, 1590–1602.

3. Jain, K.K. *Textbook of Personalized Medicine*; Springer: New York, NY, USA, 2009.

4. Meng, E.; Hoang, T. MEMS-enabled implantable drug infusion pumps for laboratory animal research, preclinical, and clinical applications. *Adv. Drug Deliv. Rev.* **2012**, *64*, 1628–1638.

5. Receveur, R.A.M.; Lindemans, F.W.; de Rooij, N.F. Microsystem technologies for implantable applications. *J. Micromech. Microeng.* **2007**, *17*, R50–R80.

6. Goldschmidtböing, F.; Doll, A.; Heinrichs, M.; Woias, P.; Schrag, H.-J.; Hopt, U.T. A generic analytical model for micro-diaphragm pumps with active valves. *J. Micromech. Microeng.* **2005**, *15*, 673.

7. Woias, P. Micropumps-summarizing the first two decades. In Proceeding of SPIE 4560, Microfluidics and BioMEMS, San Francisco, CA, USA, 22 October 2001.

8. Smits, J.G. Piezoelectric micropump with three valves working peristaltically. *Sens. Actuators A Phys.* **1990**, *21*, 203–206.

9. Kan, J.; Tang, K.; Liu, G.; Zhu, G.; Shao, C. Development of serial-connection piezoelectric pumps. *Sens. Actuators A Phys.* **2008**, *144*, 321–327.

10. Kan, J.; Tang, K.; Ren, Y.; Zhu, G.; Li, P. Study on a piezohydraulic pump for linear actuators. *Sens. Actuators A Phys.* **2009**, *149*, 331–339.

11. Geipel, A.; Goldschmidtböing, F.; Doll, A.; Jantscheff, P.; Esser, N.; Massing, U.; Woias, P. An implantable active microport based on a self-priming high-performance two-stage micropump. *Sens. Actuators A Phys.* **2008**, *145–146*, 414–422.

12. Français, O.; Dufour, I. Dynamic simulation of an electrostatic micropump with pull-in and hysteresis phenomena. *Sens. Actuators A Phys.* **1998**, *70*, 56–60.

13. Lin, Q.; Yang, B.; Xie, J.; Tai, Y.C. Dynamic simulation of a peristaltic micropump considering coupled fluid flow and structural motion. *J. Micromech. Microeng.* **2007**, *17*, 220.

14. Bourouina, T.; Grandchamp, J.P. Modeling micropumps with electrical equivalent networks. *J. Micromech. Microeng.* **1996**, *6*, 398–404.

15. Hamdan, M.N.; Abdallah, S.; Al-Qaisia, A. Modeling and study of dynamic performance of a valveless micropump. *J. Sound Vib.* **2010**, *329*, 3121–3136.

16. Oertel, H.; Böhle, M.; Dohrmann, U. *Strömungsmechanik: Grundlagen, Grundgleichungen, Lösungsmethoden, Softwarebeispiele*, 4th ed.; Springer Vieweg: Berlin, Germany, 2009.

17. Truckenbrodt, E. *Fluidmechanik*, 4th ed.; Springer Vieweg: Berlin, Germany, 2013.

18. Goldschmidtböing, F. Entwurf, Design und Experimentelle Charakterisierung von Mikro-Freistrahldispensern. Ph.D. Thesis, Universitätsbibliothek Freiburg, Freiburg, Germany, 2004. (In German)

19. Thoma, F.; Feth, H.F.; Goldschmidtboeing, F.; Woias, P. Integrated fluidic system for an artificial sphincter prosthesis. In Proceeding of 2012 IEEE Micro- and Nanoengineering in Medicine Conference, Maui, HI, USA, 3–7 December 2012

20. Biancuzzi, G.; Lemke, T.; Woias, P.; Ruthmann, O.; Schrag, H.J.; Vodermayer, B.; Goldschmidtboeing, F. Design and simulation of advanced charge recovery piezoactuator drivers. *J. Micromech. Microeng.* **2010**, *20*, 105022.

21. Ko, W.; Hynecek, J.; Boettcher, S. Development of a miniature pressure transducer for biomedical applications. *IEEE Trans. Electron Devices* **1979**, *26*, 1896–1905.

22. Yu, J.; Wang, W.; Lu, K.; Mei, D.; Chen, Z. A planar capacitive sensor for 2D long-range displacement measurement. *J. Zhejiang Univ. Sci C* **2013**, *14*, 252–257.

9

Deformation Analysis of a Pneumatically-Activated Polydimethylsiloxane (PDMS) Membrane and Potential Micro-Pump Applications

Chi-Han Chiou [1], Tai-Yen Yeh [2] and Jr-Lung Lin [2],*

[1] ITRI South Campus, Industrial Technology Research Institute, Tainan City 70955, Taiwan;
    E-Mail: prayjohn@itri.org.tw
[2] Department of Mechanical and Automation Engineering, I-Shou University, Kaohsiung City 84001,
    Taiwan; E-Mail: isu10374015b@isuo365.onmicrosoft.com

* Author to whom correspondence should be addressed; E-Mail: ljl@isu.edu.tw

Academic Editor: Joost Lötters

Abstract: This study presents a double-side diaphragm peristaltic pump for efficient medium transport without the unwanted backflow and the lagging effect of a diaphragm. A theoretical model was derived to predict the important parameter of the micropump, *i.e.*, the motion of the valves at large deformations, for a variety of air pressures. Accordingly, we proposed an easy and robust design to fabricate a Polydimethylsiloxane (PDMS)-based micropump. The theoretical model agrees with a numerical model and experimental data for the deformations of the PDMS membrane. Furthermore, variations of the generated flow rate, including pneumatic frequencies, actuated air pressures, and operation modes were evaluated experimentally for the proposed micropumps. In future, the theoretical equation could provide the optimal parameters for the scientists working on the fabrication of the diaphragm peristaltic pump for applications of cell-culture.

Keywords: larger deformation; micro-pump; peristaltic; diaphragm; operation mode

1. Introduction

In recent years, micromachining technologies have been introduced to provide a means to miniaturize microfluidic applications, such as biochemical analysis [1–3], drug delivery [4], DNA sequencing [5], nucleic acid synthesis [6], *etc.* One of the most exciting developments in microfluidic applications is the rapid evolution of biological-microelectromechanical systems (Bio-MEMS). The advantages of such applications are integrating delivery, testing, and analysis of biomedical samples, therefore dramatically reducing the required human involvement in laborious multi-step sample handling and processing, and improving data quality and quantitative analysis. Microfluidic applications also reduce the overall cost and time of measurements, and at the same time, improve the sensitivity and specificity of analyses. In addition, basic microfluidic components, such as microchannels, microvalves, micropumps, micromixers and microreactors with various novel sensors and detection platforms have been successfully incorporated in the microfluidics and Bio-MEMS fields. Among them, micropumps are the most important components because they are crucial in sample delivery and manipulation in microfluidic devices and systems. The design of an efficient micropump has been a challenging task in miniaturized biomedical systems.

Micropumps can generally be classified as either non-mechanical or mechanical based on whether their components are fixed or movable, respectively. Although non-mechanical pumps have simple structures and contain no moving parts, their performance depends on the types and surface properties of the fluids. Non-mechanical pumps transport fluids by directly converting external energy into kinetic energy. Conversely, mechanical pumps require an actuation source to provide a mechanical stroke cycle. Mechanical actuation typically drives fluids by coupling mechanical deformation of a moving boundary to fluid pressure changes. Actuation mechanics may be electrostatic [7–9], electromagnetic [10,11], pneumatic [12–14], thermopneumatic [15,16], or piezoelectric [17–19]. The actuated diaphragms are classically divided into a single reciprocating displacement and a peristaltic displacement. A single reciprocating displacement is constructed from multi-layers of material and works as the actuated diaphragm with two check valves that prevent the backflow of fluids. Alternatively, the micropumps are actuated with three diaphragms operating in series, and accordingly, are sometimes named as peristaltic micropumps. Certainly, the choice of pump diaphragm material is particularly important. Common pump diaphragm materials include silicon, glass, and plastic. Currently, PDMS elastomers are widely used for fabricating various micro fluidic devices such as passive and active structures. Because of its excellent biocompatibility, good mechanical properties and simple structural fabrication and bounding processes, the PDMS is used as both the functional membrane and the structural substrate of micropumps.

Recently, pneumatically-actuated peristaltic micropumps [20–24] have proven effective for driving fluids in microchannels and are readily integrated with bio-sensing chips. In these micropump designs, fluids are typically driven by multiple elastic membranes actuated by their corresponding pneumatic chambers. However, precise control of membrane movement is critical for pumping performance. The previous micropumps [13–15] used various vertical actuation mechanisms and fluid control valves, which increased the complexity of the structural design and the processes needed for multiple layer alignment and assembly. Therefore, the objective of this study is to design an easily fabricated, low cost micropump with fine controls needed for accurate volumetric flow rate. Although a large number of reports of PDMS micropump applications are found in the literature, and the number of papers is still increasing, the actuated working principle and membrane deformation behavior are not clearly understood. The parameters that must be considered include the geometry of the actuated structure, material properties, and applied external energy modes. All of these factors simultaneously affect micropump performance.

Using the pneumatically-actuated PDMS membrane as a flexible structure, the study investigates deformation mechanisms and the transporting performance. Theoretical and numerical models are employed to predict and evaluate experimental values for membrane deformation mechanisms. Consequently, the proposed actuated membrane design is optimized by numerical simulations. Here we demonstrate that a micro-pump with three-pair flexible structures, *i.e.*, with single- and double-sided flexible actuation, is successfully incorporated to transport the sample stream.

2. Design and Experiment

2.1. Design Principle and Fabrication

In microfluidic systems, the backflow of pumping liquid can cause cross contamination between solutions or microbial contamination [25]. Therefore, unwanted fluid backflow in the micropumps can influence the precise manipulation of fluid flow in a microfluidic system. The major contribution of this study is the design of the pneumatic side chambers that actuate the flexible structures to generate an efficient pumping performance and avoid the fluid backflow particularly for the double-side mode. The side chambers are positioned orthogonally to the fluidic microchannel, and they provide flexible structures (single- and double-side) for sample transportation (see Figure 1). Figure 2 schematically depicts how transportation by the proposed micro-pump is performed by the three-pair double-side flexible structures. The channel between the six side-chambers defines the transporting region. When atmospheric air fills the side chambers, the flexible structures are pneumatically activated by the compressed air in series to create the transporting effect. Adjusting various operation parameters, e.g., applied pressures and/or driving frequency, provides a precisely controlled pumping action.

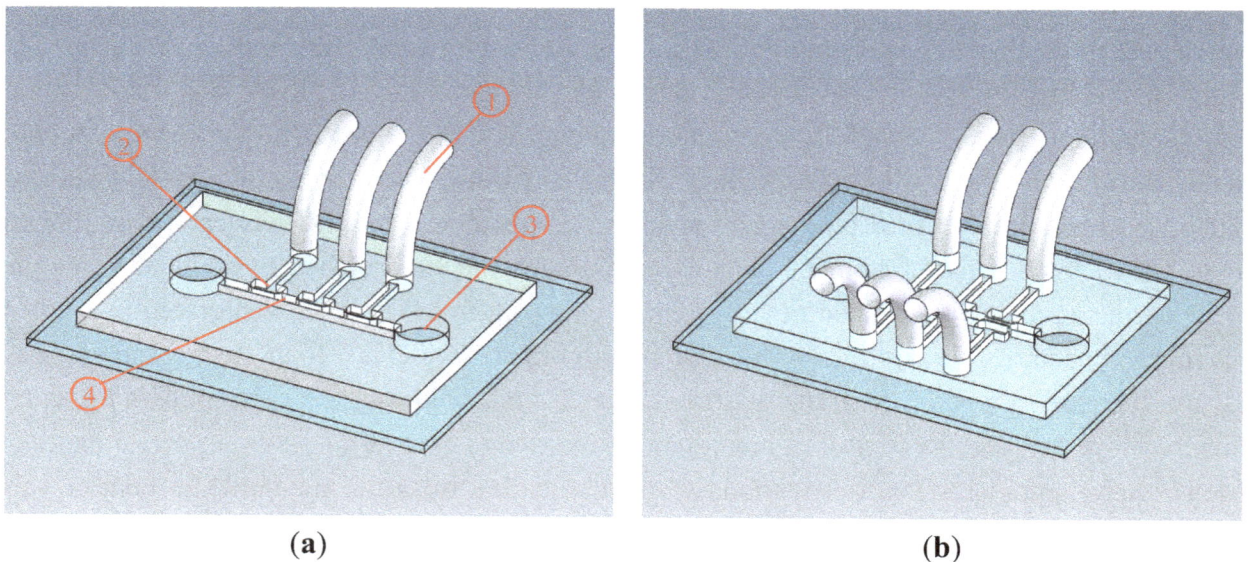

(a)   (b)

Figure 1. Schematic representation of the proposed micropump using three-pair moving structures in (**a**) single-side mode and in (**b**) double-side mode. Here, (1) is a tubing, (2) is a buried side chamber, (3) is a reservoir, and (4) is a microchannel.

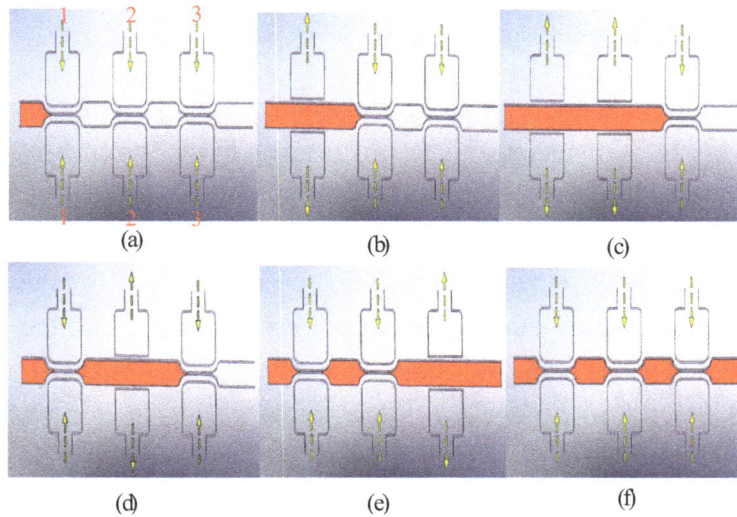

Figure 2. Schematic representations of a pumping process in a three-pair double-side mode micropump. Here, arrow direction indicates the air input/output. 1, 2, and 3 indicate the pair number. (**a**) 3 pairs simultaneously shut down to block the fluid. (**b**) 1st pair and (**c**) 2nd pair sequentially open to induce the fluid to move. (**d**) 1st pair and (**e**) 2nd pair sequentially shut down and 3rd pair open to enhance the fluid forward. (**f**) 3 pairs simultaneously shut down again to avoid the backflow.

To construct the micropump, a 250 µm-thick layer of negative photoresist SU-8 (MicroChem Corp., Newton, MA, USA) was first spun onto a silicon substrate. The SU-8 was used in the master mold in the PDMS casting process due to its excellent structural robustness, good adhesion to the silicon substrate, and suitability for producing high-aspect-ratio structures. The SU-8 master mold was formed using the standard lithography and baking processes. After fabricating the SU-8 master mold, the PDMS solution was poured into the master mold and cured at 100 °C for 10 h. The inverse structures of the SU-8 master mold were then transferred onto the PDMS chip after the de-molding process. After completing the replication process for the PDMS layer, the chip device was assembled, and an oxygen plasma treatment was used to bond the PDMS layer to the glass substrate. Figure 3a,b are close-up views of the SU-8 template. The dimensions of the pneumatic side chambers are 1500×400 µm^2 and their depths are 250 µm. The flexible structure formed between the fluidic flow channel and pneumatic side chambers is 100 µm thick. The flow channel is 150 µm wide and 250 µm deep.

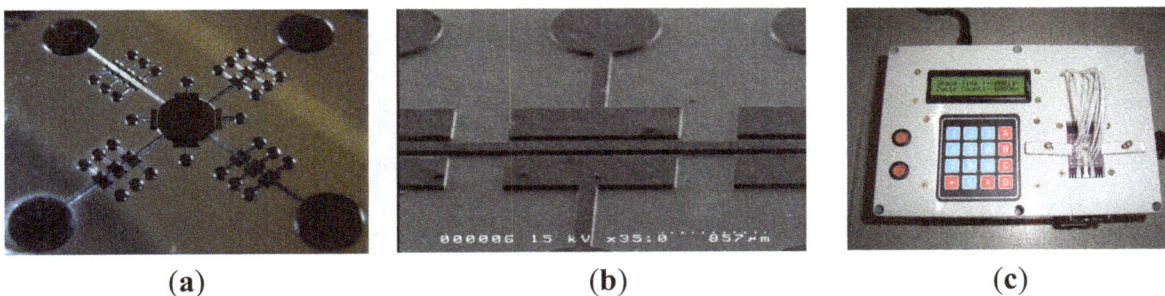

Figure 3. (**a**) Photograph of the SU-8 template for the micropump chip; (**b**) SEM image showing a close-up view of the SU-8 template; (**c**) photograph of the hand-held digital controller.

2.2. Control and Measurement System

Figure 3c is a photograph of the micropump control system, which is comprised of an air compressor (JUN-AIR Inc., MDR2-1A/11, Kawasaki-shi, Japan) for supplying compressed air to the micropump, a functional control circuit, and three electromagnetic valves (EMVs) (SMC Inc., S070M-5BG-32, Taoyuan City, Taiwan) to control the pneumatic pump. Tests were performed to measure pumping rates at various operational frequencies, pneumatic driving pressures, and operation modes. A constant current was supplied to the microflow sensor [24] throughout the tests, and the electrical signal output from the sensor was recorded by an analog-to-digital converter (ADC) (ATMEL Corp., ATMEGA8535, San Jose, CA, USA) connected to a personal computer. A syringe pump (KDScientific, KDS 100, Holliston, MA, USA) was used to calibrate the flow rate to the output voltage of the microflow sensor. The micropump was positioned under an optical microscope (Olympus, BH2-UMA, Tokyo, Japan) during testing so that fluidic motion in the microchannel could be recorded with an image capturing system (Photometrics, CoolSNAP HQ^2, Tucson, AZ, USA).

3. Theoretical and Numerical Methods

3.1. Theoretical Analysis

An approximate solution for the larger deformation of the PDMS membrane can be expressed by using the strain energy (U_{St}) method as shown below [26]:

$$U_{St} = \frac{Eh}{2(1-v^2)} \int_{-a/2}^{a/2} \int_{-b/2}^{b/2} \left\{ \left(\frac{\partial u}{\partial x}\right)^2 + \left(\frac{\partial v}{\partial y}\right)^2 + \frac{\partial u}{\partial x}\left(\frac{\partial w}{\partial x}\right)^2 + \frac{\partial v}{\partial y}\left(\frac{\partial w}{\partial y}\right)^2 \right.$$
$$+ \frac{1}{4}\left[\left(\frac{\partial w}{\partial x}\right)^2 + \left(\frac{\partial w}{\partial y}\right)^2\right]^2 + 2v\left[\frac{\partial u}{\partial x}\frac{\partial v}{\partial y} + \frac{1}{2}\frac{\partial u}{\partial x}\left(\frac{\partial w}{\partial y}\right)^2 + \frac{1}{2}\frac{\partial v}{\partial y}\left(\frac{\partial w}{\partial x}\right)^2\right] \qquad (1)$$
$$\left. + \frac{1-v}{2}\left[\left(\frac{\partial u}{\partial y}\right)^2 + 2\frac{\partial u}{\partial y}\frac{\partial v}{\partial x} + \left(\frac{\partial v}{\partial x}\right)^2 + 2\frac{\partial u}{\partial y}\frac{\partial w}{\partial x}\frac{\partial w}{\partial y} + 2\frac{\partial v}{\partial x}\frac{\partial w}{\partial x}\frac{\partial w}{\partial y}\right] \right\} dxdy$$

where *u*, *v*, and *w* are the x-, y- and z-directional displacements, respectively. Since deformations of *u* and *v* are much smaller than that of *w* ($u \ll w$ and $v \ll w$), deriving *w* is more important than deriving *u* and *v*. Moreover, the strain energy obtained by Equation (1) can be simply expressed using the following equation:

$$U_{St} = \frac{Eh}{8(1-v^2)} \int_{-a/2}^{a/2} \int_{-b/2}^{b/2} \left[\left(\frac{\partial w}{\partial x}\right)^2 + \left(\frac{\partial w}{\partial y}\right)^2\right]^2 dxdy \qquad (2)$$

here, *E* and *h* are Young module and membrane thickness, respectively. The expressions clearly satisfy the boundary conditions but contain several arbitrary parameters, which must be determined by applying virtual displacement principles. To illustrate the methods, consider a uniformly loaded rectangle membrane with the dimensions *a* (transversal width) and *b* (axial length) (Figure 4a). The steady-state solution *w*(*x*, *y*) for the larger membrane deformation problem is assumed by a cosine functions regarding of x- and y-directions. The solution has to be satisfied by the boundary conditions in Figure 4a:

$$\frac{\partial w(\bar{c}/2,y)}{\partial x}=0; \quad w(b/2,y)=0; \quad \frac{\partial w(x,0)}{\partial y}=0; \quad w(x,a/2)=0 \tag{3}$$

The displacement w can be obtained by:

$$w=\begin{cases} w_0\cos\dfrac{\pi y}{a} & 0\le x\le c/2 \\[2mm] w_0\cos\dfrac{\pi y}{a}\cos\dfrac{\pi(x-\bar{c}/2)}{(b-\bar{c})} & \bar{c}/2\le x\le b/2 \end{cases} \tag{4}$$

here w_o is the maximum value of w, c is varied with the y-axial direction and is assumed as below:

$$c=\bar{c}\cos\frac{\pi y}{a} \tag{5}$$

By substituting Equations (4) and (5) into Equation (1), the strain energy can be integrated as:

$$U_{St}=\frac{Eh\pi^4 w_o^4}{512(1-v^2)a^2}\left[\frac{128}{5\pi}\frac{\bar{c}}{a}+\frac{9(b-\bar{c})}{a}+\frac{2a}{(b-\bar{c})}+\frac{9a^3}{(b-\bar{c})^3}\right] \tag{6}$$

Equation (6) can be rewritten as another form as below:

$$U_{St}=\frac{Eh\pi^4 w_o^4}{512(1-v^2)a^2}\left[\frac{\alpha b}{a}+(9-\alpha)\lambda+\frac{2}{\lambda}+\frac{9}{\lambda^3}\right] \tag{7}$$

here $\alpha=\dfrac{128}{5\pi}(\approx 8.15)$ and $\lambda=\dfrac{(b-\bar{c})}{a}$.

By deriving $\dfrac{\partial U_{St}}{\partial\lambda}$ (implied that U_{St} achieved the maximum value as the $(b-\bar{c})$ attained the minimum value), the solution obtained for λ is:

$$\lambda=\frac{(b-\bar{c})}{a}=\sqrt{\frac{1+\sqrt{1+27(9-\alpha)}}{(9-\alpha)}}=2.63 \tag{8}$$

here, λ is a constant value ($\sim$2.63). The $\lambda/2$ defines the dimensionless curved-shaped length, which is dependent on the shorter side length of the rectangle. The outcome can be verified by the latter numerical calculations and experimental observations. The external pneumatic energy (U_{Pn}) produced by air pressure loading is integrated by:

$$U_{Pn}=\int_{-a/2}^{a/2}\int_{-b/2}^{b/2}Pwdxdy=\frac{2a^2 Pw_o}{\pi^2}\left(\frac{\pi^2 b}{4a}+(2-\frac{\pi^2}{4})\lambda\right) \tag{9}$$

Since the strain energy (U_{St}) is equal to the external pneumatic energy (U_{Pn}), is performed by:

$$\frac{Eh\pi^4 w_o^4}{512(1-v^2)a^2}\left[\frac{\alpha b}{a}+(9-\alpha)\lambda+\frac{2}{\lambda}+\frac{9}{\lambda^2}\right]=\frac{2a^2 Pw_o}{\pi^2}\left(\beta\frac{b}{a}+\gamma\lambda\right) \tag{10}$$

here $\beta=\dfrac{\pi^2}{4}(\approx 2.47)$ and $\gamma=2-\dfrac{\pi^2}{4}(\approx -0.47)$.

(a)

(b)

(c)

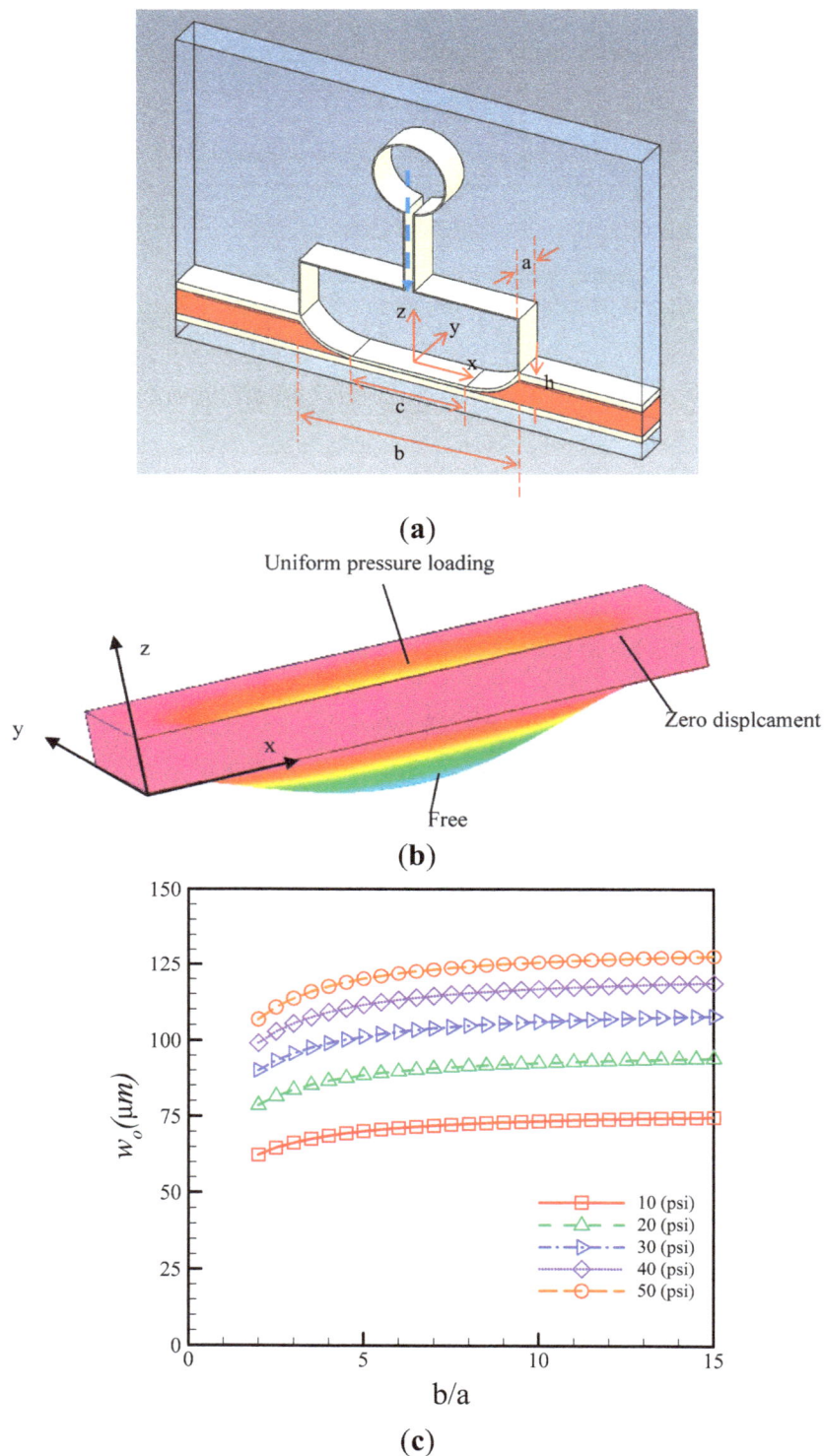

Figure 4. (**a**) Schematic illustration of the flexible membrane, side chamber, and microchannel; (**b**) The boundary conditions of 3-D numerical simulation for the larger deformation; (**c**) Deformed profiles of flexible membrane related with the axial/transversal length (b/a) ratio under different air pressures.

During deforming, the membrane satisfies the principles of energy minimization in achieving equilibrium. Hence, the dimensionless maximum deformation w_o/a caused by pressure loading can be estimated as:

$$\frac{w_o}{a} = \left(\frac{4}{\pi}\right)^2 \left[\frac{(1-v^2)\left(\beta\frac{b}{a}+\gamma\lambda\right)}{4\cdot\left(\frac{\alpha b}{a}+(9-\alpha)\lambda+\frac{2}{\lambda}+\frac{9}{\lambda^3}\right)}\frac{Pa}{Eh}\right]^{\frac{1}{3}}$$

(11)

Equation (11) can be easily calculated for the maximum deformation of the rectangle membrane for the pneumatically-driven air pressures. If $\lambda = 1$ (*i.e.*, $a = b$ and $c = 0$) implied that the membrane is square, Equation (11) can be simplified as:

$$\frac{w_o}{a} = 0.474\left[(1-v^2)\frac{Pa}{Eh}\right]^{\frac{1}{3}}$$

(12)

Equation (12) was also used to predict the maximum deformation for the square membrane. These equations can be used to design the optimum sizes of the actuated membrane of the proposed micropump.

3.2. Numerical Modeling and Optimizations

To investigate the transporting performance of the developed micro-pump, the design was optimized by using flexible structures, pneumatically activated by side air chambers. A numerical simulation was performed to design the micro-pumps and to investigate deformation of the PDMS membrane. The deformation was simulated numerically using a commercial code (CFD-ACE+, CFD-RC, Huntsville, AL, USA). Enhanced first order brick elements were recommended for introduction to the mechanism of PDMS membrane [27]. The moving boundary condition was simulated by using both the stress module and the deformation module. The moving boundary of the membrane was discretely separated to ensure smooth motion. The deformation grid of the moving boundary was constructed using the auto-remesh function in the deformation process. Dense grids were used in moving boundary regions where deformation was induced by the side chamber. The density (ρ), Young module (E), and Poisson ratio (v) of the PDMS are 970 kg/m^3, 1.4 MPa [28], and 0.5, respectively. A stringent residuals criterion (less than 10^{-8}) and a nonlinear stress residuals criterion (less than 10^{-4}) were used between each iterative solution step to ensure convergence of the solution. Figure 4b shows the numerical geometry and boundary conditions. A uniform pressure loading was applied to the top of X-Y plane. The bottom of the X-Y plane was set as the free module. The surroundings of X-Z and Y-Z planes were set to zero for the displacements. As we know, when the PDMS membrane is deflected by the applied pressure as if $b \approx a$, the deformation profile will present symmetry curved-shaped profile, or, if $b >> a$, that will exhibit two segments: A curved-shaped and a flat-shaped profile. A flat-shaped membrane increases the stroke volume, which in turn increases the flow rate. In double-side actuation mode, a flat-shaped profile for the flexible structure in the rectangular membrane was clearly desirable for increasing stroke volume. The theoretical calculations were employed to optimize the b/a (axial length/transversal width) for the flexible element. The calculation results showed the relationship profile of maximum deformation and b/a value for the membrane under the different air pressures as shown as Figure 4c. The profile of the maximum deformation initially increases along with the increase in b/a ratio. However, when the ratio b/a is larger than 6.0, the variations of maximum deflections slightly attain saturation values. Considering the difficulty in fabricating a micropump with high aspect ratio, the optimum design is $b/a = 6.0$, *i.e.*, a 1500 μm axial length and a 250 μm depth.

4. Results and Discussion

4.1. Estimation of Membrane Deformation

Figure 5a shows the deformation of the flexible membrane under applied air pressures of 10, 20, 30 and 40 psi. As expected, the maximum deformation increases with increases in applied pressure. The experimental results show that increasing applied pressure increases the deformation of the flexible membrane structure and results in an increase in the stroke volume. In addition, experimental observation shows that the fluid flow was difficult to interrupt while the membrane is completely deflected down to another side at a pressure of 40 psi or even higher. Therefore, high pressure applied in the microvalves is unable to shut the fluid flow in the single mode. This results in fluid back flow that causes potential sample contamination. The deformation of the flexible membrane structure was numerically investigated under different operating pressures as shown in Figure 5b. The numerical representation of the deformation contours of x-z cross-sectional and y-z cross-sectional surfaces are shown under the air pressures of 10, 20, 30, and 40 psi, respectively. As was observed in experiments, the numerical deformation profiles of x-z plane (longer side) also exhibit two segments: Flat-shaped and curve-shaped profiles. Contrarily, the deformation of y-z plane (shorter side) demonstrated a symmetric curve-shaped profile. Figure 5c shows the deformation profiles for the numerical calculation and experimental data. The comparison results show that the flexible membrane deformation profiles obtained by numerical calculations generally agree with the measurement data, although the deformations obtained by numerical calculations are larger than that of the experimental ones in the flat-shaped profile. Notably, the curved-shaped profiles occur within 300 μm from the edges. In contrast, the distance calculated in theoretical mode is about 329 μm. Figure 6 shows the numerical deformations of the x-directional and y-directional displacements under the pressure of 50 psi. The x- and y-directional displacements of PDMS membranes are limited to about 8 μm and 0.1 μm, respectively. The results demonstrate the x- and y-directional displacements are much less than the z-directional displacement. Therefore, these results verify the assumptions in the theoretical analysis, *i.e.*, the x- and y-direction displacements are much shorter than the z-direction displacement.

Figure 7a shows the present analytical model (Equation (12)) compared with the experimental measurements [29]. The results demonstrated that the prediction of the present model is slightly lower than that of experimental data. However, the trend of the calculations are reasonable in agreement with experimental observations. The maximum deviations of the results are about 17.5%. Figure 7b compares the relationship between maximum deformation and applied pressures according to the theoretical analysis, numerical calculations and experimental measurements. The theoretical and numerical models are reasonably consistent with the experimental measurements. The numerical results are more precise than the theoretical analysis results, however, the numerical model requires longer time to calculate membrane deformation by using FEM method. Figure 7 also shows that maximum deformations are clearly divided into two regions: A small deformation region (implied that maximum deformations less than the membrane thickness, *i.e.*, $w_o < h$) described by the linear relationship, and a larger deformation region (implied that maximum deformations larger than the membrane thickness, *i.e.*, $w_o > h$) described by the non-linear relationship. The PDMS membranes are affected by the elasticity properties in the smaller deformation, but exhibit plasticity properties in the larger deformation. If a flexible membrane is working under a high applied pressure that results in large deformation, the membrane will be elastically fatigued, and the lifetime will also be reduced.

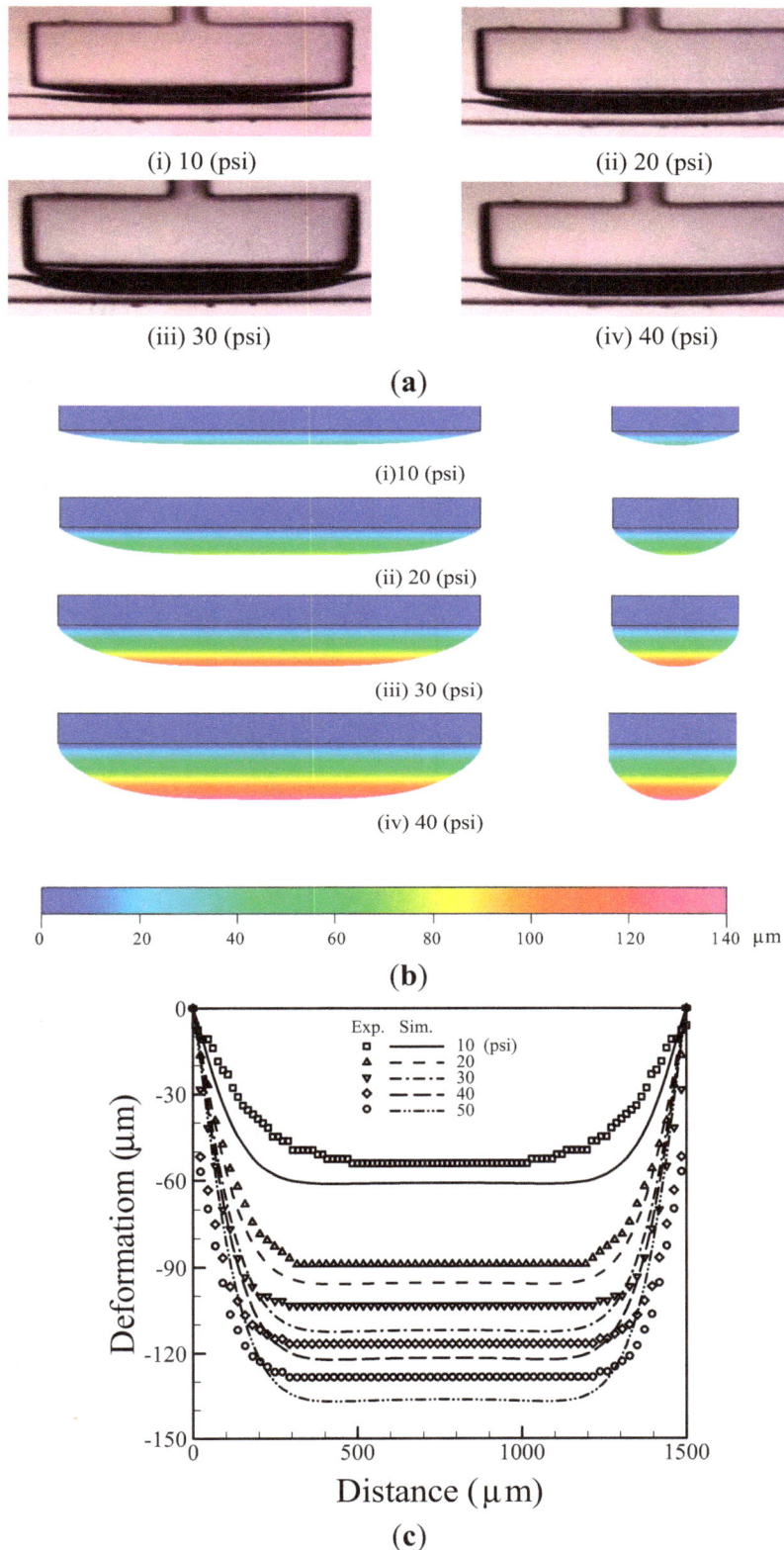

(i) 10 (psi)   (ii) 20 (psi)

(iii) 30 (psi)   (iv) 40 (psi)

(a)

(i)10 (psi)

(ii) 20 (psi)

(iii) 30 (psi)

(iv) 40 (psi)

0   20   40   60   80   100   120   140 μm

(b)

(c)

Figure 5. (a) Experimental and (b) numerical deformation of the flexible membrane structure (c) comparison of numerical calculations and experimental measurements at various air pressures.

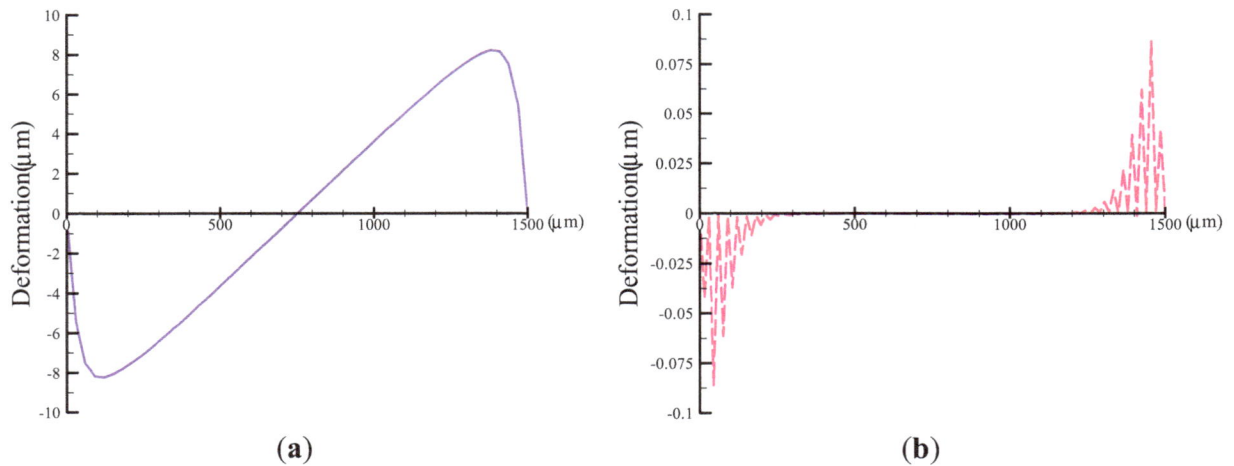

Figure 6. Numerical deformations of (**a**) the *x*-directional; and (**b**) *y*-directional displacements at 50 psi.

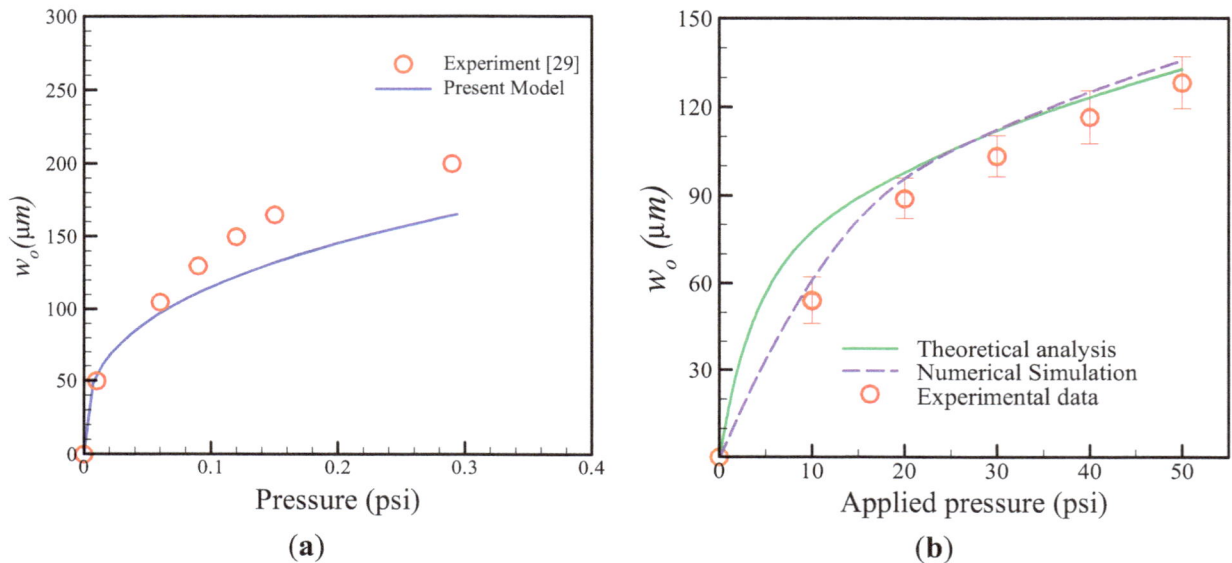

Figure 7. (**a**) Comparison of the present model and experimental data for maximum deformation; (**b**) Comparison of theoretical analysis, numerical simulation and measured data for maximum deformation of membrane at various air pressures.

4.2. Evaluation of Pumping Rate

The volumetric flow rate in the proposed medium pumping scheme is determined predominantly by the applied pneumatic pressure and its frequency. To establish a quantitative link between these parameters, the pumping rate was measured at various applied pneumatic pressures (10, 20, 30 and 40 psi) and frequencies (5–70 Hz). An observation in Figure 8a shows the liquid pumping rates correlate with pneumatic pressures. This phenomenon can be reasonably explained by the fact that the increased pneumatic pressure increases membrane deformation and forces liquids through the microchannel. Figure 8a also reveals that, under the experimental conditions investigated, the pumping rate performance shows a similar profile of an initial increase with an increase in applied frequency and then followed by a decline in the saturation flow rate when frequency reached 45–50 Hz. The decline in the liquid pumping rate observed in the investigations results mainly from the lagging mechanical response of PDMS membranes at high frequencies. This phenomenon has also been reported previously. The lag results from the longer time needed to re-actuate the flexible membranes and the time needed to switch on/off the EMVs. Because the membranes do not have enough time to reopen completely, less fluid passes through. Within the experimental conditions explored, 40 psi pneumatic pressure, which obtained the best liquid pumping performance, provided a maximum flow rate approximating 28.0 $\mu L \cdot min^{-1}$. To minimize the lagging response and the fluid backflow effect, performance was measured in double-side actuation mode (Figure 8b). The performance tests showed that, at lower operation frequencies (<50 Hz), pumping behavior resembled that in the single-side actuation mode. The flow rates correlated positively with the applied pressure and the operated frequency. However, no declines occurred at higher frequencies. The reason is that because of its shorter deflection distance, the PDMS membrane does not activate the lagging response mechanism. Notably, compared to the single-side actuation mode, the double-side mode can afford maximum pumping rate of 30.0 $\mu L/min$ at 20 psi that is half of the total pressure energy (40 psi) required in the single-side mode to attain similar pumping rate (28.0 $\mu L \cdot min^{-1}$). Most importantly, the design of the double-side actuated mode can prevent the backflow of the fluids particular to microfluidic cell culture systems. In conclusion, the double-side actuated mode not only enhances pumping performance, it also decreases a hysteresis effect on the membrane.

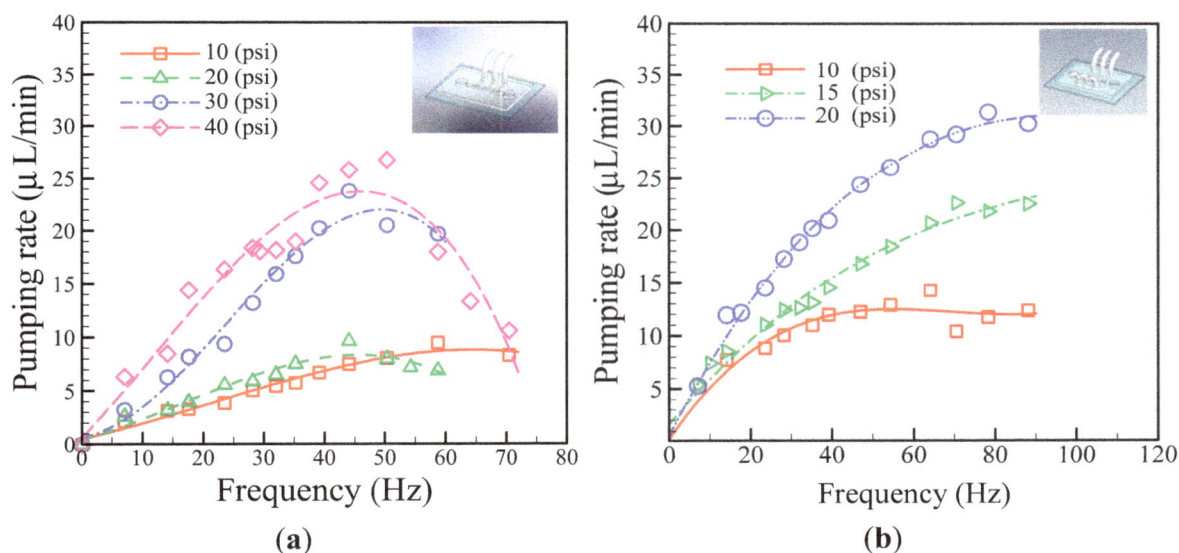

Figure 8. Relationship between the pumping rate and the driving frequencies at various operating air pressures. (**a**) the single-side; and (**b**) double-side actuated mode micropumps.

5. Conclusions

In this paper, we derived a theoretical model that can describe the deformation behavior of a PDMS membrane in a microfluidic micropump. With this theoretical model, numerical simulations are able to predict larger deformation mechanisms. The results of theoretical and numerical calculations are in reasonable agreements with the experimental results for the deformation of PDMS membrane. The theoretical and numerical results were used to facilitate optimum design of the actuated membrane of the proposed micropump. Performance tests to evaluate pumping efficiency also showed that the double-side actuated mode not only reduces the required applied pressure, it also avoids the lagging response effect of the PDMS membrane and prevents the backflow of fluids. Pumping rates observed in the performance tests reached as high as 30.0 μL/min. The integration of these newly designed micropumps can accelerate the development of miniaturized systems and fully-automated microfluidic systems.

Acknowledgments

The authors would like to thank the National Science Council of the Republic of China, Taiwan for financially/partially supporting this research under NSC102-2221-E-214-031 and ISU100-INT-09.

Author Contributions

Tai-Yen Yeh and Jr-Lung Lin conceived and designed the experiments; Tai-Yen Yeh and Chi-Han Chiou performed the experiments; Jr-Lung Lin analyzed the data; Chi-Han Chiou contributed reagents/materials/analysis tools; Jr-Lung Lin wrote the paper.

Conflicts of Interest

The authors declare no conflict of interest.

References

1.   Laser, D.J.; Santiago, J.G. A review of micropumps. *J. Micromech. Microeng.* **2004**, *14*, R35–R64.
2.   Fu, A.Y.; Chou, H.-P.; Spence, C.; Arnold, F.H.; Quake, S.R. An integrated microfabricated cell sorter. *Anal. Chem.* **2002**, *74*, 2451–2457.
3.   Fu, A.Y.; Spence, C.; Scherer, A.; Arnold, F.H.; Quake, S.R. A microfabricated fluorescence-activated cell sorter. *Nat. Biotechnol.* **1999**, *17*, 1109–1111.
4.   Maillefer, D.; Gamper, S.; Frehner, B.; Balmer, P. A high-performance silicon micropump for disposable drug delivery systems. In Proceedings of The 14th IEEE International Conference on Micro Electro Mechanical Systems, Interlaken, Switzerland, 25 January 2001; pp. 413–417.
5.   Hong, J.W.; Studer, V.; Hang, G.; Anderson, W.F.; Quake, S. A nanoliter-scale nucleic acid processor with parallel architecture. *Nat. Biotechnol.* **2004**, *22*, 435–439.
6.   Liu, J.; Enzelberger, M.; Quake, S.R. A nanoliter rotary device for polymerase chain reaction. *Electrophoresis* **2002**, *23*, 1531–1536.
7.   Bourouina, T.; Bosseboeuf, A.; Grandchamp, J.P. Design and simulation of an electrostatic micropump for drug-delivery applications. *J. Micromech. Microeng.* **1997**, *7*, 186–188.
8.   Saif, M.T.A.; Alaca, B.E.; Sehitoglu, H. Analytical modeling of electrostatic membrane actuator for micro pumps. *J. Microelectromech. Syst.* **1999**, *8*, 335–345.
9.   Francais, O.; Dufour, I.; Sarraute, E. Analytical static modeling and optimization of electrostatic micropumps. *J. Micromech. Microeng.* **1997**, *7*, 183–185.
10. Gong, Q.L.; Zhou, Z.Y.; Yang, Y.H.; Wang, X.H. Design, optimization and simulation on microelectromagnetic pump. *Sens. Actuators A* **2000**, *83*, 200–207.
11. Capanu, M.; Boyd, J.G.; Hesketh, P.J. Design, fabrication, and testing of a bistable electromagnetically actuated microvalve. *J. Microelectromech. Syst.* **2000**, *9*, 181–189.

12. Jeong, O.C.; Konishi, S. Fabrication and drive test of pneumatic PDMS micro pump. *Sens. Actuators A* **2007**, *135*, 849–856.

13. Unger, M.A.; Chou, H.-P.; Thorsen, T.; Scherer, A.; Quake, S.R. Monolithic microfabricated valves and pumps by multilayer soft lithography. *Science* **2000**, *288*, 113–116.

14. Thorsen, T.; Maerkl, S.J.; Quake, S.R. Microfluidic large-scale integration. *Science* **2002**, *298*, 580–584.

15. Chou, H.P.; Unger, M.A.; Quake, S.R. A microfabricated rotary pump. *Biomed. Microdevice* **2001**, *3*, 323–330.

16. Jeong, O.C.; Yang, S.S. Fabrication and test of a thermopneumatic micropump with a corrugated p^+ diaphragm. *Sens. Actuators A Phys.* **2000**, *83*, 249–255.

17. Van de Pol, F.C.M.; van Lintel, H.T.G.; Elwenspoek, M.; Fluitman, J.H.J. Therpneumatic micropump based on microengineering techniques. *Sens. Actuators A Phys.* **1990**, *21*, 198–202.

18. Spencer, J.G. Piezoelectric micropump with three valves working peristaltically. *Sens. Actuators A Phys.* **1990**, *21–23*, 203–206.

19. Morris, C.J.; Forster, F.K. Optimization of a circular piezoelectric bimorph for a micropump driver. *J. Micromech. Microeng.* **2000**, *10*, 459–465.

20. Husband, B.; Bu, M.; Evans, A.G.R.; Melvin, T. Investigation for the operation of an integrated peristaltic micropump. *J. Micromech. Microeng.* **2004**, *14*, S64–S69.

21. Lin, Q.; Yang, B.; Xie, J.; Tai, Y.-C. Dynamic simulation of a peristaltic micropump considering coupled fluid flow and structural motion. *J. Micromech. Microeng.* **2007**, *1*, 220–228.

22. Wang, C.H.; Lee, G.B. Automatic bio-sampling chips integrated with micropumps and microvalves for multiple disease detection. *Biosens. Bioelectron.* **2005**, *21*, 419–425.

23. Tseng, H.Y.; Wang, C.H.; Lin, W.Y.; Lee, G.B. Membrane-activated microfluidic rotary devices for pumping and mixing. *Biomed. Microdevices* **2007**, *9*, 545–554.

24. Wang, C.H.; Lee, G.B. Pneumatically-driven peristaltic micropumps utilizing serpentine-shape channels. *J. Micromech. Microeng.* **2006**, *16*, 341–348.

25. Lin, J.L.; Wang, S.S.; Wu, M.H.; Oh-Yang, C.C. Development of an integrated microfluidic perfusion cell culture system for real-time microscopic observation of biological cells. *Sensors* **2011**, *11*, 395–411.

26. Timoshenko, S.; Woinosky-Krieger, S. *Theory of Plates and Shells*, 2nd ed.; McGraw-hill: New York, NY, USA, 1959; pp. 415–420.

27. Fan, B.; Song1, G.; Hussain, F. Simulation of a piezoelectrically actuated valveless micropump. *Smart Mater. Struct.* **2005**, *14*, 400–405.

28. Fuard, D.; Tzvetkova-Chevolleau, T.; Decossas, S.; Tracqui, P.; Schiavone, P. Optimization of poly-di-methyl-siloxane (PDMS) substrates for studying cellular adhesion and motility. *Microelectron. Eng.* **2008**, *85*, 1289–1293.

29. Sim, W.; Kim, B.; Choi, B.; Park, J.-O. Theoretical and experimental studies on the parylene diaphragms for microdevices. *Microsyst. Technol.* **2005**, *11*, 11–15.

10

Optimized Simulation and Validation of Particle Advection in Asymmetric Staggered Herringbone Type Micromixers

Eszter L. Tóth [1,2,]*, Eszter G. Holczer [2], Kristóf Iván [1] and Péter Fürjes [2]

[1]   Faculty of Information Technology and Bionics, Pázmány Péter Catholic University, Práter utca 50/a, H-1083 Budapest, Hungary; E-Mail: ivan.kristof@itk.ppke.hu
[2]   Research Centre for Natural Sciences, Institute for Technical Physics and Materials Science, Hungarian Academy of Sciences, Konkoly Thege M. út 29-33, H-1121 Budapest, Hungary; E-Mails: holczer@mfa.kfki.hu (E.G.H); furjes@mfa.kfki.hu (P.F.)

*   Author to whom correspondence should be addressed; E-Mail: toth.eszter@itk.ppke.hu

Academic Editor: Yong Kweon Suh

Abstract: This paper presents and compares two different strategies in the numerical simulation of passive microfluidic mixers based on chaotic advection. In addition to flow velocity field calculations, concentration distributions of molecules and trajectories of microscale particles were determined and compared to evaluate the performance of the applied modeling approaches in the proposed geometries. A staggered herringbone type micromixer (SHM) was selected and studied in order to demonstrate finite element modeling issues. The selected microstructures were fabricated by a soft lithography technique, utilizing multilayer SU-8 epoxy-based photoresist as a molding replica for polydimethylsiloxane (PDMS) casting. The mixing processes in the microfluidic systems were characterized by applying molecular and particle (cell) solutions and adequate microscopic visualization techniques. We proved that modeling of the molecular concentration field is more costly, in regards to computational time, than the particle trajectory based method. However, both approaches showed adequate qualitative agreement with the experimental results.

Keywords: staggered herringbone micromixer (SHM); concentration distribution; particle trajectory; finite element modeling; experimental validation

1. Introduction

The precisely controlled manipulation of fluids in microanalytical systems or microchemical reactors is a key issue in terms of the use of these devices. In addition to fluidic transport, integrated functional microfluidic elements (pumps, mixers, separators, *etc.*) are also essential building blocks of sample preparation systems. The reliable modeling of the microscale fluidic processes in these systems is of critical importance to their economical design and development. Accordingly, the finite-element modeling of these microstructures is of current interest, with respect to their use in analytical devices [1,2]. The reliable modeling of biological fluids, with a precise estimation of particle motion, is a significant challenge due to the non-Newtonian behavior of the media and the special geometric and physical properties of the cells [3,4]. Although fast and cost effective, the prediction of the functional performance of preliminary analysis of the increasingly complex microfluidic systems is problematic considering that a coarse mesh resolution can deteriorate the resulting solutions. Our aim was to compare different modeling strategies to suggest a truly economical way to simulate demonstrative microfluidic systems, and to find a compromise between detailed solutions and fast analyses.

One of the basic functions of microfluidic systems integrable into microanalytical devices such as microchemical reactors is the dilution and complete mixing of the analyte with an adequate buffer solution or different reactants to ensure homogeneous concentration distribution over the critical active area of the architecture [5,6]. However, the mixing possibilities are highly limited at the microscale since turbulent flow cannot build up due to the dominant viscous forces. Evolving transversal (or chaotic) advection [7] could be a promising and effective mixing method in case of microfluidic applications considering stable and laminar flow in low Reynolds regime [2,8]. We have chosen chaotic advection as a complex process to demonstrate the performance of the different modeling approaches. Micromixers could be the key components in sample preparation units of proposed lab-on-a-chip systems providing efficient mixing of sample and analyte fluids or reagents in spite of the laminar flow present at the microscale.

1.1. Mixing at the Microscale

Mixing can be explained as a gradient driven transport process increasing entropy and decreasing global inhomogeneity of particle concentration, temperature and phases. Various effective mixing strategies were demonstrated at the macroscale such as molecular diffusion, turbulent diffusion, advection and Taylor-dispersion. In contrast to macroscale fluid dynamics where turbulent flows are easily achieved, microscale flow processes cannot generate turbulence due to the dominant viscosity of the system. Thus, the efficient mixing possibilities are limited. This is because the small characteristic dimensions of the microfluidic systems liquid flows are laminar and characterized by low Reynolds number. In such a laminar fluidic environment, the species can be transported by molecular diffusion between the component streams, where a dynamically diffusive interface is created with predictable geometry. Advection is defined as a transport process generated by the fluid flow causing a nonlinear or even chaotic distribution of molecules. This phenomenon can be realized using peculiar geometries. The Taylor-dispersion can be explained as a special advection generated by the sheared flow with mixing efficiency two orders of magnitude greater than strategies based on molecular diffusion [9].

1.2. Chaotic Advection

The transversal or chaotic advection refers to a transport process where advection, *i.e.*, particle transport is generated by the fluid flow. Even in case of a simple laminar flow, velocity distribution can lead to a chaotic motion of particles without the need for turbulence. In chaotic advection, the flow velocity components are constant over space and time in a steady flow whereas in a turbulent system these are considered to be random. In a three-dimensional steady state advection flow, the representative streamlines of the system or the particle trajectories in a particle tracing setup can cross each other, thus fulfilling the requirements of minimum complexity for chaotic behavior [10]. Advection can become chaotic in a two-dimensional unsteady flow (unsteady meaning time-dependent perturbations), and in three-dimensional steady or unsteady flows [11]. The importance of chaotic advection, therefore, lies in the possibility to enhance mixing in the laminar flow regimes by orders of magnitude compared to molecular diffusion, thus enabling the use of smaller microfluidic devices with slightly more complex channel geometries (such as staggered herringbone type double layered channels) to greatly outperform straight channels in mixing and sample preparation tasks known to be of critical importance at the microscale.

2. Experimental Section

2.1. Applied Geometries

To demonstrate and compare different modeling strategies for mixing processes, various staggered herringbone microstructures were designed with different parameters. The microchannel cross-section dimension was set to 100 μm (height) × 200 μm (width) with the groove depth of 100 μm (Figure 1). The grooves are patterned at a 45° angle to the x axis asymmetrically to divide the channel into one third and two thirds as suggested by Stroock *et al.* [7]. One mixing unit was defined as two sets of consecutive herringbone grooves with two different cusp positions as presented in Figure 1. The grooves comprising the same half unit are intended to build transversal advection from the same side of the channel via the generation of counter-rotating vortices. The total length of the mixer channel was fixed at 6000 μm. For detailed modeling and experimental verification of the advective processes and mixing performances, different microfluidic geometries were applied using the following parameters: width of the grooves varied between 30 and 40 μm, the number of grooves per half unit varied between 4 and 6 and the number of mixing units varied between 6 and 4 as summarized in Table 1.

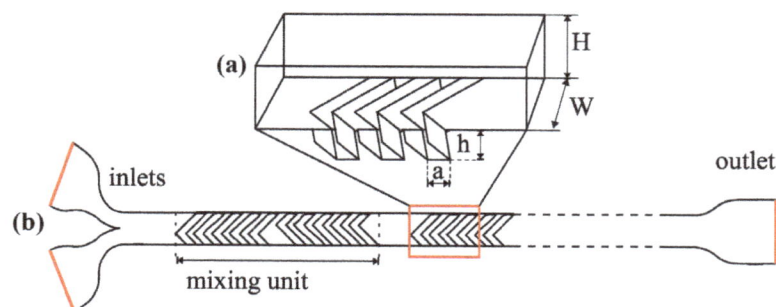

Figure 1. Parameters of the Staggered Herringbone Mixer (SHM): (**a**) width (W) and height (H) of the channel, depth (h) and width of the herringbones (a), and the schematic structure of the overall microfluidic system (**b**) including inlets and outlet and one mixing unit.

Table 1. Structural parameters of the applied micromixers. Width of the herringbone grooves, number of herringbones per half unit, and the number of mixing units were considered in our study.

Herringbone groove width	Number of herringbones per half unit	Number of mixing units	Name
30 μm	4	6	30/4/6
30 μm	6	4	30/6/4
35 μm	4	5	35/4/5
35 μm	5	4	35/5/4
40 μm	4	5	40/4/5
40 μm	5	4	40/5/4

2.2. Modeling

This paper focusses on two different approaches regarding the modeling of microscale mixing. The concentration-based method calculates the diffusion of a reagent and estimates the concentration field as a result. The particle tracing-based method calculates particle trajectories with information about the path of each particle along the channel. Both aspects are based on a pre-calculated stationary velocity field as a result of solving the Navier-Stokes equation [2] numerically with finite element method. Models were built in COMSOL Multiphysics version 4.4 [12]. Laminar inflow boundary condition was applied on the inlet with 0.002 m/s average velocity in case of each numerical study of this series and zero pressure was set at the outlet. On the channel walls, no slip boundary conditions were defined for the laminar flow model as summarized in Table 2. The properties of room temperature water (density: 1000 kg/m^3, kinematic viscosity: 10^{-6} m^2/s) were applied as material parameters. The maximal 0.0112 Reynolds number was calculated by the simulator solver considering 0.002 m/s average inlet velocity. The average Reynolds number was estimated to be 0.00244.

Table 2. Summary of boundary conditions for the model.

Boundary	Model	Boundary condition	Value
Inlet	CFD	Laminar inflow with average flow velocity	0.002 m/s
Inlet	Trajectory	Particle inlet	6000 particles with uniform density
Inflow$_1$	Concentration	Concentration	100 mol/m^3
Inflow$_2$	Concentration	Concentration	0 mol/m^3
Channel wall	CFD	No slip	-
Channel wall	Trajectory	Bounce (for the trajectory model)	-
Channel wall	Concentration	No flux	-

In case of the concentration-based modeling, the concentration field of the diluted molecular size particles can be described by stationary scalar transport. In the finite element model, a step function initial concentration distribution was defined at the inlet of the mixer channel according to the influx of the particle solution. The concentration values at the mesh nodes were set to 100 mol/m^3 on the left side and zero on the right side of the inlet section. At the channel walls and the outlet, no flux and outflow boundary conditions were defined, respectively. Diffusion coefficient of the model molecule was set to 10^{-10} m^2/s. The mesh resolution was refined to avoid numerical diffusion effects [13], which could be a key issue in modeling molecular diffusion processes and typically induces false mixing effects in the channel at low mesh resolutions (see Table 3 for more details).

The particle tracing based model calculates, follows and depicts the individual particle trajectories according to the hydrodynamic drag force described by Stokes' law. The lift force was not considered due to the low Reynolds number and the small difference between the particle and water density. For recording the individual particle trajectories a total number of 6000 spherical particles were released from the left half of the inlet plain of the mixer channel after the steady-state velocity field had been obtained. Properties of the model particles were set to be in correspondence with the principal properties of red blood cells (density: 1100 kg/m^3, particle diameter: 6 μm [14,15]). Spherical particle geometry was used as an approximation of the cell geometry and this approach was in accordance with the experimental methods, as well. Particle trajectories follow the fluid streamlines established by prior hydrodynamic simulations, thus reducing the need for higher mesh resolution and higher computational demand as compared to the concentration-based model.

Tetrahedral mesh was applied for both the convergence study and the Finite Element Modeling of the microfluidic mixing processes in the different geometries described in Table 1, as well. Mesh element size for the convergence study is listed in Table 3. For the resulting concentration calculations, we used approximately 9,000,000 elements and for the particle trajectory calculations we used approximately 1,000,000 mesh elements.

Table 3. Mesh parameters for the convergence study on the 40/5/4 geometry.

Mesh name	Number of elements	Average element size (μm)	Standard deviation of element size (μm)
Fine	11,828,167	2.3752	1.8527
Medium	4,922,845	3.7606	2.4241
Low	1,788,920	5.4645	3.2021
Coarse	121,404	13.8776	8.5472

2.3. Fabrication

Polydimethylsiloxane (PDMS) [16] soft lithography is a conventionally applied fabrication method for polymer microfluidic structures. Microfabrication into PDMS needs moulding and polymerising on the photoresist replica, then removal by peel-off from the moulding form as schematically demonstrated in Figure 2. Micropatterned photoresist structures such as SU-8 are ideally suitable for such moulding forms. For our proposed mixer structure, a multi-layered SU-8 epoxy based negative photoresist was used [17–19] as a master replica for PDMS molding. Lin *et al.* reported substrate fabrication for 3D herringbone mixer by femtolaser direct writing [20] as an alternative fabrication method. An improved 3D SU-8 multilayer fabrication process was developed to be applicable for formation of microfluidic systems with high aspect ratio sidewalls and advanced functional parts. SU-8 layers were patterned by subsequent spin-coatings (Brewer Science Cee 200CBX spin-coater [21]), lithographic exposures (Süss MicroTech MA6 mask aligner [22]) and a final development step, as illustrated in Figure 2a. The layers of the SU-8 replica contained different functional components of the micromixer. The main microfluidic channel and reservoirs were fabricated in the first layer while the inlets and outlets as well as the herringbone structures were created in the second layer. PDMS prepolymer was poured onto the developed replica and polymerized in two days in room conditions. At the final step of the fabrication process, PDMS was sealed to glass by low temperature bonding after oxygen plasma treatment applying 200 W plasma power, 100 kPa chamber pressure and 1400–1900 sccm oxygen flow (Terra Universal Plasma Preen Cleaner/Etcher [23]). Hence, stable covalent bonds were developed between the surfaces.

These materials have excellent structural, mechanical, optical and technological properties which make them suitable for application in microtechnology [24]. PDMS is a silicon-based organic polymer: $(H_3C)_3SiO[Si(CH_3)_2O]_nSi(CH_3)_3$ which has become a versatile material to realize microfluidic test structures due to its transparency, flexibility, reliable geometry transfer, price, and biocompatibility. This material provides an excellent solution for fast prototyping of simple microfluidic test structures, and can also be convenient for more complex demonstrator applications, although its applicability for large scale production is not yet proven considering its long term chemical and biological stability.

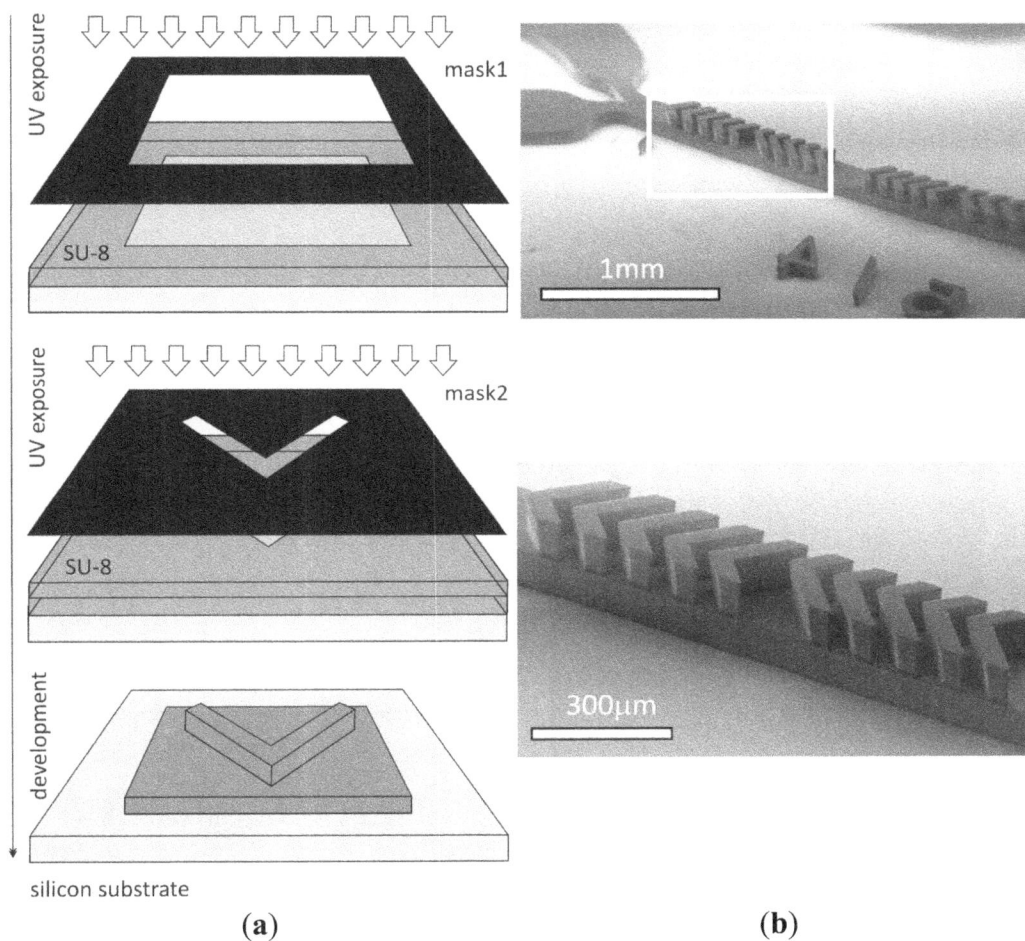

(a)   **(b)**

Figure 2. (**a**) SU-8 multilayer lithography steps and (**b**) the 3D SU-8 molding replica imaged by Scanning Electron Microscopy (SEM).

2.4. Measurement

To validate the model results, both modeling strategies were tested in the microfabricated fluidic systems. The concentration-based mixing process was characterized by simultaneous injection of diluted yellow and blue food color dye (molecular weight is approximately 500 g/mol) [25], and the mixing of the two solutions was observed. We used an upright microscope (Zeiss AxioScope A1 [26]) with bright field color imaging applying adequate flow parameters corresponding to the modeled values (4×10^{-11} m^3/s inlet flow rate).

To validate the particle-based modeling method, yeast cells were diluted in room temperature, with the water being of a similar volume as set in the simulation. The size distribution of the applied yeast cells was verified by Diatron Abacus Junior 30 hematology analyzer [27] and compared to the size histogram of human red blood cells (RBC) as presented in Figure 3. The RBC concentration was around $4.2–5.4 \times 10^{12}$ L^{-1} (woman) and the applied yeast cell concentration was 7.12×10^{10} L^{-1} to avoid clogging and multiple scattering during the measurements. The hematology analyzer uses 25 μL sample volume for one test. The spherical shape used in the model calculations is a good approximation for the yeast cells. To enhance and visualize cell trajectories and the resulting mixing states in the microchannel, dark field imaging was used. This imaging method facilitates the recording of the light scattered from the cells crossing the light beam, and we estimated the local cell concentration from the lateral distribution of the scattered light intensities, *i.e.* from the local brightness levels of the image. A relative light intensity was calculated by transforming the recorded intensities to follow the following conservation law:

$$\frac{\int_0^w I(x_0, y)dy}{w} = cont. = \frac{\int_0^w I(0, y)dy}{w} = \frac{\int_0^{w/2} 1dy + \int_{w/2}^w 0dy}{w} = \frac{1}{2} \tag{1}$$

where x_0 and y represents the local coordinates in the channel at a given cross-section and w denotes the width of the channel. This calibration ensures that the integral of the relative scattered light intensity is constant and equal to the initial value (0.5 at the inlet).

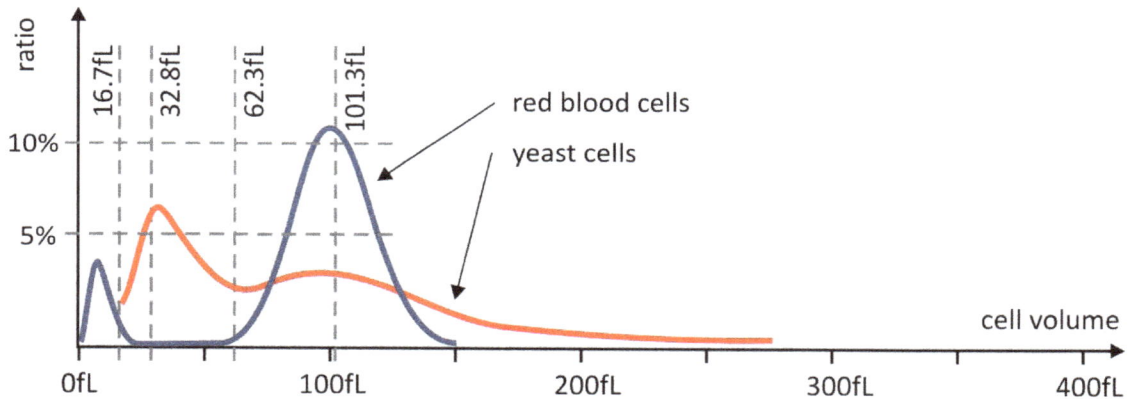

Figure 3. Size distribution of yeast cells and red blood cells. The two cell types have a similar range of cell diameters making yeast cells a well suited model of human cells in our experiments. The significant peaks in the volume distribution of the RBCs and the yeast cells are around 100 fL, and the spherical equivalent diameter belonging to this value is 5.76 μm. In the size distribution of the yeast cell, there is another peak at around 30 fL with an equivalent diameter of 4.16 μm. The total number of counted yeast and human red blood cells was 1.78×10^6 and 1.25×10^8, respectively.

3. Results and Discussion

3.1. Modeling Results

At first, the laminar flow field was modeled as a basis for estimation of both the diffusion-based transport of diluted molecular species and the advection-based particle trajectories. Detailed analysis and visualization of the velocity field in Figure 4 highlights the rotating streamlines due to anisotropic fluid resistance generated by the secondary channel grooves.

The applied mesh resolution is a key issue in concentration-based modeling of micromixers as proven by our work. The numerical diffusion demonstrated a significant role during modeling the mixing process with lower mesh resolutions (mainly above 5 μm mesh element size). Figure 5 clearly shows the effect of numerical diffusion in the case of the herringbone micromixer after one mixing unit. In this case, the numerical diffusion may induce significant error in the estimation of the resulting concentration field as demonstrated by the vanishing phase stratification in the lowest resolution numerical solution (Figure 5d).

Mesh convergence study was implemented for the trajectory model as well and the results were summarized in Figure 6. As a demonstrative model parameter, the mixing efficiencies were calculated in case of different mesh resolutions in the 40/5/4 geometry after one mixing unit considering the ratio of the cells on the right side and the left side of the channel. A slight and saturating improvement in the modeling results could be experienced, but the effect was significantly smaller compared to the diffusion model.

To observe the evolution of the mixing states four planes of interest were defined locally: one at the inlet, one after the first and one after the second mixing unit and one at the outlet (Figure 7). Concentration fields in these planes (Figure 7a–d) show the mixing effect of the rotation considering the additional diffusion that occurs on the increased interface between the two fluids due to transversal advection.

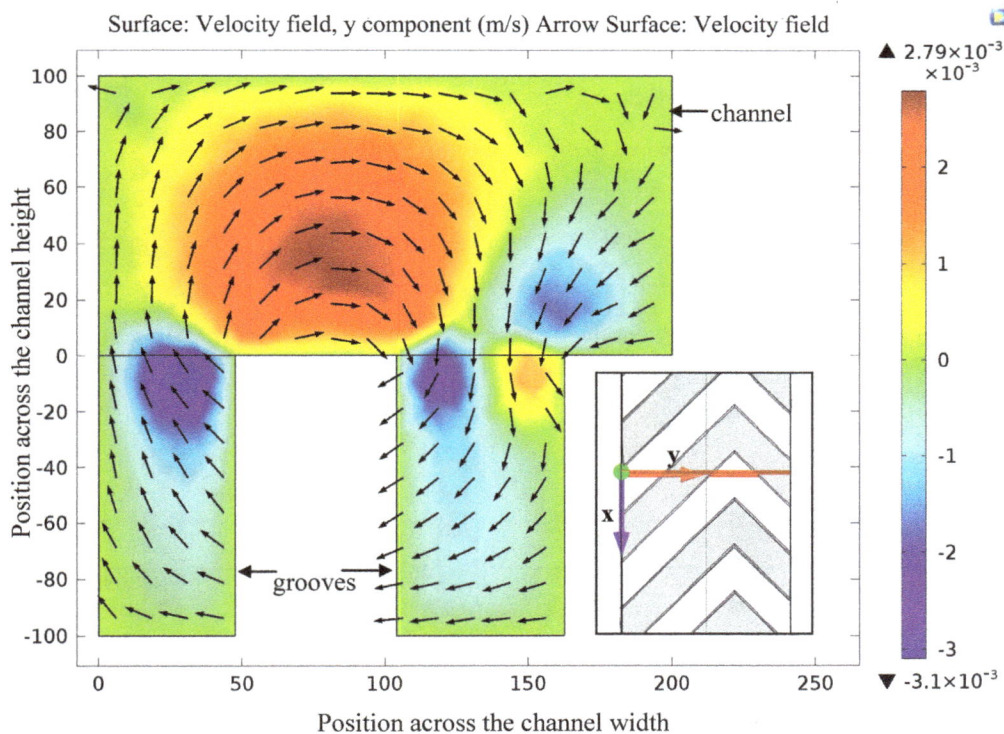

Figure 4. Velocity field in the cross section of the channel and grooves. Coloring and arrows denote the amplitude (blue: left direction, yellow to red: right direction) and the direction of the *y* component of the local flow velocity vector, respectively. Rotating effect of the herringbone shaped grooves is clearly observable due to the flow direction modification effect of the grooves.

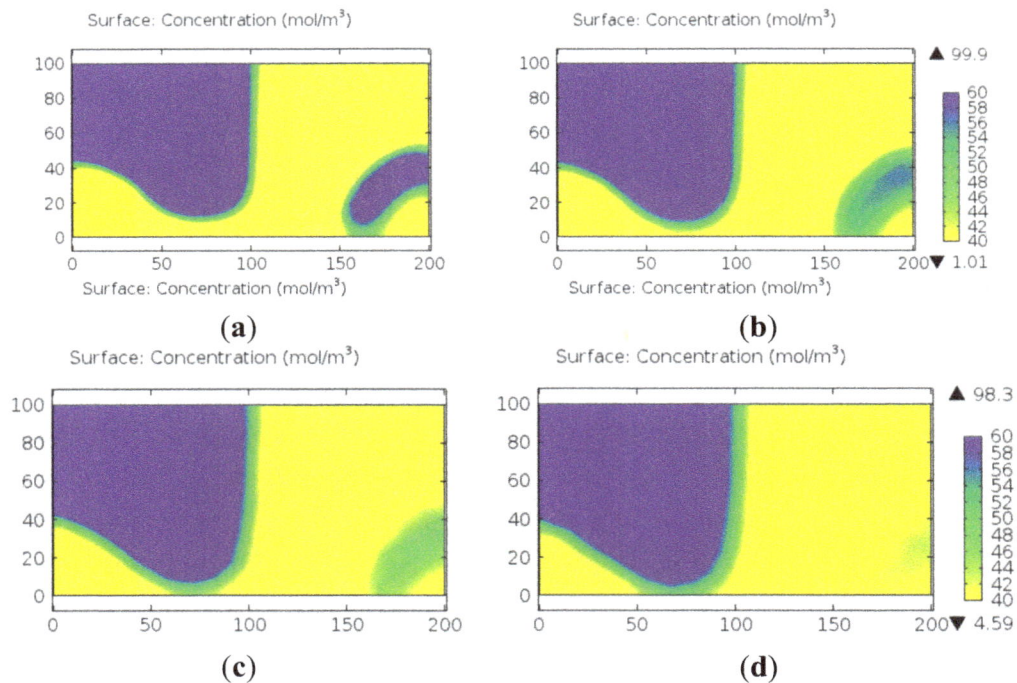

Figure 5. Effect of mesh element size on the quality and reliability of the numerical solution in case of the concentration based model. Concentration fields after the first mixing unit with fine (**a**), medium (**b**), low (**c**) and coarse (**d**) mesh resolutions. The numerical diffusion caused by the poor mesh resolution is well observable on the more extensive green colored areas on (**b**) and (**d**) compared to the respective areas on (**a**) and (**c**).

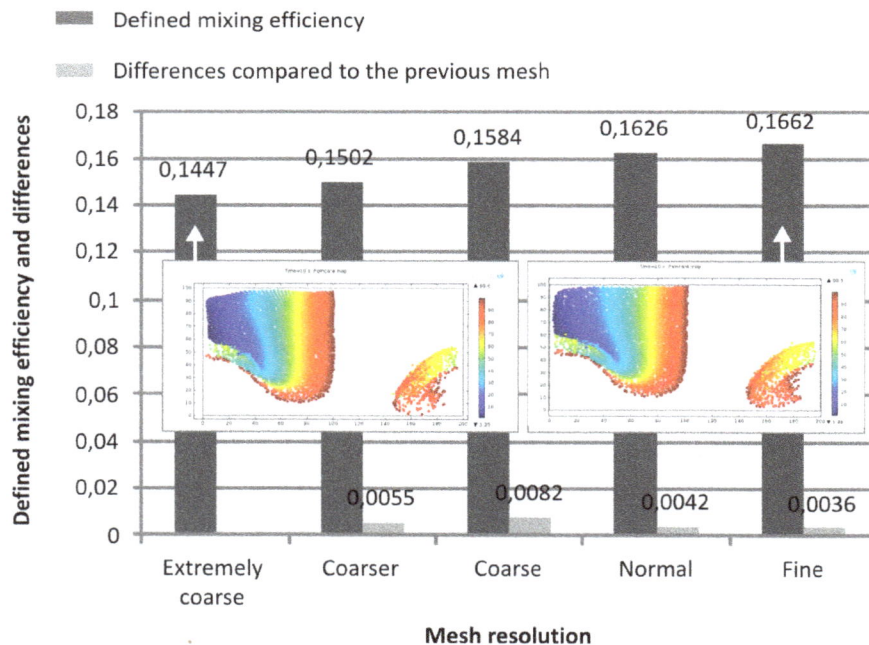

Figure 6. Mesh convergence study for the trajectory model, by calculating the ratio of the number of transferred and remained particles. The difference between values is calculated from to the actual and the previous mesh resolution results (depicted with grey bars). Poincaré maps at the outlet for the extremely coarse and the fine mesh show minor differences in the particle trajectories compared to the diffusion model.

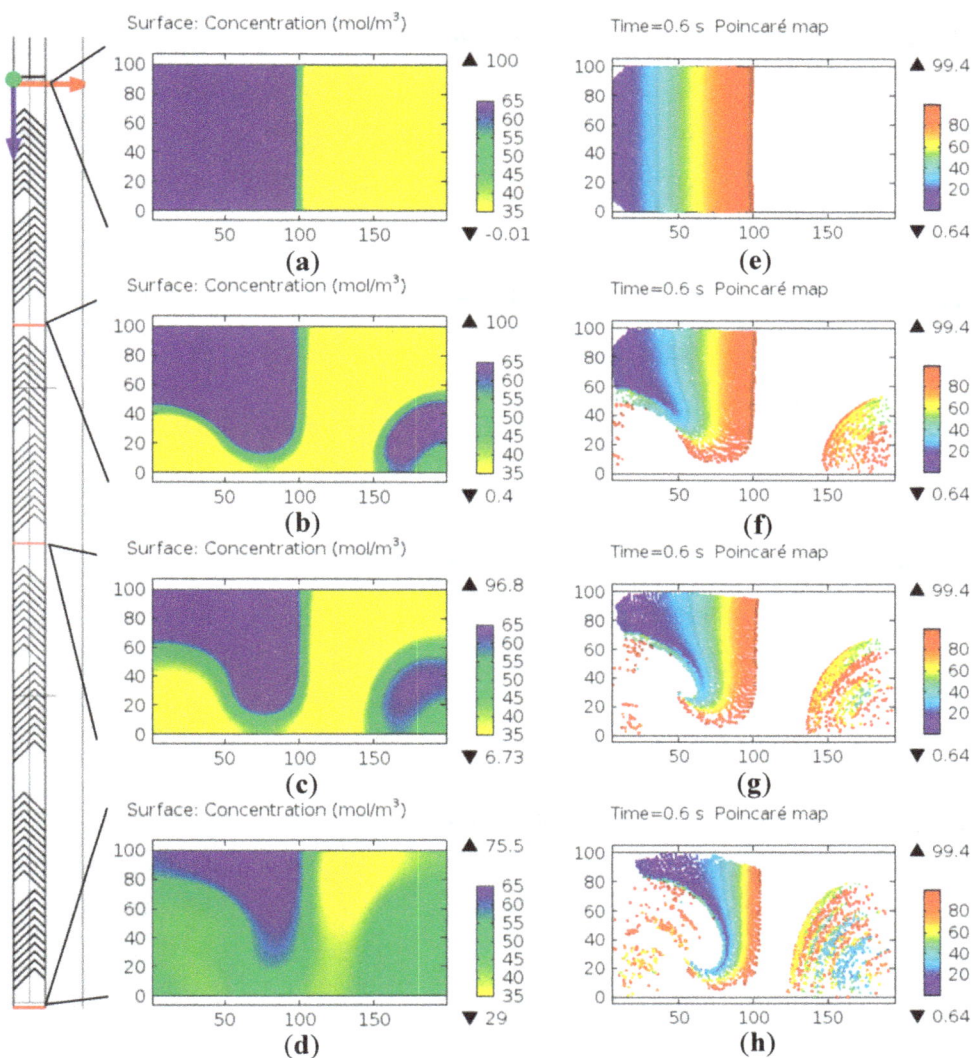

Figure 7. Mixing evolution of two different solutions modeled by concentration based method (**a–d**) and particles modeled by trajectory based method (**e–h**) along the microchannel. The mixing distributions are qualitatively similar which is observable on the similar distributions of shapes on (**a–d**) compared to (**e–h**), respectively.

For modeling cell (or particle) distribution in the studied micromixer the particle tracing module of COMSOL Multiphysics was applied. Poincaré maps (Figure 7e–h) show the actual location of individual particles at the selected planes. The different points correspond to the plain sections of trajectories of the monitored particles. The color of the particles denotes their initial location along the y axis (blue: left, red: right) of the channel to support visualization of the cell displacements relative to each other. As revealed in Figure 7h, stratified particle "shells" have evolved on the right side of the channel due to the transversal advection effect. On the left side of the channel, the particles also show a layered structure with the buffer solution.

To determine the effects of the selected geometric parameters of the proposed herringbone mixers (see parameters in Table 1), the mixing efficiencies were characterized and compared to each other in the case of trajectory-based models. By creating the outline of the Poincaré sections at the outlet plane of the channel, the change of its area estimates the number of transferred particles from the initial channel-side to the other. We defined a mixing efficiency to be better if it corresponds to a larger transferred particle shell area relative to the cross section of the channel. Efficiency values were calculated by considering the ratio of the number of particles on the left (initial) side of the channel and the number of particles on the right side of the channel. Figure 8 and Table 4 show that mixing effect improves with increased width of the grooves as well as by applying more mixing units rather than more grooves per unit. The effect of the depth of the grooves has been characterized previously by Fürjes et al. [2] and it was found that saturating efficiency can be achieved when the depth of the grooves are similar in range to the channel depth.

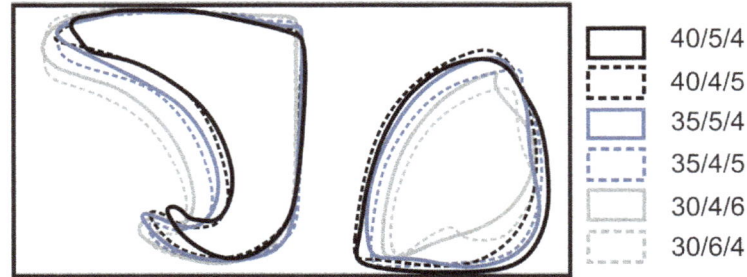

Figure 8. Outlines of the mixed particles at the outlet plane of the mixer channel. The color of the lines denote the width of the grooves, line type indicates the number of grooves. Shrinkage of the area with large number of particles at their initial side (left) and expansion of the area rich in particles on the right side is well observable, which results in the narrowing of the unmixed region in the middle of the channel.

Table 4. Mixing efficiencies for the proposed geometries.

Geometry	40/5/4	40/4/5	35/5/4	35/4/5	30/4/6	30/6/4
Efficiency	0.4526	0.4513	0.3655	0.2998	0.2105	0.1984

3.2. Experimental Validation

To validate the applied numerical simulation methods, micromixer structures were fabricated and characterized with bright field microscopy for the concentration-based and dark field microscopy for the particle-based approach. Figure 9 shows a qualitative comparison of the modeled and the measured concentration fields near the channel outlet after three mixing units. Blue and yellow regions indicate high concentration levels of the unmixed food color dies while green color denotes the mixed regions. The deducted relative concentration curves recorded along the line section of the mixer channel by ImageJ [28] represent the mixing efficiency of the herringbone mixer.

When mixing the yeast cell solution in our device, the layered structure of the flow is well observable in accordance with the particle trajectory model results (Figure 10). The cell depleted and cell enriched regions (dark and bright regions in the channel) correspond to the modeled particle advections that evolved in the micromixer structure.

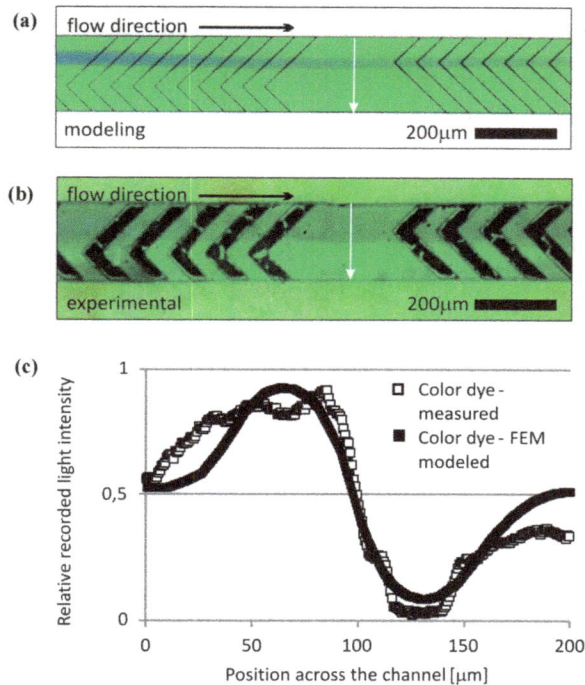

Figure 9. Modeled and measured concentration fields in case of diffusion based approach. (**a**) Result of finite element concentration modeling. (**b**) Mixing of food dies in the microchannel monitored by bright field microscopy imaging. (**c**) The recorded histograms represent the calculated and measured relative concentrations along a line section perpendicular to the channel direction.

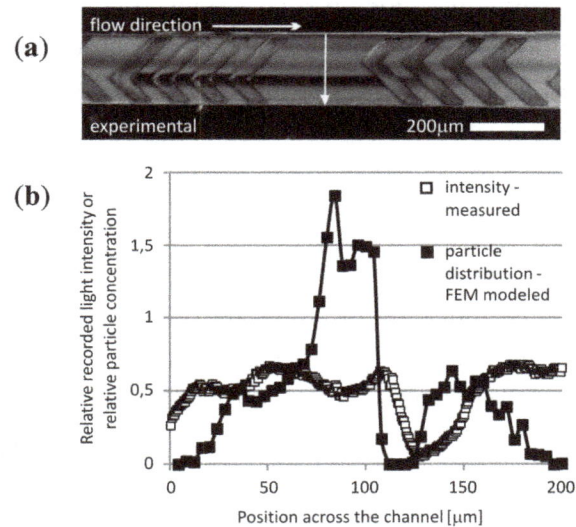

Figure 10. Relative scattered light intensity distribution corresponding to the local yeast cell concentrations recorded by dark field microscopy. (**a**) The cell-depleted region is clearly observable and coinciding with the food color dye concentration field presented in Figure 9. (**b**) The histogram extracted from the intensity field along a line section perpendicular to the channel direction is in good agreement with the modeled Poincaré map evolved at the corresponding plane.

Both modeling methods demonstrated the impressive characteristics of the herringbone micromixer. The multilayer streamline structure developed by the rotating effects and the evolving transversal advection due to the special microchannel geometry were clearly visible in both cases. The enhanced mixing by molecular diffusion on the additional dynamic surfaces was well captured, and concentration fields showed a good qualitative agreement between simulated and observed results, proving the reliability of these modeling approaches. The particle trajectory model provides accelerated qualitative information on the functional behaviour of the proposed microfluidic systems. However, for a precise description and prediction of concentration dependent processes (e.g., in microreactors), the application of high resolution concentration-based models is required. The results obtained with COMSOL Multiphysics provide a solid basis for further development and testing of more complex microfluidic systems with cost-efficient numerical modeling.

4. Conclusions

Numerical modeling of complex microfluidic systems is a cost-efficient way of developing and testing novel structures. However, models have to be verified and set up properly to obtain reliable results. In this study, we compared two significantly different modeling approaches and verified the results experimentally to characterize the applied strategies regarding their mesh resolution sensitivity, reliability, performance and computational demand. For this purpose, the staggered herringbone micromixer was selected due to its high mixing efficiency and complex 3D microstructure. The mixing process was numerically modeled applying both concentration-based and particle trajectory-based approaches.

The concentration-based method showed a high sensitivity to mesh resolution having a significant effect on false numerical diffusion in model results. Therefore, a high resolution mesh with high computational cost was needed to obtain accurate results. Application of the particle tracing approach resulted in a much lower computational cost (approximately 20 times lower) and makes an extensive geometric parameter optimization more feasible. Both concentration and particle trajectory-based models were evaluated by representative measurements, and adequate accordance in the resulted mixing states was experienced. Comparing the estimated mixing efficiencies of the various structures, we verified that the performance of the mixer can be improved by increasing the width of the grooves and by increasing the number of herringbones per mixing unit.

Cell and particle manipulation is a key issue in the development of Lab-on-a-chip applications (e.g., CTC (circulating tumor cell) separation and counting [29]). Therefore, the quality of model simulations is critical in the effective development of these new devices. Our test structure—the staggered herringbone micromixer—was integrated into a microfluidic sample preparation system developed for polymer-based photonic biosensor applications in the P3SENS project [30].

Acknowledgments

The supports of the European Commission through the seventh framework program FP7-ICT4-P3SENS (248304), the National Development Agency through TÁMOP grants (recipients: Eszter L. Tóth and Kristóf Iván, project numbers: TÁMOP-4.2.1.B-11/2/KMR-2011-0002 and TÁMOP-4.2.2/B-10/1-2010-0014), Pázmány University KAP grant (recipient: Kristóf Iván), Hungarian NAP grant No.: KTIA-NAP 13-1-2013-0001, and the MedinProt fellowship of the Hungarian Academy of Sciences (recipient:

Péter Fürjes) are gratefully acknowledged. The significant efforts of Magda Erős, Margit Payer and Anna Borbála Tóth are highly appreciated.

Author Contributions

In this work Eszter L. Tóth was responsible for numerical modeling and optimization of the mixer structures. The layout level design and the microfabrication of the microfluidic systems were implemented by Eszter G. Holczer and Péter Fürjes. The experimental validation of the mixer structures were also implemented by Eszter L. Tóth, Eszter G. Holczer and Péter Fürjes. The theoretical explanation of the modeling and experimental results was supported by Kristóf Iván and Péter Fürjes.

Conflicts of Interest

The authors declare no conflict of interest.

References

1. Erickson, D. Towards numerical prototyping of labs-on-chip: Modeling for integrated microfluidic devices. *Microfluid. Nanofluid.* **2005**, *1*, 301–318.
2. Fürjes, P.; Holczer, E.G.; Tóth, E.; Iván, K.; Fekete, Z.; Bernier, D.; Dortu, F.; Giannone, D. PDMS microfluidics developed for polymer based photonic biosensors. *Microsyst. Technol.* **2014**, *2014*, 1–10.
3. Morenoa, N.; Vignalb, P.; Li, J.; Calo, V.M. Multiscale modeling of blood flow: Coupling finite elements with smoothed dissipative particle dynamics. *Proced. Comput. Sci.* **2013**, *18*, 2565–2574.
4. Shah, S.; Liu, Y.; Hu, W.; Gao, J. Modeling Particle Shape-Dependent Dynamics in Nanomedicine. *J. Nanosci. Nanotechnol.* **2011**, *11*, 919–928.
5. Capretto, L.; Carugo, D.; Mazzitelli, S.; Nastruzzi, C.; Zhang, X. Microfluidic and lab-on-a-chip preparation routes for organic nanoparticles and vesicular systems for nanomedicine applications. *Adv. Drug Deliv. Rev.* **2013**, *65*, 1496–1532.
6. Khan, I.U.; Serra, C.A.; Anton, N.; Vandamme, T.F. Production of nanoparticle drug delivery systems with microfluidics tools. *Expert Opin. Drug Deliv.* **2014**, *2014*, 1–16.
7. Stroock, A.D.; Dertinger, S.K.W.; Ajdari, A.; Mezić, I.; Stone, H.A.; Whitesides, G.M. Chaotic Mixer for Microchannels. *Science* **2002**, *295*, 647–651.
8. De Mello, A.J. Control and detection of chemical reactions in microfluidic systems. *Nature* **2006**, *442*, 394–402.
9. Nguyen, N.-T. *Micromixers: Fundamentals, Design and Fabrication*; William Andrew Publishing: Norwich, NY, USA, 2011.
10. Aref, H. Stirring by chaotic advection. *J. Fluid Mech.* **1984**, *143*, 1–21.
11. Aref, H. The development of chaotic advection. *Phys. Fluids* **2002**, *14*, 1315–1325.
12. COMSOL Multiphysics. Available online: http://www.comsol.com (accessed on 30 September 2014)
13. Lantz, R.B. Quantitative Evaluation of Numerical Diffusion (truncation Error). *Soc. Pet. Eng. J.* **1970**, *11*, 315–320.
14. Density of Blood. Available online: http://hypertextbook.com/facts/2004/MichaelShmukler.shtml (accessed on 15 November 2014).

15. Kutz, M. *Standard Handbook of Biomedical Engineering and Design*; McGraw-Hill: New York, NY, USA, 2003.

16. Dow Corning Corp. Available online: http://www.dowcorning.com (accessed on 30 September 2014).

17. MicroChem Corp. Available online: http://www.microchem.com (accessed on 30 September 2014).

18. Del Campo, A.; Greiner, C. SU-8: A photoresist for high-aspect-ratio and 3D submicron lithography. *J. Micromech. Microeng.* **2007**, *17*, 81–95.

19. Mata, A.; Fleischman, A.J.; Roy, S. Fabrication of multi-layer SU-8 microstructures. *J. Micromech. Microeng.* **2006**, *16*, 276.

20. Lin, D.; He, F.; Liao, Y.; Lin, J.; Liu, C.; Song, J.; Cheng, Y. Three-dimensional staggered herringbone mixer fabricated by femtosecond laser direct writing. *J. Opt.* **2013**, *15*, 025601.

21. Brewer Science Inc. Available online: http://www.brewerscience.com (accessed on 30 September 2014).

22. SÜSS MicroTec AG. Available online: http://www.suss.com (accessed on 30 September 2014).

23. Terra Universal Inc. Available online: http://www.terrauniversal.com (accessed on 30 September 2014).

24. Gervais, L.; Delamarche, E. Toward one-step point-of-care immunodiagnostics using capillary-driven microfluidics and PDMS substrates. *Lab Chip* **2009**, *9*, 3330–3337.

25. Material Safety Data Sheet Tartrazine MSDS. Available online: http://www.vinayakcorporation.com/tarmsdc.htm (accessed on 30 September 2014).

26. ZEISS International. Available online: http://www.zeiss.com (accessed on 30 September 2014).

27. Diatron MI PLC. Available online: http://www.diatron.com (accessed on 30 September 2014).

28. ImageJ. Available online: http://imagej.nih.gov/ij/index.html (accessed on 30 September 2014).

29. Dong, Y.; Skelley, A.M.; Merdek, K.D.; Sprott, K.M.; Jiang, C.; Pierceall, W.E.; Lin, J.; Stocum, M.; Carney, W.P.; Smirnov, D.A. Microfluidics and Circulating Tumor Cells. *J. Mol. Diagn.* **2013**, *15*, 149–157.

30. P3SENS Project. Available online: http://www.p3sens-project.eu/ (accessed on 22 November 2014).

11

Laser Controlled Synthesis of Noble Metal Nanoparticle Arrays for Low Concentration Molecule Recognition

Enza Fazio [1], **Fortunato Neri** [1], **Rosina C. Ponterio** [2], **Sebastiano Trusso** [2], **Matteo Tommasini** [3] **and Paolo Maria Ossi** [4,*]

[1] Dipartimento di Fisica e di Scienze della Terra, Università di Messina, v.le F. Stagno d'Alcontres 31, 98166 Messina, Italy; E-Mails: enfazio@unime.it (E.F.); fneri@unime.it (F.N.)
[2] IPCF-CNR, Istituto per i Processi Chimico Fisici, Consiglio Nazionale delle Ricerche, v.le F. Stagno d'Alcontres 37, 98158 Messina, Italy; E-Mails: ponterio@ipcf.cnr.it (R.C.P.); trusso@its.me.cnr.it (S.T.)
[3] Dipartimento di Chimica, Materiali e Ingegneria Chimica, "G. Natta", Politecnico di Milano, P.zza L. da Vinci 32, 20133 Milano, Italy; E-Mail: matteo.tommasini@polimi.it
[4] Dipartimento di Energia & Center for NanoEngineered Materials and Surfaces-NEMAS, Politecnico di Milano, via Ponzio 34-3, 20133 Milano, Italy

* Author to whom correspondence should be addressed; E-Mail: paolo.ossi@polimi.it

External Editors: Maria Farsari and Costas Fotakis

Abstract: Nanostructured gold and silver thin films were grown by pulsed laser deposition. Performing the process in an ambient gas (Ar) leads to the nucleation and growth of nanoparticles in the ablation plasma and their self-organization on the substrate. The dependence of surface nanostructuring of the films on the deposition parameters is discussed considering in particular the number of laser pulses and the ambient gas nature and pressure. The performance of the deposited thin films as substrates for surface-enhanced Raman spectroscopy (SERS) was tested against the detection of molecules at a low concentration. Taking Raman maps on micrometer-sized areas, the spatial homogeneity of the substrates with respect to the SERS signal was tested.

Keywords: laser ablation of solids; morphology of thin films; optical properties of thin films; SERS

1. Introduction

Lasers are widely used for macro- and micro-machining in industrial applications, ranging from electronics to automotive, astronautical and biomedical. The extension to laser nanomachining is a fascinating and relatively new field, where basic research and advanced applications have moved at the same pace since the first report in 1987 that ultrafast lasers are feasible for materials processing. The sharp ablation of polymethyl methacrylate was demonstrated, using a 160-femtosecond UV excimer laser [1]. Two impressive results were reported that, namely, a heat affected zone (HAZ) almost does not form in the irradiated target and that the ablation threshold is significantly lower than using a nanosecond excimer laser. A considerable expansion of the field resulted in the 1990s from the introduction of the chirped-pulse amplification technique in Ti: sapphire regenerative amplifiers that allow producing high energy fs pulses avoiding damage or nonlinear effects in the amplification medium. The absence of HAZ makes it possible to extend nanomachining to materials that are challenging due to their brittleness and hardness, such as insulators and semiconductors [2]. A relevant feature of ultrafast laser absorption is its nonlinearity; thereby, multiphoton absorption (MA) results in important absorption in transparent materials [3]. This opened the way to three-dimensional internal nanomachining of glass and polymers that found application in optical waveguide writing in glass [4] with an immediate rise of industrial interest. A prominent characteristic of MA is the combination of non-linearity with the strong lowering of the fluctuation associated with the optical breakdown threshold when an ultrafast laser is used. This allows overcoming the diffraction limit of the laser light, which is the crucial bottleneck to laser nanomachining [5]. The introduction, around the year 2000, of a stable and robust chirped pulse amplifier, as well as of a high-power laser pumped by a solid state diode, starting from the 2010s, made the use of (few) picosecond lasers in industry for nanomachining realistic. Among the most impressive achievements of ultrafast laser irradiation were surface nanoripple structures with periodicity consistently lower than the laser wavelength produced on several materials at intensities around the ablation threshold [6]. Furthermore, irradiating with fs pulses of a Si laser in an oxidizing atmosphere (e.g., Cl_2), spectacular arrays of self-aligned nanocones with unique anti-reflecting and infrared (IR) absorbing properties were obtained [7]. All of the above examples demonstrate the potentialities of pico- and femto-second lasers, where an ultra-short pulse width and very high peak intensity are available for nanomachining, where surface and volume material modification can be performed in a top-down approach; for a recent review on the topic, the reader may refer to Sugioka [8].

Noble metal nanoparticles exhibit a wide variety of spectroscopic, electronic and chemical properties that largely depend on their size, shape and spatial arrangement. By fs pulsed laser irradiation, highly randomly stacked arrangements of irregularly-shaped nanoparticles (NPs) with a narrow size distribution are obtained [9]. However, although presenting different advantages, this technique is very expensive and requires a high degree of precision in the choice of the process parameters. From an alternative viewpoint, an easy, but simple, bottom-up approach using nanosecond pulsed laser ablation can be adopted to prepare the surface-enhanced Raman spectroscopy (SERS) substrate of metal nanoparticles, as described in this contribution. Hence, controlled NP synthesis is problematic, besides being expensive. From an alternative viewpoint, we have developed an easier to control bottom-up approach to laser nanoprocessing, using ns pulsed laser ablation to prepare quasi-two-dimensional noble metal NP arrays with morphology tailored *ad hoc*, deposited on suitable substrates. This is the subject of the present contribution.

The deposition is performed using a nanosecond excimer UV laser, ablating a target in an inert gas atmosphere through which the ablation plume propagates. Depending on the subtle interplay among the ablated mass per laser pulse, the gas mass and the gas pressure, NPs of different size are synthesized. The fine control of the NP morphology results in the controlled tailoring of their optical properties. As is well known, silver and gold NPs show optical properties that are different from the corresponding bulk materials. In the presence of an external electromagnetic (EM) field, the conduction electrons give rise to collective oscillations on the surface of the NPs. When the wavelength of the incident light is larger than the NP typical size, surface electrons oscillate in phase with the EM field. Such collective oscillations are know as localized surface plasmons (LSP). When LSP and the external field are in resonance (LSPR), the induced electric field about the NP can be much more intense than the field associated with the incident radiation. One of the most striking consequences of such an EM enhancement is the so-called surface-enhanced Raman scattering effect. If a Raman active molecule comes into contact with a point on the NP surface where the EM field is amplified, the intensity of its Raman features will be enhanced, as well. As a first approximation, when the incident radiation is in resonance with the LSP, the Raman intensity is:

$$I_{SERS} = \left| \frac{E(r, \omega_o)}{E_0(\omega_0)} \right|^4 \tag{1}$$

where $E(r, \omega_0)$ and $E_0(\omega_0)$ are the field at the molecule position and the incident radiation, respectively [10]. A parameter that accounts for the enhancement relative to a particular system is the enhancement factor (EF) defined as the ratio between the observed intensities of the SERS signal and the normal Raman one:

$$EF = \frac{I_{SERS}/N_{SERS}}{I_{Raman}/N_{Raman}} \tag{2}$$

N_{SERS} and N_{Raman} being the number of molecules that contribute to the SERS and to the normal Raman intensities, respectively. In specific cases, EF values beyond a factor of 10^{10} were observed, allowing the detection of Raman features from a single molecule [11].

Yet, the magnitude of the SERS EF was, for a long time, part of the self-perpetuating controversies in the field, with quoted values that could differ by several orders of magnitude for similar experimental conditions. The myth of SERS EFs as large as 10^{14}, which originates in the pioneering single molecules-SERS studies from an incorrect normalization of the SERS intensity with respect to a non-resonant Raman signal, has long been (and to some extent still is) a hindrance to progress in the field. In fact, the EF of SERS active substrates critically depend on their morphology, *i.e.*, on the size and shape of the NPs, besides their spatial disposition. A careful control of such properties is thus of paramount importance to optimize SERS substrates for their application in biosensing, catalysis, cultural heritage and materials science [12].

2. Experimental Section

As mentioned above, pulsed laser deposition (PLD) of thin films is based on the vaporization of a small quantity of a target material by a high energy pulsed laser beam. The vaporized material gives rise to the formation of a plasma plume that expands in the direction normal to the surface of the target material. Material growth occurs on the surface of substrates placed in front of the target. Although conceptually simple, the process is controlled by several experimental parameters: laser fluence (F), *i.e.*, the energy density deposited at the target surface J·cm^{-2}, the target to substrate distance (d_{T-S}), ambient gas nature and pressure (P_g) and ablated mass per pulse (m_p), the latter quantity depending on the choice of fluence and laser spot area. All of these parameters play a crucial role in the resulting optical, morphological and structural properties of the growing film. In most of the experiments discussed below, we kept fixed some of the above parameters, namely fluence F = 2.0 J·cm^{-2}, target to substrate distance d_{T-S} = 35 mm and nature of ambient gas, Ar. Ambient gas pressure and the number of laser pulses were changed to investigate their role on the properties of noble metal (silver and gold) films. Target ablation was performed using a KrF excimer laser (Compex 205, Lambda Physik, Coherent Inc., Santa Clara, CA, USA) that provides light pulses at a wavelength of 248 nm with a pulse time width of 25 ns. The energy of the laser pulses can be tuned from 500 mJ down to a few mJ using both the laser high voltage value and an external optical attenuator. The laser beam was focused with a quartz lens (focal length f = 35 cm) at the target surface, the latter being high purity Au or Ag plates (K.J. Lesker, Goodfellow). Targets were positioned on a rotating holder inside a vacuum chamber (base pressure lower than 10^{-4} Pa) to minimize surface damage in case of long lasting depositions. Experiments were performed using different deposition conditions. The laser fluence (J·cm^{-2}) was deduced from the energy of the laser pulse and the measured irradiated spot area, by suitably changing the lens to target distance, with an optical microscope equipped with a motorized table. The pulse energy was measured with an energy monitor placed inside the deposition chamber. This procedure guarantees properly taking into account all causes of energy loss (*i.e.*, energy attenuator, 45° mirror, circular mask, focusing lens, entrance window of the deposition chamber). Different substrates were used depending on the experimental analysis to perform on the sample. Crystalline silicon and amorphous-carbon covered Cu grids were used for scanning and transmission electron microscopy, respectively (SEM Supra 40 Field Ion Emission, TEM Leo 91, Zeiss, Jena, Germany). Corning 7059 glass was used for SERS and UV-Vis absorption spectroscopy. Raman spectra were acquired using an HR800 micro-Raman spectrometer (Horiba Jobin-Yvon, Grenoble, France), at an excitation wavelength of 632.8 nm from a He-Ne laser line. To test the SERS activity, rhodamine 6G and organic pigments were dispersed in aqueous solutions at concentration levels between 10^{-10} and 10^{-3} M. The substrates were soaked in the solutions for 1 h and left to dry in air. Raman spectra were acquired choosing power density and integration times, such that the signal-to-noise ratio was maximized and sample degradation was avoided. UV-Vis absorption spectra were collected using a UV/VIS/NIR Lambda 750 spectrometer (Perkin Elmer) in the 190–1100 nm range. Measurements of the ablated mass per pulse were performed using an Alpha Step 500 profilometer following the procedure detailed in the next section.

3. Results and Discussion

In this section, we discuss the role of different process parameters to the morphology of thin films made of noble metal NP arrays, thus to their optical properties and, in turn, to their performance when used as SERS substrates. We initially focus on the determination of the amount of target material ablated by a single pulse m_p, since this quantity critically affects both the expansion dynamics of the ablation plume and the development of the film nanostructure.

3.1. Determination of Ablated Mass per Pulse

Measurements of m_p were performed by producing on the surface of a freshly polished target a number of craters using a different number of laser shots. In a typical experiment, four or more craters are produced on the target surface. Several craters of increasing depth result upon increasing the laser shot number, at fixed laser fluence. We report here as an example the ablation of a silver target at the fixed fluence of 2.0 J·cm^{-2}. Four craters were produced using 10, 50, 100 and 200 laser shots. The corresponding crater volumes were measured by scanning the crater area with a stylus profilometer. Two representative crater depth profiles are shown in Figure 1. At low laser pulse number N, craters are practically free of rims and well defined, while, at higher N values, craters are surrounded by much re-deposited material. In all cases, the aspect ratio of the crater (i.e., the depth-to-diameter ratio) is below unity, thus minimizing any effect of the deep-hole confinement on the estimate of the ablation rate per pulse.

Figure 1. 3D representation of the craters produced on the surface of a silver target at 2.0 J·cm^{-2} by (**a**) 50 and (**b**) 200 laser pulses. At the bottom are the corresponding contour plots.

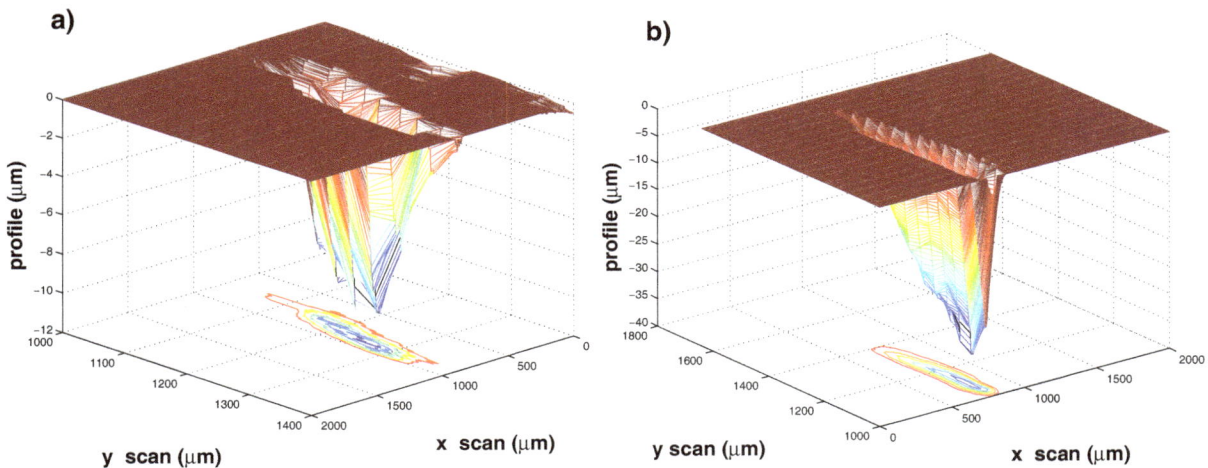

Considering the photo-thermal activation of the material removal in open air, the excitation energy from the laser propagates mainly within the target material due to the poor thermal contact between the target surface and the surrounding air: this results in the considerable depth of the craters as compared to their width. At higher N, the strong heating of the target surface combined with the instabilities arising within the melted layer at the target surface heated up to the critical temperature may produce a roughening of the crater surface that we observe experimentally. The presence of rims around the craters can be also explained by the strong temperature rise of the target surface that, in turn, leads to hot and dense ablation plasmas. The latter decouple from the liquefied target surface with a large acceleration oriented normal to it, thus exerting a strong recoil pressure onto the melted layer. Some data treatment was performed leading to the crater pictures in Figure 1: target surface leveling along the XY plane (see the flat area in Figure 1) and removal of any protruding features for which $Z > 0$. The volume of such redeposited and then melted material removed from the crater is subtracted from the total volume of the crater. The estimated crater volumes are reported in Table 1. We notice from Figure 2 that the number of laser shots and the ablated mass are linearly correlated with each other. The data were fitted to the linear relation $M = m_p N$, M being the mass removed by N laser pulses. The results of the fit are reported in Table 1. In the considered case, about 60.0 ng, corresponding to about 3.35×10^{14} silver atoms, are removed from the target by a single laser pulse. With an estimated error for each volume value (taking into account the proper error on x, y and z measurements), less than 3%, the ultimate error on m_p is about 5%, as evaluated from the fitting procedure. Determining m_p is of critical relevance when the morphologies of samples deposited by PLD in the presence of an ambient gas are investigated. Film properties, as will be discussed in the next sections, sensitively depend on the interaction between the ablation plasma and ambient gas via the interplay of plasma mass m and gas density ρ.

Table 1. Measurement of crater volume on an irradiated silver target. Data were fitted to the linear relation: $M = m_p N$, M is the total mass removed; the mass and number of silver atoms ablated per single pulse are: $m_p = 60.0$ ng and $N_{at} = 3.35 \times 10^{14}$ atoms, respectively.

Laser Pulse Number (N)	Volume (mm^3)
10	$(0.61 \pm 0.0167) \times 10^{-4}$
50	$(3.25 \pm 0.0343) \times 10^{-4}$
100	$(5.42 \pm 0.1490) \times 10^{-4}$
200	$(1.07 \pm 0.2940) \times 10^{-3}$

Figure 2. Crater volumes on the surface of a silver target as a function of the laser shot number. The line is a linear fit to the data.

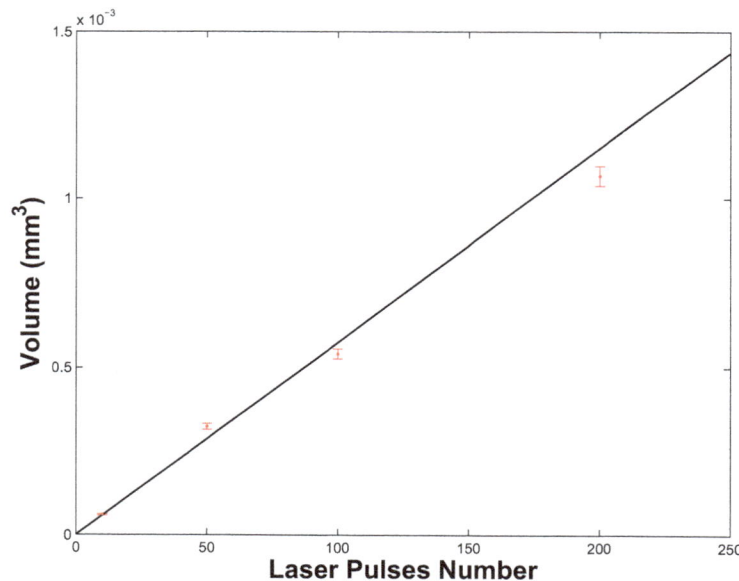

3.2. Surface Morphology

The optical properties of gold and silver nanostructured thin films are sensitively affected by their surface morphology. For isolated silver and gold NPs with a typical size ranging from a few nm up to

30 nm, the LSPR lies at about 400 and 520 nm, respectively. Moving to NP arrangements, self-organized or engineered at the nanoscale, both the position and the width of the LSPR peak depend on the size, shape, number density, mutual distance and aggregation features of the NPs [13]. The control of these properties then results in the control of the optical properties and, ultimately, of the SERS activity of the assembly. Among the several process parameters that affect the morphology of PLD synthesized films, two appear to play a major role: laser pulse number and gas pressure. While the first controls the degree of the NP number density and aggregation on the substrate, the latter mostly influences the interaction between plasma and ambient gas. In Figure 3, we show the surface of silver samples grown under identical conditions ($F = 2.0$ J·cm^{-2}, $P_g = 70$ Pa of Ar, $d_{T-S} = 35$ mm), but changing the laser pulse number between 1000 and 3×10^4.

Figure 3. SEM images of the surface morphology of silver samples deposited at 70 Pa of Ar with: (**a**) 1000, (**b**) 10,000, and (**c**) 3×10^4 laser pulse.

We see that, at a low pulse number (see Figure 3a), the surface is covered by isolated, nearly spherical NPs. As N increases, the NP number density also increases, and deviations towards non-spherical geometry, with increasing NP size, are evident at some locations, together with incipient particle coalescence. Finally, at the highest N value (3×10^4; see Figure 3c), the extensive coalescence among adjacent NPs occurs, thereby islands with an irregular shape and smooth edges form. Notably, such a surface morphology is characterized by the presence of inter-island channels with a defined average length and width. Couples or triplets of such channels meet together at junction points.

To the different morphologies correspond different optical properties of the films. The trend of the UV-Vis absorption spectra as a function of the different morphologies is shown in Figure 4. Increasing N, keeping fixed all other deposition parameters, a red-shift and a broadening of the FWHM of the LSPR peak is observed. A similar trend of change of the film surface morphology and, hence, of the optical properties of the film is observed if N is kept fixed, while the Ar pressure is changed. The effect of Ar pressure on film properties can be understood if the role of the gas on the expanding plasma is considered. Just after the end of the laser pulse, the ablated material starts to expand through the ambient gas, the expansion dynamics being driven by collisions: in the initial expansion stage, intra-plume collisions are dominant, but soon, collisions between plasma species and gas atoms become more and more effective. The collision rate depends on the gas density ρ and increases with ρ. The higher the collisional rate, the less the amount of material that reaches the substrate. At the same time, a higher collision rate promotes the formation and growth of NPs in the plasma. Such a complex process leads to the observed morphology dependence on N and on ambient gas density [14]. Thus, it is possible to tailor the optical properties of deposited films through the control of two easily adjustable deposition parameters (laser pulse number and Ar pressure), provided the relevant process parameters, in particular the laser fluence F, are kept fixed. The latter, given by the ratio of the laser pulse energy E_L to the irradiated spot area A, can be kept fixed by changing appropriately both E_L and A. Fixing F does not ensure that the ablated mass remains fixed as well. If we irradiate a larger area with a higher energy pulse, we can keep constant the fluence value, but a different m_p value results. Referring to the above discussion, for expansions through an ambient gas, the interaction with gas atoms depends on the plasma mass. We have investigated this point growing a set of samples by changing m_p, still keeping fixed F [15]. Two Ag samples were grown at 70 Pa of Ar, at $d_{T-S} = 35$ mm, with 10^4 laser shots, $F = 2.8$ J·cm^{-2}. To explore the role of m_p, two focusing conditions were chosen, both corresponding to the same F value: $E_1 = 10$ mJ, $A_1 = 0.011$ cm^2 and $E_2 = 30$ mJ, $A_1 = 0.037$ cm^2. For the two conditions, different m_p values result, namely $m_{p1} = 7.0$ ng and $m_{p2} = 16.4$ ng, as measured according to the procedure outlined in Section 3.1.

Figure 4. UV-Vis absorption spectra of Ag samples deposited at 70 Pa of Ar with different numbers of laser shots.

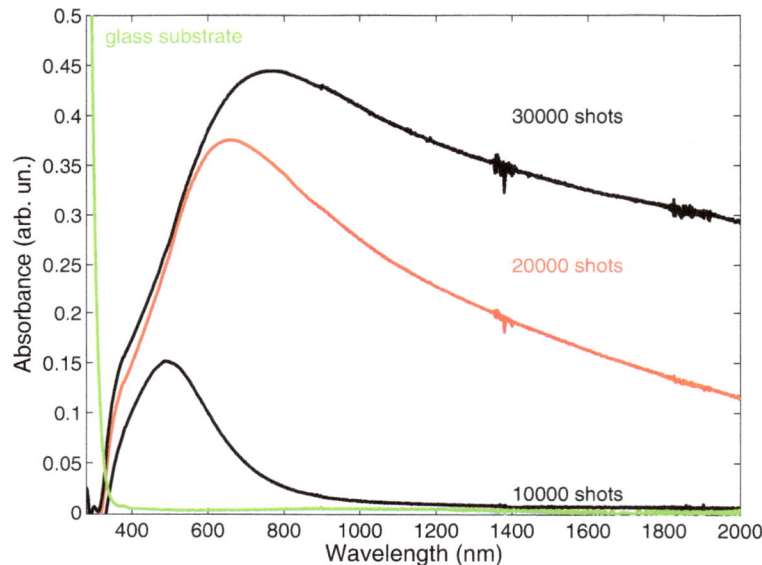

In Figure 5 are shown the TEM pictures of the surface morphology of the two Ag films. The difference between the two morphologies is impressive. The surface of the sample deposited at lower m_{p1} (7.0 ng) is characterized by the presence both of small, nearly spherical NPs and of larger ones with the shape progressively more and more irregular, up to islands that result from the coalescence of several NPs. Looking at the surface of the sample deposited at higher m_{p2} (16.4 ng), we see a nearly percolated structure made of larger islands interconnected by a network of channels. In this sense, m_p has the same effect on the evolution of sample surface morphology as lowering Ar pressure or increasing the laser pulse number. It is worth outlining that two well-differentiated surface morphologies were obtained adopting the same fluence. Thus, in a report on a PLD experiment, the fluence value is not enough to make it reproducible, if laser pulse energy E_L and irradiated area A are not reported, as well, because these two parameters determine m_p.

In Figure 6a are displayed the UV-Vis absorption spectra of the above discussed samples. The different m_p values lead to key differences in their optical properties.

Figure 5. TEM images of the surface morphologies of two Ag samples deposited at a laser fluence of 2.8 J·cm^{-2} with ablated mass per pulse of (**a**) 7.0 ng and (**b**) 16.4 ng.

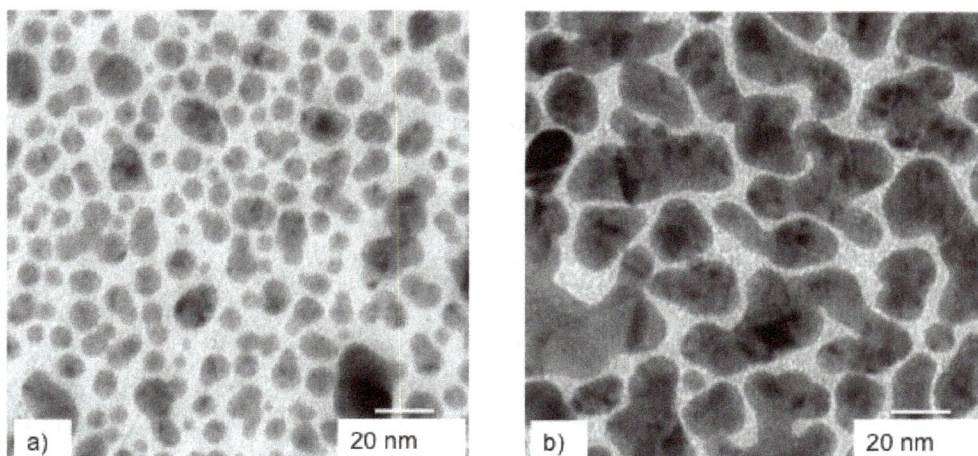

Figure 6. (**a**) UV-Vis absorption spectra of Ag samples deposited at 70 Pa of Ar, 2.8 J·cm^{-2}, 10^4 laser shots and $m_p = 7.0$ ng or $m_p = 16.4$ ng. (**b**) Raman spectra acquired on the surface of the two samples soaked in an aqueous solution of rhodamine 6G (R6G) at a concentration $c = 10^{-4}$ M.

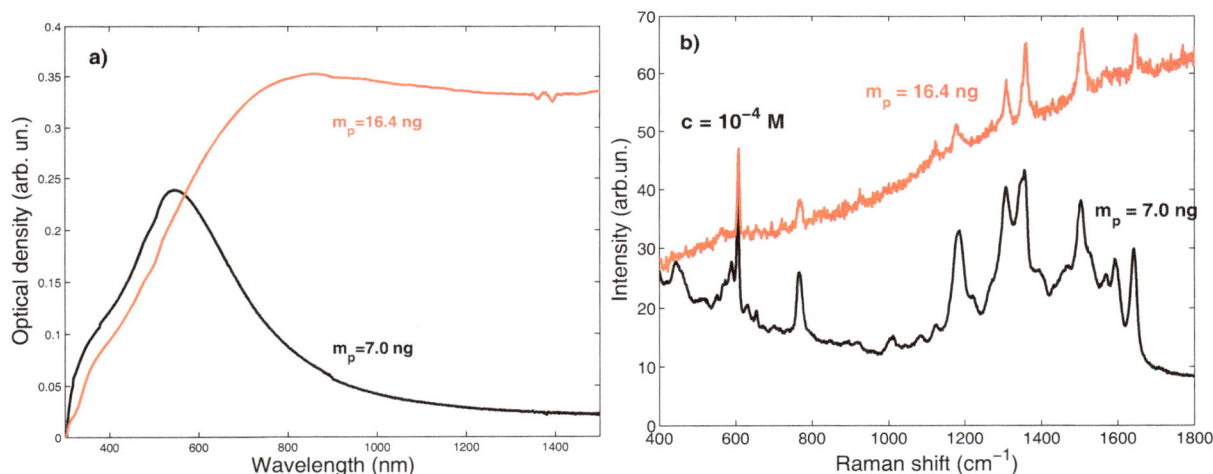

We see that the LSPR considerably red-shifts and its FWHM increases, as expected once the morphology of the two samples is taken into account. Moreover, the large FWHM increase observed in the film deposited with the higher m_p value (compare the film morphologies in Figure 5) points out the setting up of a composite system made of the metallic film and the supporting dielectric substrate (glass). In this system, the optical transitions are detuned via dipole-dipole interactions that broaden the LSPR peak [16,17].

The Raman spectra acquired on the surface of the two samples after being soaked for 1 h in an aqueous solution of rhodamine 6G (R6G) at a concentration of 10^{-4} M (Figure 6b) show a dramatic difference between the SERS response of the two films. The sample deposited at the lower m_{p1} shows a notably higher SERS efficiency. Thus, tuning the morphology and optical properties of nanostructured noble metal thin films is imperative to obtain highly SERS active substrates. Such a tuning cannot disregard detailed knowledge of the influence of all parameters on the deposition process, as exemplified by the study of the ablated mass per pulse.

3.3. SERS Activity of Pulsed Laser Ablated Silver and Gold Substrates

In this section, we report on some applications of Ag and Au SERS active substrates grown by PLD, whose properties were properly optimized from a parametric analysis [18,19]. We observed that Ag films deposited at 70 Pa of Ar, at $F \approx 2.0$ J·cm^{-2} ($A = 2 \div 3 \times 10^{-3}$ cm^2), $d_{T-S} = 35$ mm, with 3×10^4 laser shots show the best SERS activity, so far. Highly SERS active Au substrates were grown under identical conditions, but the Ar pressure was 100 Pa.

In Figure 7, we report the SERS spectra acquired on Ag and Au samples soaked in aqueous solutions of R6G, at the lowest concentrations we tested. The corresponding laser excitations were 632.8 nm for Ag and 785 nm for Au to achieve the maximum EF by matching the corresponding SPR absorption peak positions, observed near 550 and 800 nm, respectively [20,21].

Figure 7. Raman spectra acquired on silver and gold nanostructured substrates soaked in aqueous solutions of R6G at concentrations $c = 5 \times 10^{-8}$ and $c = 10^{-10}$ M respectively.

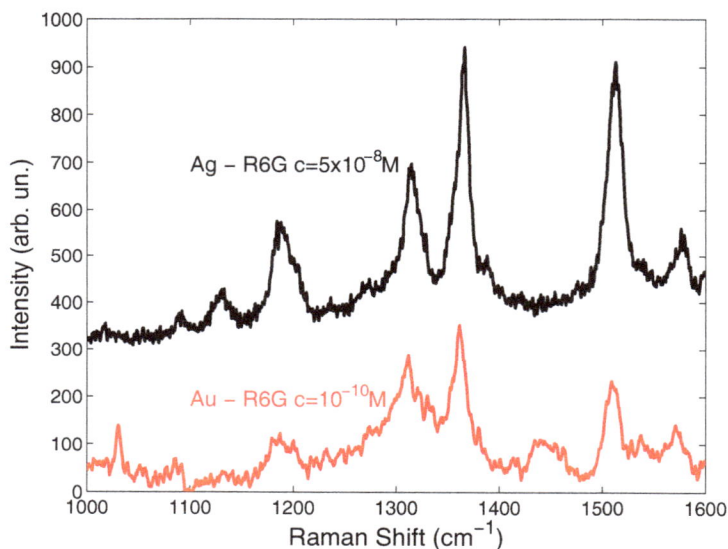

The R6G peaks lying at 1189, 1314, 1366 and 1513 cm^{-1} assigned to the stretching modes of aromatic C bonds are clearly visible. Typical integration times varied between a few seconds and 240 s, pointing out that detection might be possible for this reference analyte at even lower concentration levels.

We believe that the detection at such a low concentration level is due to the formation of hot spots, whose number density strictly depends on film surface morphology, in turn controlled by an easily accessible deposition parameter, i.e., the laser pulse number (see Figure 2.). It is noteworthy that compared to other deposition methods, like thermal evaporation and sputtering, PLD allows synthesizing films without any post-deposition thermal treatment. Moreover, the process is performed at room temperature, thus making it possible to use every kind of substrate material. Finally, PLD is a physical deposition technique that does not require any chemical precursor, different from routes based on the chemical reduction of silver and gold salts to obtain NPs: these have undesired chemical residuals, often introducing disturbing features in SERS spectra.

The observed SERS enhancement was so high, since we choose the excitation wavelength (632.8 nm), such that the surface plasmon absorption band (see Figure 4) lies between the wavelengths of the exciting and the Raman scattered light; indeed, according to the electromagnetic theory, the optical properties of the metallic NPs set the choice of the excitation wavelength in SERS.

Raman spectroscopy is one of the most employed techniques in the cultural heritage field. Identification of dyes is a relevant issue in the field from the historical, conservation and restoration view points. Raman scattering presents some advantages with respect to other techniques, like X-ray fluorescence, UV-Vis absorption and Fourier transform infrared spectroscopy. The most relevant are its non-destructive character and its selectivity. With the advent of portable Raman microscopy apparatuses, measurements can be performed *in situ* on micrometer-sized areas of the artwork. SERS applied to this kind of investigation opens the way to the detection of exiguous amounts of substances employed in the realization of the work of art [22]. In Figure 8, we report as an example SERS measurements performed on a silver substrate soaked in a 10^{-4} M concentrated aqueous solution of alizarine. Alizarine, with its hydrolyzed counterpart, purpurine, are the two chromophores that characterize the organic garanza lake dye extracted from the root of the *Rubia tinctorium* plant. Red lake has been used since ancient Egyptian times until today. Its detection and the relative concentration of the two chromophores can allow identifying the origin and authenticity of the work of art or the presence of restoration works.

Figure 8. (a) SERS spectrum acquired on a silver substrate soaked in an aqueous solution of alizarine at a concentration $c = 10^{-4}$ M. (b) Intensity of the Raman peak at 900 cm^{-1} as a function of the position on the substrate surface over an area of 10×12 μm^2. Detection point spacing, 1 μm.

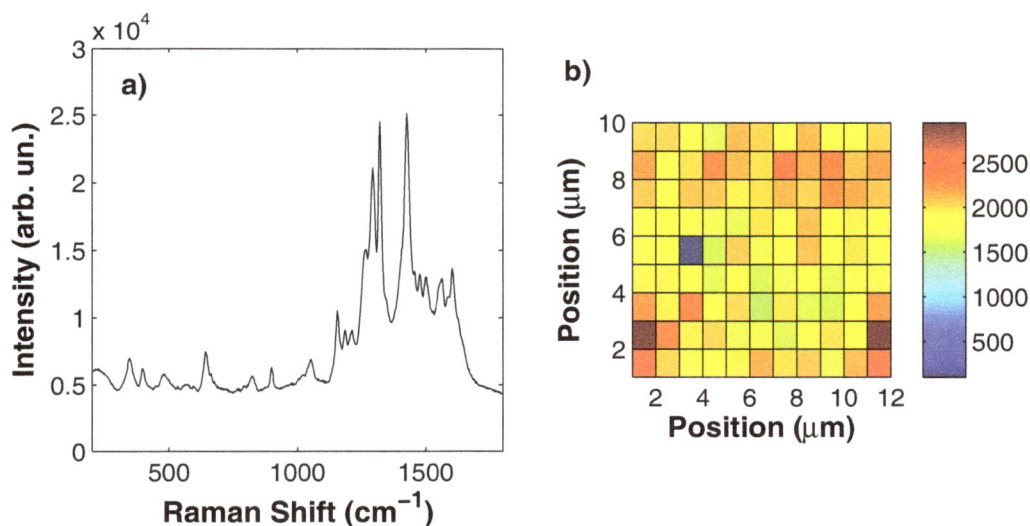

All the alizarine Raman features were detected at 658, 823, 900, 1065, 1156, 1186, 1211, 1269, 1293, 1320, and 1425 cm^{-1}. One of the most important requirements for a SERS substrate is the spatial homogeneity; indeed, Raman spectra should be reproducible over micron-sized areas with intensity variations within 20%. In Figure 8b, we report the results of a Raman mapping experiment performed over an area of 10×12 μm^2 on an Ag substrate soaked in an aqueous solution of alizarine at a 10^{-4} M concentration. Raman spectra were acquired with a spatial resolution of about 1 μm, both along x and along y. In the Raman map (see Figure 8b), we report the intensity of the Raman peak at 900 cm^{-1} as a function of the position. We chose this peak because it is well separated from other alizarine features, thus making it easier to evaluate its intensity over the background. The mean intensity of the peak was of 2003 counts with a standard deviation of 318 counts, corresponding to a variation of about 15%. A companion experiment was performed on a gold substrate soaked in a purpurine aqueous solution at the same 10^{-4} M concentration. The results are reported in Figure 9. Purpurine Raman features are located at 620, 650, 820, 904, 970, 1065, 1313, 1440 and 1470 cm^{-1}. The investigated area to test the substrate spatial homogeneity was $10 \times 8 \ \mu m^2$ and the intensity fluctuations of the peak at 1065 cm^{-1} as a function of the position on the considered surface was about 11% of the average peak value.

Figure 9. (**a**) SERS spectrum acquired on a silver substrate soaked in an aqueous solution of purpurine at a concentration $c = 10^{-4}$ M; (**b**) intensity of the Raman peak at 1065 cm^{-1} as a function of the position on the substrate surface over an area of $10 \times 8 \ \mu m^2$. Detection point spacing, 1 μm.

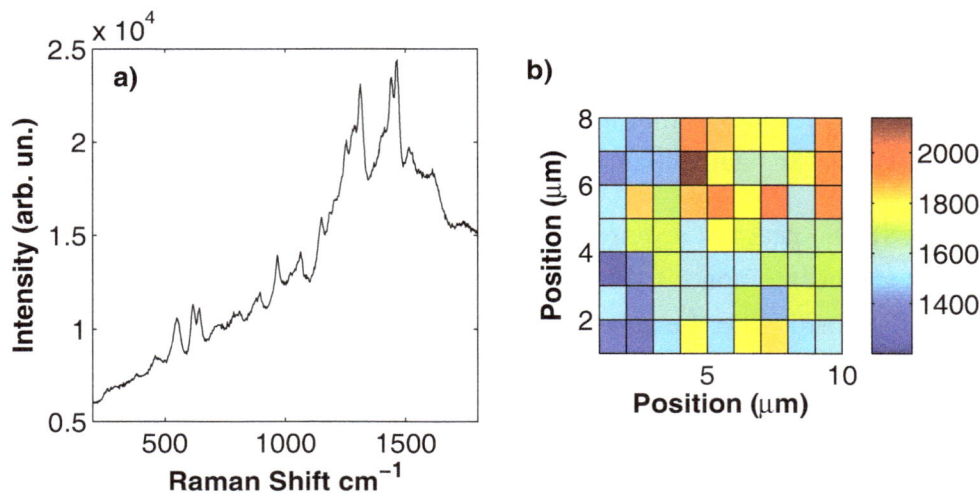

4. Conclusions

In conclusion, nanosecond laser ablation in an inert gas atmosphere at high pressure allows depositing NP arrays of noble metals with a finely controlled nanostructure. By such a kind of bottom-up laser nano-machining, metallic substrates with optical properties tailored *ad hoc* were prepared and utilized for the selective recognition of analytes in a small concentration via surface-enhanced Raman spectroscopy.

Acknowledgments

Matteo Tommasini and Paolo Maria Ossi acknowledge support from Polisocial Award 2014, Project "Controllare l'epilessia nei Paesi in via di sviluppo" (Controlling epilepsy in Developing Countries).

Author Contributions

Fortunato Neri, Sebastiano Trusso and Paolo Maria Ossi conceived and designed the experiments; Enza Fazio, Rosina C. Ponterio and Sebastiano Trusso performed the experiments; Enza Fazio and Rosina C. Ponterio analyzed the data; Fortunato Neri and Matteo Tommasini contributed reagents/materials/analysis tools; Sebastiano Trusso and Paolo Maria Ossi wrote the paper.

Conflicts of Interest

The authors declare no conflict of interest.

References

1. Srinivasan, R.; Sutcliffe, E.; Braren, B. Ablation and etching of polymethylmethacrylate by very short laser pulses. *Appl. Phys. Lett.* **1987**, *51*, 1285–1287.

2. Bärsch, N.; Körber, K.; Ostendorf, A.; Tönshoff, K.H. Ablation and cutting of planar silicon devices using femtosecond laser pulses. *Appl. Phys. A* **2003**, *7*, 237–242.

3. Küper, S.; Stuke, M. Ablation of polytetrafluoroethylene (Teflon) with femtosecond UV excimer laser pulses. *Appl. Phys. Lett.* **1989**, *54*, 4–6.

4. Davis, K.M.; Miura, K.; Sugimoto, N.; Hirao, K. Writing waveguides in glass with a femtosecond laser. *Opt. Lett.* **1996**, *21*, 1729–1731.

5. Kawata, S.; Sun, H.B.; Tanaka, T.; Takada, K. Finer features for functional microdevices. *Nature* **2001**, *412*, 697–698.

6. Reif, J.; Varlamova, O.; Uhlig, S.; Varlamov, S.; Bestehorn, M. On the physics of self-organized nanostructure formation upon femtosecond laser ablation. *Appl. Phys. A* **2014**, *117*, 179–184.

7. Carey, J.E.; Crouch, C.H.; Shen, M.; Mazur, E. Visible and near-infrared responsivity of femtosecond-laser microstructured silicon photodiodes. *Opt. Lett.* **2005**, *30*, 1773–1775.

8. Sugioka, K. Ultrafast Laser Micro- and Nano-processing of Glasses. In *Lasers in Materials Science*; Castillejo, M., Ossi, P.M., Zhigilei, L.V., Eds.; Springer: Berlin, Germany, 2014; pp. 359–380.

9. Nöel, S.; Hermann, J.; Itina, T. Investigation of nanoparticle generation during femtosecond laser ablation of metals. *Appl. Surf. Sci.* **2007**, *253*, 6310–6315.

10. García-Vidal, F.J.; Pendry, J.B. Collective Theory for Surface Enhanced Raman Scattering. *Phys. Rev. Lett.* **1996**, *77*, 1163–1166.

11. Nie, S.; Emory, S.R. Probing Single Molecules and Single Nanoparticles by Surface-Enhanced Raman Scattering. *Science* **1997**, *275*, 1102.

12. Halas, N.J.; Moskovits, M. Surface-enhanced Raman spectroscopy: Substrates and materials for research and applications. *MRS Bull.* **2013**, *38*, 607–611.

13. Mulvaney, P. Surface plasmon spectroscopy of nanosized metal particles. *Langmuir* **1996**, *12*, 788–800.

14. Bailini, A.; Ossi, P.M.; Rivolta, A. Plume propagation through a buffer gas and cluster prediction. *Appl. Surf. Sci.* **2007**, *253*, 7682–7685.

15. Spadaro, M.C.; Fazio, E.; Neri, F.; Ossi, P.M.; Trusso, S. On the influence of the mass ablated by a laser pulse on thin film morphology and optical properties. *Appl. Phys. A* **2014**, *117*, 137–142.

16. Shalaev, V. Electromagnetic properties of small particle composites. *Phys. Rep.* **1996**, *272*, 61–137.

17. Micali, N.; Neri, F.; Ossi, P.M.; Trusso, S. Light scattering enhancement in nanostructured silver thin film composites. *J. Phys. Chem. C*, **2013**, *117*, 3497–3502.

18. D'Andrea, C.; Neri, F.; Ossi, P.M.; Santo, N.; Trusso, S. The controlled pulsed laser deposition of Ag nanoparticle arrays for surface enhanced Raman scattering. *Nanotechnology* **2009**, *20*, 245606.

19. Fazio, E.; Neri, F.; D'Andrea, C.; Ossi, P.M.; Santo, N.; Trusso, S. SERS activity of pulsed laser ablated silver thin films with controlled nanostructure. *J. Raman Spectrosc.* **2011**, *42*, 1298–1304.

20. Agarwal, N.R.; Fazio, E.; Neri, F.; Trusso, S.; Castiglioni, C.; Lucotti, A.; Santo, N.; Ossi, P.M. Ag and Au nanoparticles for SERS substrates produced by pulsed laser ablation. *Cryst. Res. Technol.* **2011**, *46*, 836–840.

21. Merlen, A.; Gadenne, V.; Romann, J.; Chevallier, V.; Patrone, L.; Valmalette, J.C. Surface enhanced Raman spectroscopy of organic molecules deposited on gold sputtered substrates. *Nanotechnology* **2009**, *20*, 215705.

22. Leona, M. Microanalysis of organic pigments and glazes in polychrome works of art by surface-enhanced resonance Raman scattering. *Proc. Natl. Acad. Sci. USA* **2009**, *106*, 14757–14762.

sBCI-Headset—Wearable and Modular Device for Hybrid Brain-Computer Interface

Tatsiana Malechka [1,2,†], **Tobias Tetzel** [2,†], **Ulrich Krebs** [1,†], **Diana Feuser** [1,†] **and Axel Graeser** [1,†,*]

[1] Institute of Automation (IAT), University Bremen, Otto Hahn Allee NW1, 28359 Bremen, Germany; E-Mails: tatsiana.malechka@googlemail.com (T.M.); krebs@iat.uni-bremen.de (U.K.); diana.feuser@uni-bremen.de (D.F.)

[2] Friedrich Wilhelm Bessel Institute gGmbH, Otto Hahn Allee NW1, 28359 Bremen, Germany; E-Mail: tetzel@embedded-design-team.com

[†] These authors contributed equally to this work.

[*] Author to whom correspondence should be addressed; E-Mail: ag@iat.uni-bremen.de

Academic Editor: Dean Aslam

Abstract: Severely disabled people, like completely paralyzed persons either with tetraplegia or similar disabilities who cannot use their arms and hands, are often considered as a user group of Brain Computer Interfaces (BCI). In order to achieve high acceptance of the BCI by this user group and their supporters, the BCI system has to be integrated into their support infrastructure. Critical disadvantages of a BCI are the time consuming preparation of the user for the electroencephalography (EEG) measurements and the low information transfer rate of EEG based BCI. These disadvantages become apparent if a BCI is used to control complex devices. In this paper, a hybrid BCI is described that enables research for a Human Machine Interface (HMI) that is optimally adapted to requirements of the user and the tasks to be carried out. The solution is based on the integration of a Steady-state visual evoked potential (SSVEP)-BCI, an Event-related (de)-synchronization (ERD/ERS)-BCI, an eye tracker, an environmental observation camera, and a new EEG head cap for wearing comfort and easy preparation. The design of the new fast multimodal BCI (called sBCI) system is described and first test results, obtained in experiments with six healthy subjects, are presented. The sBCI concept may also become useful for healthy people in cases where a "hands-free" handling of devices is necessary.

Keywords: brain computer interface; assistive systems control; eye tracker; environmental control; human machine interface

---

1. Introduction

The achieving of autonomy is very important for disabled people, as it means improving the quality of their lives. Ambient assistive living, environmental control systems and supporting devices like the assistive robot FRIEND (**F**unctional **R**obot arm with user-fr**IEN**dly interface for **D**isabled people) [1] are designed to restore the autonomy of disabled users in All Day Living (ADL) scenarios and in professional life. To ensure economic feasibility, the support systems have to provide user independence from care personal for several hours. The user needs then complete control over the Human Machine Interface (HMI), which itself depends completely on the remaining mental and physical capabilities of the user. Additionally, a gradual or sudden change in motion capabilities may happen depending on the disability. HMI concepts that are helpful at a specific point in time, like systems that use head motion capabilities, may later become insufficient. That holds, e.g., for patients with muscular dystrophy, multiple sclerosis or amyotrophic lateral sclerosis (ALS) where motion based communication capabilities may decrease over time. Steady-state Visual Evoked Potential Brain Computer Interfaces (SSVEP-BCI) and Motion imagination Event-related (de)-synchronization (ERD/ERS) BCI are then becoming important.

Depending on the disability of the user, many HMI methods to issue commands are possible [2], however all have some disadvantages. Often BCI is mentioned as a specific HMI for a large class of disabled users, esp. for users with very limited motion capabilities. The feasibility of an SSVEP-BCI to control FRIEND has been shown in [3] and was demonstrated also on several exhibitions and conferences (CEBIT Hannover 2008, RehaCare Duesseldorf 2008, ICORR 2007). Figure 1 shows control of FRIEND with a SSVEP-BCI in an All Day Living laboratory experiment (ADL) [2]. The control of different assistive devices like an internet radio, the assistive robot FRIEND, and other environmental systems serve here as use cases for BCI application. However, BCI use is time consuming and has an error rate that cannot be neglected, which is the reason to design a new hybrid BCI and research its capabilities.

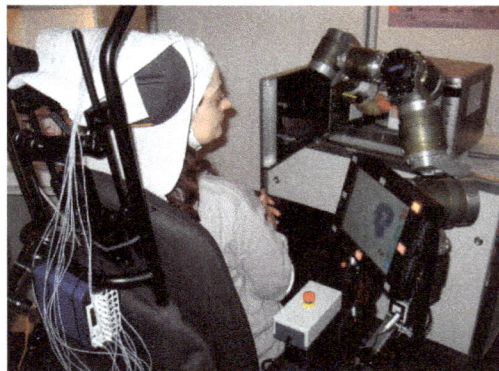

Figure 1. Control of FRIEND with a SSVEP-BCI in an ADL-test bed. SSVEP Diodes are located at the frame of the screen.

To simplify preparation, improve the comfort of the user, extend the user group, and enable research for optimized HMI, a new electroencephalography (EEG)-cap was designed and integrated with an SSVEP-BCI, an ERD/ERS-BCI, an eye-tracker, and an environmental camera. The integration of signal processing for both BCI methods with image processing for an eye tracker and environmental camera in one software package is also accomplished and improves software development capabilities in research and future application of sBCI.

In a first setup, we tested the control of an internet radio, a microwave, and a fridge with six healthy participants. Signal processing for the BCI signals is carried out with a Bremen-BCI software package, which was described [4–6] including results of European Project BRAIN [7]. In this paper, we are reporting first test results. The BCI tests aim mainly on feasibility of the sBCI concept and on a comparison with BCI results which we achieved with a standard EEG cap.

We focus here on the sequential use of the sBCI components whereby the eye tracker or the ERD/ERS-BCI acts as a selector for systems in the environment and the SSVEP-BCI enables the control of the selected device. Five subjects used the eye tracker and one subject who was trained in ERD/ERS–BCI used that one for selection. In both cases, the SSVEP-BCI facilitated the control of the selected device. All participants succeeded in performing five requested tasks with a good performance: the subjects who used eye tracker spent 3.9 s on average on the selection of the target device; however, 20 s on average were needed with ERD/ERS-BCI; a peak information transfer rate (ITR) of 73.9 bit/min was achieved with the SSVEP-BCI; a mean ITR of 41.2 bit/min and an accuracy of 96.3%.

The paper is organized as follows: In Section 2, we consider two use cases for BCI—control of home appliances and control of an Internet radio. Section 3 discusses the state of art in EEG-based BCI. In Section 4, we discuss the layout of the sBCI system and design decisions. Section 5 describes the user interface and control methods chosen so far. Section 6 gives details about the subjects, the data acquisition and the results. The paper closes in Section 7 with a discussion of the results and the lessons learned.

2. Use Cases and General Design Decisions for sBCI

The use cases are inspired by the support of users with tetraplegia in an ADL scenario. Simple devices like a fridge or microwave and more complex ones like an internet radio have to be operated. Typically, the user has to select a device and then issue commands to the selected device. For a fulfillment rate of the initialized tasks larger than 80%, the user has to support the automation system. The user can interrupt it if a problem in task execution arises and issue corrective commands. After the problem is solved, control is handed back to the automation system.

2.1. Use Case 1: Control of Simple Home Appliances

Here we consider an ADL scenario. The user sits in a wheelchair with a robot arm and would like to drink. The user may first select the fridge and then send an "open" command to the actuator that opens the door of the fridge.

Once the door is open, the user chooses the robot arm and issues a high level command "grasp bottle". The robot arm which is mounted on the wheelchair performs an autonomous picking up of the object in the fridge. The control of the robot arm is realized as a combination of vision-based object recognition and advanced path planning as described in [1].

2.2. Use Case 2: Operation of Complex Devices

For a device like an internet radio, a sequence of inputs and commands has to be generated to start a specific task. It may also be necessary to intervene in task execution to support the automation algorithms. Typical commands after choosing the internet radio are: (a) turn radio on/off; (b) select channel; (c) set volume up/down; (d) operate playback functions (pause, resume fast forward/backward *etc.*). The user has to select the HMI and then issue one or more commands.

2.3. General Design Requirements

General requirements for a BCI that is used by a disabled user as an HMI are: (a) preparation of the user should be possible within 5 min, (b) wearing comfort has to allow uninterrupted usage for several hours, (c) the user must be able to switch the system or components on and off without external support.

3. State of Art for EEG-Based BCI

The BCI system establishes a direct communication channel between the human brain and a control or communication device. BCIs detect the human intention from various electrophysiological signal components, such as steady-state visual evoked potentials (SSVEPs) [8,9], P300 potentials [10] and sensorimotor rhythms (SMR) and translate it into commands. The brain signals are recorded from the scalp using electroencephalography (EEG). The movement-related modulation of mu (7–13 Hz) and beta (13–30 Hz) sensorimotor rhythm induced by the imagination of limb movements [11,12] has gained considerable interest as a more natural paradigm for the non-invasive BCIs. Compared to those BCIs based on evoked potentials such as SSVEP and P300, motor imagery BCIs do not need external stimulation. SMR modulation patterns in the form of event-related de-synchronization (ERD) and event-related synchronization (ERS) are independent in terms of any stimulation and allow the user to freely decide when they wish to generate a control signal. Nevertheless, ERD/ERS-BCIs are generally even more demanding in the usage and more complex in the implementation. ERD/ERS-BCIs require extensive user training and adaptive signal processing algorithms tailored to the mental states (fatigue, workload and emotion) and learning rates of each subject [13,14], although advanced signal processing methods and well-designed training interfaces are used in BCI research. However, the output of the ERD/ERS-based BCI systems is still less reliable and the interaction speed is much lower in comparison to other paradigms and still far away from mainstream interaction modalities such as joysticks or mice. Currently, the SSVEP approach provides the fastest and most reliable paradigm for non-invasive BCI system implementation and requires little or no training. Steady-state visual evoked potentials are brain responses elicited by presenting repetitive visual stimulation above 5 Hz. Exposing the user to flickering lights in frequencies between 5 and 20 Hz evokes SSVEPs with comparatively large amplitude [15] but it is inconvenient and tiring for the user. Moreover, during the non-control state (NC state), when the user does not want to generate any command, the continuing flickering could induce false positive classifications if the BCI system mistakenly declares the NC state as an intentional control state.

As explained in the next chapter, the requirements for control of complex devices lead to the development of a BCI with hybrid architecture (hBCI). A hybrid BCI (hBCI) is usually defined as a BCI combined with at least one other interface system or device. Such a system or device might be a

BCI channel relying on other brain patterns, the output of an external assistive device (chin-joystick, switch, *etc.*) or other bio signals (heart rate, eye gaze, muscular activities, *etc.*) [16]. Recently multimodal interfaces have demonstrated promising results towards more reliability, flexibility and faster interaction. Some of them are based on multiple brain signals [17–21]. Others combine brain and additional bio signals [22,23]. A multimodal interface including eye tracking to determine the object of interest and a brain-computer interface to simulate the mouse click is presented in [24,25]. Eye movements are detected with a remote eye tracker that is mounted opposite of the user and uses an infrared camera to observe gaze direction. A limitation of such a remote system is that it can only be used with screen-based interfaces.

A further important issue in non-invasive BCI research is to make the EEG acquisition system more comfortable and suitable for daily use. Basically, two trends of development can be identified for this design aspect. Firstly, minimizing the number of electrodes in order to reduce hassle and setup time and, secondly, the development of dry [26–28] or water-based electrodes [29]. Several commercial non-medical EEG headsets following these trends have been released over the last few years. The commercially available BCI headsets promise ease of use, low cost, short setup time, as well as mobility. The headsets are wireless and vary in the number of electrodes: NeuroSky's MindSet and MindWave (http://www.neurosky.com) use a single EEG electrode positioned on the user's forehead; ENOBIO [30] provides four channels of bioelectric signals; Emotiv's EPOC headset [31] measures electrophysiological signals using 14 saline non-dry sensors placed over the user's head. However, the data acquired by such consumer devices contains neither event related potentials nor sensorimotor rhythms. These commercial-graded systems focus on gaming, entertainment, and biofeedback training based on the attention level and meditation. Based on this state of art analysis, the system layout and design decisions are discussed in the next chapter.

4. System Layout and Design Decisions

The capacity of EEG based BCIs must be carefully reflected in order to integrate a BCI with the ambient intelligence environment and control strategy of assistive systems for a disabled user. A detailed analysis of the support tasks carried out for FRIEND and other devices in the environment of the disabled user lead to the following principal actions that had to be carried out with a BCI:

- Start and stop the BCI completely or only a specific BCI mode;
- Selection of a specific support device (e.g., robot, gripper, microwave, fridge or radio);
- Selection of a specific action of the selected device (e.g., open fridge, grasp bottle, open/close gripper, chose moving direction of robot arm, move arm a specific distance in chosen direction);
- Switch to the HMI screen;
- Navigate within the HMI screen (left, right, up, down);
- Start a specific action;
- Stop a running action.

In this paper, a new EEG head cap for BCI in combination with an eye tracking system is introduced. We will mainly focus on the sequential hybrid BCI, where the eye tracker or ERD/ERS-BCI acts as a selector and the SSVEP-BCI generates the control commands.

The sBCI system includes a lightweight, ergonomically designed headset which integrates multi-channel EEG equipment, an eye tracker system with environmental observation and an integrated visual stimulator for SSVEP-BCI. The full integration of the SSVEP stimulator, SSVEP electronic and BCI software with the eye tracker was the main reason to design an eye tracker instead of using a commercially available one. An EEG amplifier Porti 32 (Twente Medical Systems International, Oldenzaal, The Netherlands) connected to a standard notebook is used to measure the brain activity, while an LED pattern generator produces flicker frequencies. A programmable remote controller sends the control commands to the controlled devices. The universal structure of the sBCI interface allows the easy integration of additional devices and further functionalities into one common graphical user interface. The following sections describe the system layout and design criteria.

4.1. Gaze-Controlled Interface

An overview about eye tracking technology can be found in [32]. The sBCI system described here is from a Video-Occulography (VOG) type. It uses a mobile head mounted eye tracker, which is specifically developed for use with the BCI-headset and SSVEP-BCI. Wearable eye trackers which were available on the market did not allow a seamless integration into the sBCI multimodal headset. The provided software is also not open source and the integration into a single hybrid system is rather difficult, especially if additional information (e.g., the pupil diameter) is required. The sBCI eye tracking system detects the user's intention to interact with a specific device in its environment. For this, three cameras are used. Two cameras are for tracking of left and right eye, the remaining one is for monitoring the environment. Figure 2 shows the principal layout, with one of the eye cameras and the environmental camera of the tracking system DeLock USB CMOS Cameras 95,852, 1.3 Megapixel are used in all cases. The resolution of the DeLock cameras is set to 1280 × 1024 pixels for environmental images and to 640 × 480 pixels for eye tracking. The user's eyes are illuminated by Infrared (IR)-LEDs (LD271, OSRAM Opto Semiconductors GmbH, Regensburg, Germany) P_{max} = 2.7 mW) through a dichroic mirror (Edmund Optics NT62-630, Barrington, NJ, USA).

The eye cameras are mounted in parallel to the IR-LEDs and have modified lenses without the mandatory IR blocking filter. Instead, two IR pass filters (LUXACRYL-IR 1698) in front of the lenses suppress most of the visible environmental light. The light of the IR-LEDs in combination with the two infrared filters in front of the eye cameras give a stable illumination of the eyes. This significantly eases the process of detecting the pupils and estimating the gaze direction. Figure 3 shows the characteristic image of an eye camera and the result of pupil recognition.

The dichroic mirrors allow mounting the eye cameras outside the user's field of vision. It reflects the IR light coming from the LEDs into the user's eyes and back into the eye cameras while being fully transparent for visible light. Additionally, the mirror holders can be used for the SSVEP stimuli LEDs. Based on the estimation of the gaze direction, the object of interest is determined. The object recognition may follow different approaches. The object may be recognized in the image of the environmental camera by using specific colors that are not immediately recognizable as marker, by using SIFT (Scale Invariant Feature Transform) or a comparable algorithm to recognize known objects or by specific markers, e.g., chosen from ARToolKit [33]. Figure 4 shows typical markers of the different categories. Based on the estimation of the gaze direction the object of interest is determined.

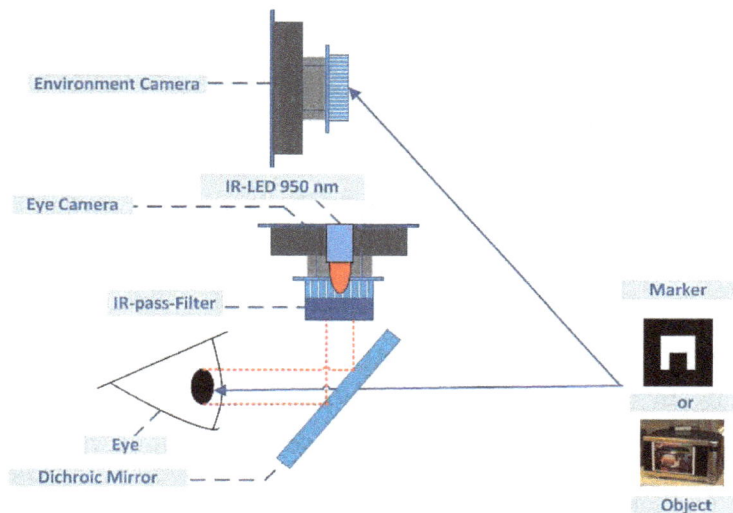

Figure 2. Layout of eye tracker and environmental camera to recognize gaze direction and objects to be controlled.

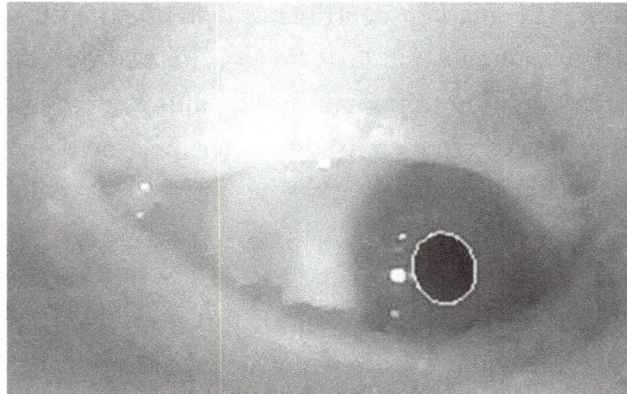

Figure 3. Eye and pupil detection (large white circle) shown in eye camera image.

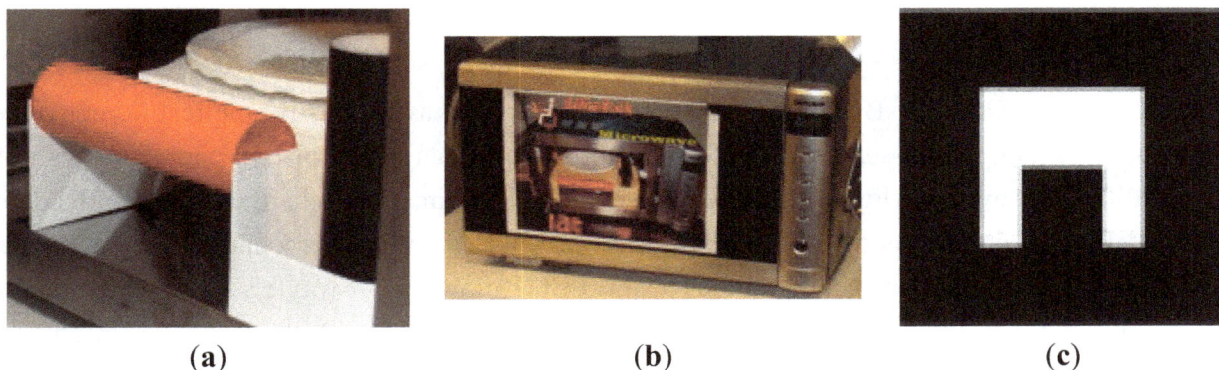

(a)   **(b)**   **(c)**

Figure 4. Marker layout used to distinguish between home devices, object with specific color and size as marker (**a**), object to be recognized by SIFT (**b**) and artificial ARToolKit marker (**c**).

Although very powerful SIFT algorithms have been developed for object recognition [34,35] the recognition approach in this paper is based on specific ARToolKit markers which are detected in the live video stream [36]. This decision was made to decouple sBCI and SIFT related research projects and to be able to show the feasibility of BCI usage in complex control tasks without image processing overhead. The ARToolKit allows extracting the 3D position of the marker in the world coordinate system while knowing the exact marker dimensions.

To determine the gaze direction, it is necessary to extract the pupil centers from both eye camera images. The following standard image processing methods are used:

- Calculating the histogram;
- Finding first distinctive maximum;
- Threshold the image based on that maximum;
- Erode and dilate the image with a circle element;
- Use edge detection and calculate best fitting ellipse (regarding its roundness and size).

Before using the eye tracker system, a calibration is required. The eye tracker calibration specifies the relationship between both eye cameras and the scene camera. Each subject is instructed to fixate upon a calibration marker shown in front of him/her for a short time while changing the head position and the viewing angle. Simultaneously, the eye tracker records the pupil centers and marker coordinates for each fixation. The marker coordinates and the pupil centers are grouped into a constraint matrix that is used to compute a transformation matrix via singular value decomposition. After calibration, this matrix can be used to transform the pupil centers to gaze coordinates relative to the scene image. The gaze coordinates are then used to determine the object in the environment where the user is concentrating on. This object can then be selected as the one to where future input commands are directed.

For the selection procedure, the eye tracker or the ERD/ERS-BCI can be used. A recognized object may be considered to be selected if the user looks at it for more than a predefined fixation interval. To achieve a fast object selection while avoiding unwanted selection actions, the selection interval should be chosen carefully. Best results are achieved with a selection time between 1.5 and 3 s. Usage of ERD/ERS for selection is described in the next section.

4.2. sBCI-Headset

The primary goal of the sBCI newly developed head set was an easily applicable, convenient, wearable, and appealing multi-sensor device, which allows easy data fusion of BCI paradigms with other input modalities like eye tracking. The sBCI, shown in Figure 5, includes a hard case cap with adapters for up to 22 EEG-electrodes, two eye cameras, one camera for environment monitoring, and a miniature SSVEP stimulator with four surface-mounted (SMD) LEDs allowing a 4-way SSVEP interaction.

The EEG-electrodes are placed at the pre-defined positions according to the extended international 10-10 system of EEG measurement. The electrodes which are integrated in the headset are commercially available conventional Ag-AgCl EEG gel electrodes. The electrode assembly with the holder and spiral spring provides many degrees of freedom for good adaptation on the user's head and also offers the possibility of gel injection and abrading the skin for reducing the impedance. The CAD-models of the headsets are developed with the help of 3D-models of adult heads and manufactured by using a rapid prototyping technology. The sBCI-headset is built in three different sizes based on head circumferences in order to fit the heads to a large number of users.

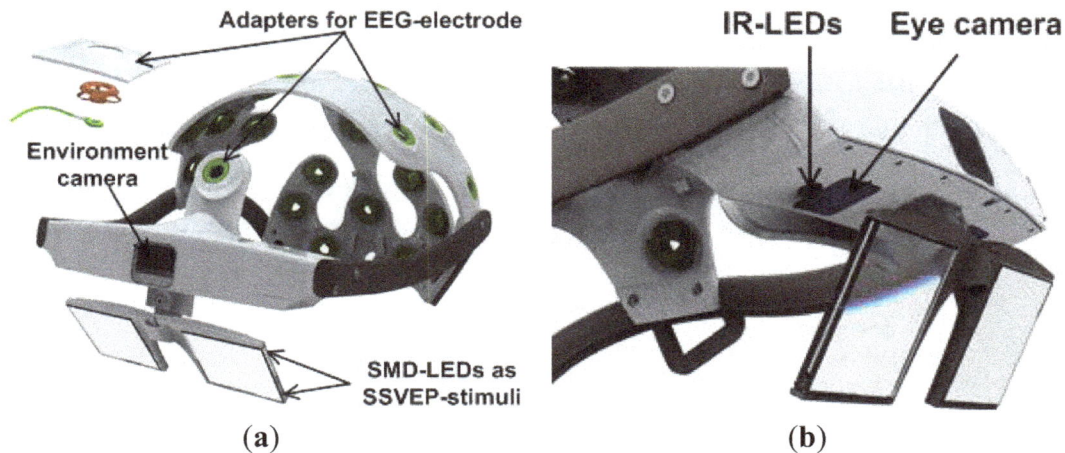

Figure 5. sBCI headset which includes a hard case cap with 22 EEG electrode positions (**a**) two eye cameras, one camera for observation of the environment and four SSVEP-stimuli at the edges of the infrared mirrors (**b**).

4.3. Brain Computer Interface

4.3.1. SSVEP-BCI

Eight EEG electrodes are used to record neural activity from the occipital region of the scalp. The visual stimulation is provided using four LEDs in SMD 0805 size (surface mounted device, 2.0×1.25 mm). The flicker frequency and brightness of the LEDs is controlled by a dedicated LED controller. This device uses a PIC18F4550 microcontroller (Microchip, Chandler, AZ, USA) as a communication interface and master timing generator and eight PIC16F690 microcontrollers as independent brightness controllers for up to eight output channels. The flicker frequency is adjustable from 0.2 to 1000 Hz with a timing resolution of 50 µs. The brightness can be set between 0% and 100% with a resolution of 1%. All brightness and timing parameters are controlled by software and can be adjusted at runtime from a host computer via USB interface. The LED controller is configured by software specially written for this purpose which allows easy access to the complete controller functionality. The four stimuli LEDs are placed at the corners of the eye tracker's dichroic infrared mirrors, as shown in Figure 6. To simply the navigation among LEDs for the user, two different colors are used: green LEDs are placed on the top and red LEDs at the bottom of the mirrors. The LED controller also allows a different assignment of frequencies to the four LED positions. This arrangement of LEDs allows focusing on a stimulus while simultaneously observing the environment and/or the controlled device.

The distance between the lights and the user's eye is about 3 cm. The chosen stimulation frequencies are: 13 Hz (LED on left-bottom), 14 Hz (LED on left-top), 15 Hz (LED on right-top) and 16 Hz (LED on right-bottom). With the LED controller, it is easy to change frequencies of all LED as well as the on/off intervals in future consideration. The signal processing module uses the Minimum Energy Combination (MEC) for spatial filtering, signal power estimation, and normalization for each stimulation frequency [4]. The spatial filter re-adjusts the input channels in order to minimize background activity and noise. The next step of signal processing is the estimation of power in each stimulation frequency based on a 2 s data window. Spatial filtering and power estimation is executed every 125 ms. The power values are normalized to convert the absolute values into relative probabilities. The probability in a stimulus frequency has to exceed a predefined threshold in order to be classified. After classification of a nonzero class, an idle period of 2 s is introduced, in which no further classification is made.

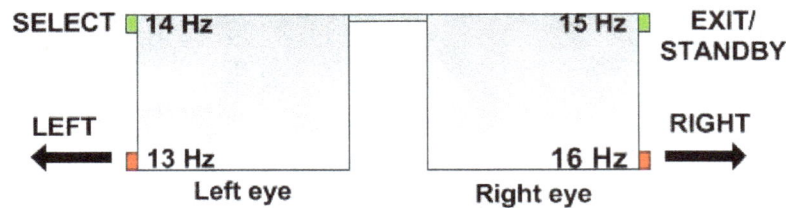

Figure 6. Location of four flickering LEDs at the dichroic infrared mirrors and possible meaning in a specific application.

4.3.2. ERD/ERS-BCI

Motion imagination (MI) causes characteristic changes in EEG signals (ERD/ERS) which can be measured as power changes in specific frequency bands. The topography and time course of the changes depend on several factors, including the type of MI, the contents of the imagery, the subject training and experience. Before the ERD/ERS-BCI can be efficiently used, the user has to undergo a calibration procedure performed in a cued paradigm. Three types of motion imagination are used as intentional control (IC) states: left hand (LH), feet (F) and right hand (RH). The data recorded during the calibration session was used to adapt the classifier to the individual user. This is ensured by identification of individual frequency bands for μ and β rhythms, estimation of spatial filters using the Common Spatial Pattern (CSP) algorithm [37], extraction of features and training of two-stage classifiers. As first, logistic regression classifier is trained to discriminate between non-control (NC) and IC states. If the data represents a potential IC state, it is passed to the second stage classification procedure, which performs the discrimination between three IC states [7]. In the operational phase, the participants used the asynchronous ERD/ERS-BCI to scroll through the device submenu and select a device of interest. The EEG subject-specific features extracted from the 2 s long sliding window were exploited as inputs to the created classifier. The classifier output, which is updated every 125 ms, is passed on the circular buffer containing the classifier outputs for the last 5 s. The control signal is forwarded to the sBCI interface if six uniform IC classifications were collected. After the execution of each command, an idle period of 2 s is introduced.

5. User Interface and Control Method

In order to consider feasibility of the sBCI system for the control of complex systems, a test bed is designed which contains the HMI for the operation of an internet radio, a fridge and a microwave. The internet radio is used as a complex device while the fridge and microwave are two examples for simple devices which are used together with FRIEND in an ADL scenario. The control of the internet radio with sBCI has a similar complexity as the control of an assistive robot. A robot can be used in a study only after additional safety measures are introduced in order to avoid endangering of the user. Ensuring user safety requires extensive study of failure modes of sBCI and the development of a safety concept [38]. The safety study would exceed the content of this publication.

The user interface takes the classification from the BCI and gaze commands from the eye tracker as an input and sends commands to the selected devices. Depending on the subject's abilities and preferences, either the eye tracker or the ERD/ERS-based BCI can be set up to select the target device. The SSVEP-BCI facilitates the control of the selected device.

Figure 7 shows the graphical user interface (GUI) for the internet radio. It gives access to the main functionalities and informs the user about the currently selected position inside the hierarchical structure of the interface. sBCI user interface presently contains three external home devices: internet radio (Grundig Cosmopolit 7), fridge and microwave. The interface is similar for all devices and is divided in two parts: The left side allows the device selection while the right side changes dynamically and gives access to the respective functionalities of the chosen device. All icons of this interface originate from www.iconarchive.com. Each command coming from the BCI or the eye tracker is accompanied by an audible and visual feedback to the user.

The following devices and main functionalities are currently implemented:

- **Internet radio**: On/off, channel selection from 0 to 9, music source selection (internet radio, FM radio, music archive), menu configuration (right, up, left, down, and OK), volume up/down and standard playback functions (pause, resume and end playback, select previous or next track).
- **Fridge:** open and close door.
- **Microwave:** Cooking time selection, start cooking and release door.

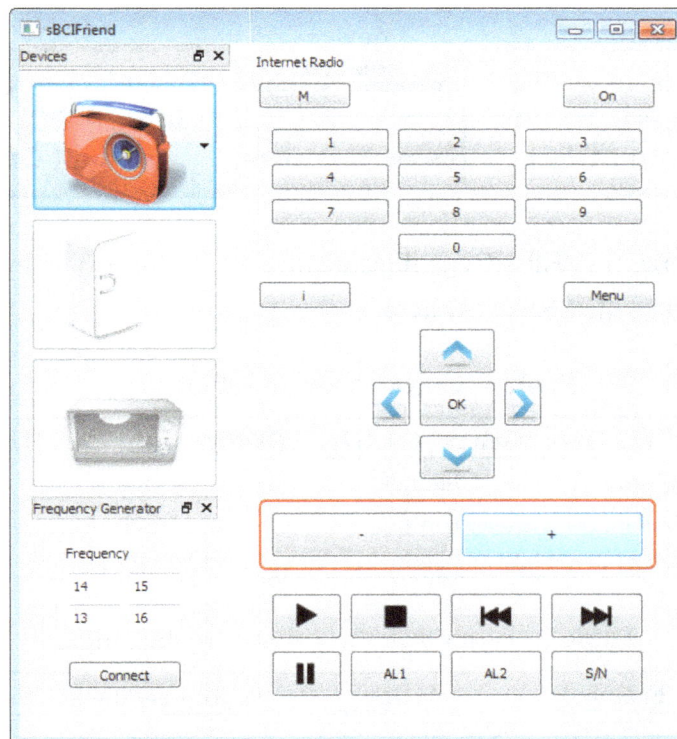

Figure 7. GUI for the internet radio.

5.1. Gaze-SSVEP Interface

The combined control interfaces of the headset allow the user to select a specific device with the eye tracker by simply focusing on it for a predefined selection time. After selection, the device is operated using the SSVEP-BCI. When the user has selected a specific device with the eye tracker, several buttons appear on the GUI showing different functionalities which are related to this device, as shown in Figure 8. The SSVEP stimuli start flickering and the SSVEP-BCI automatically starts analyzing the EEG signals.

The hierarchical menu structure of the interface implies a two-level control:

Level 1: All functionalities of a specific device are grouped together in sub-menus. This level enables the navigation through the functionalities' sub-menus (13 and 16 Hz), offering access to the sub-menu actions (14 Hz) and stopping the SSVEP processing and returning to the primary BCI-system by using the standby-command (15 Hz) (see Table 1). Furthermore, the first level enables the direct access to several sub-menus in order to accelerate navigation.

Level 2: The sub-menu provides the user several buttons related to the different actions. The user can navigate through the available actions (13 and 16 Hz), execute the action by using the select-command (14 Hz), or quit this menu level (go back to level 1) (15 Hz).

To turn down the volume on the radio, the user has to produce the command sequence shown in Table 2. A two-level menu structure has great advantages, especially for devices with a large number of functionalities, e.g., the internet radio or the microwave or the robot arm. There is no need to have the same menu for the fridge application because only the door can be chosen for a control command. However, to avoid inconsistencies in the user interface, the same hierarchical structure has been applied to all considered devices.

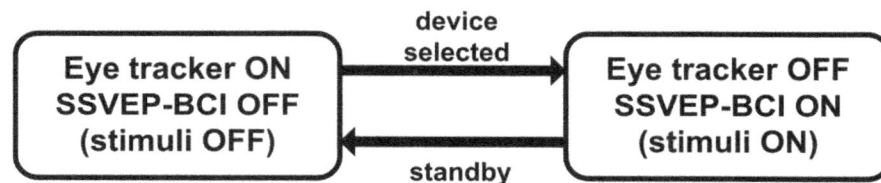

Figure 8. Hybrid Gaze-SSVEP BCI with sequential processing. The eye tracker acts as a selector which activates the SSVEP system.

Table 1. Quick access to submenus.

Device	"Left" (13 Hz)	"Select" (14 Hz)	"Exit/Stand-by" (15 Hz)	"Right" (16 Hz)
Internet radio	Volume	Channel	Mode	Menu
Fridge	Door	Door	Door	Door
Microwave	Timer	Timer	Operation	Operation

Table 2. Turn down the volume on the radio.

Number	Frequency	Action	Audit. Feedback	Command
1	13Hz	Volume (level 1)	Volume	-
2	14 Hz	Select (level 1)	Increase	-
3	13Hz	Left (level 2)	Decrease	-
4	14 Hz	Select (level 2)	Select	Turn Down
5	15 Hz	Exit (level 2)	Volume	-
6	15 Hz	Standby (level 1)	Standby	-

5.2. ERD/ERS-SSVEP Interface

The ERD/ERS acts in this mode as a selector which activates the SSVEP system. Figure 9 shows the principle operation. In the current study, sensorimotor rhythms related to the imagery of left hand, feet, and right hand movements are detected, thus providing three control commands.

Left and right hand imagery are used to browse available devices. The select command is carried out by feet imagery. At the beginning of each experiment the "Internet Radio" button is highlighted. This indicates the starting point for navigating through the device submenu. Figure 10 shows the operation of the hybrid ERD/ERS-BCI.

In order to, e.g., select the device "Microwave", the subject has to imagine the motion sequence right hand, right hand, feet. As the menu wraps at each end, the sequence left hand, feet gives the same result. After a selection command, the cursor automatically moves back to the initial position. "Internet Radio" button is chosen in this layout as initial position.

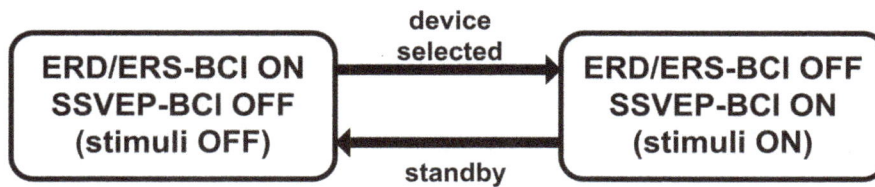

Figure 9. Hybrid ERD/ERS-SSVEP BCI with sequential processing. The ERD/ERS-BCI acts as a selector which activates the SSVEP system.

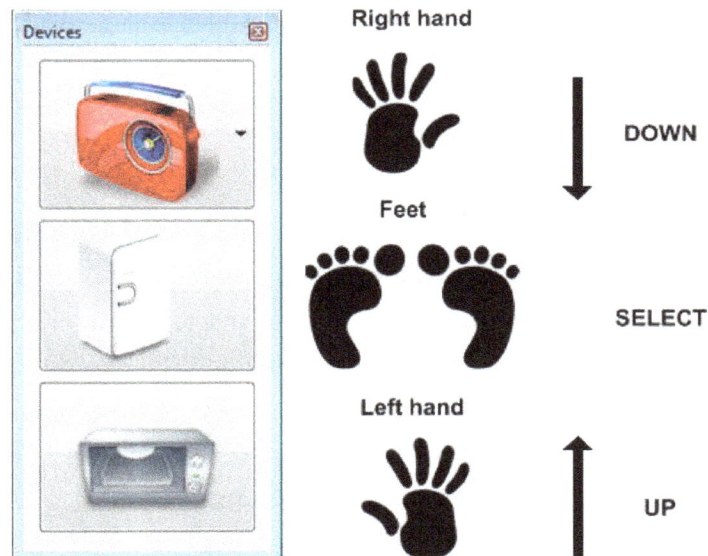

Figure 10. Multiclass self-paced ERD/ERS-BCI is designed to discriminate three MI-classes. Left/right hand and feet motor imagery are used to scroll down/up and to select a target device.

6. Subject and Data Acquisition and Results

A preliminary evaluation of the sBCI-SSVEP and a comparability test of the sBCI-ERD/ERS system is carried out in order to prove the feasibility of the concept and compare sBCI results with previous ones that use the same software packages. A total of six able-bodied subjects were recruited. All participants used the sBCI-SSVEP system. Five participants (subjects A–E, aged 30.8 ± 7.4; 3 female and 2 male) used the eye tracker system to select a target device. One participant for the comparability test (subject F, 31 years old, female) used the ERD/ERS-BCI as the primary system of the sBCI. Subject F has been trained regularly for 12 months on the ERD/ERS-BCI. She also attended the study published in [7] but was the only trained subject available at the time of the sBCI study. In all cases the SSVEP based BCI was used to control the selected device. All subjects had normal or corrected to normal vision. According to self-reports, none of the participants had a previous history of neurological and psychiatric diseases that may have influenced the experimental results. The subjects were sitting in a comfortable chair. The markers were placed on physical devices at a distance of about 3 m from the subject. 8 channels (Pz, PO3, PO4, PO7, Oz, PO8, O9, O10) were used for SSVEP detection, and 14 channels (FC3, FCz, FC4, C5, C3, C1, Cz, C2, C4, C6, CP3, CPz,CP4, Pz) covering the sensorimotor area were passed to the ERD/ERS signal processing modules. All channels were grounded to the AFz which was placed on the forehead. The impedance was kept below 5 kΩ. An EEG amplifierPorti32 (Twente Medical Systems International, Oldenzaal, The Netherlands) was used to capture the EEG data. The Porti32amplifier records unipolar inputs, configured as the reference amplifier, *i.e.*, all channels are amplified against the average of all connected inputs. The signals were high-pass filtered at 0.1 Hz and digitized with a sampling rate of 256 Hz. To reduce power line noise, a 50 Hz notch filter was applied. The hybrid sBCI system was realized using the BCI 2000 general software framework [39]. The signal processing module implemented the methodology presented in Sections 4.3.1 and 4.3.2 using the Matlab interface of BCI2000. The user interface (GUI) and gaze detection software were implemented as external modules in C++. The BCI2000 was configured to stream its output to external modules via UDP (User Datagram Protocol) in real time.

6.1. Experimental Protocol

Each experiment starts with an Eye tracker calibration or an ERD/ERS calibration.

(1) Eye tracker calibration: Eye tracker calibration is required to establish mapping between the eye cameras and the environmental camera for a particular user. During a short fixation interval, the system records the 2D pupil center and the corresponding 3D coordinates of the marker. The participants were asked to fixate the marker (placed *ca.* 3 m away) from nine different points of view. This procedure was repeated for two further distances (*ca.* 2.5 m and 2 m) in order to maintain accurate gaze detection even if the markers were not in the same plane.

(2) ERD/ERS calibration: Before the ERD/ERS system could be used to select a target device, the classifier needs to be trained in the participant's EEG patterns. During the calibration phase, a total number of 120 trials consisting of 40 randomly distributed trials for each motor imagery were conducted. The user was instructed to perform one of the three imagery movements (left Hand, right Hand, Foot-motion imagination) which were indicated on the screen. Motion imagination was expected for 4 s at an interval of 5 s. During this calibration session, the subject was not provided with any feedback.

After the calibration phase, five tasks covering all three devices (Table 3) were introduced to the subjects. Each subject was free to familiarize with the hybrid interface before the actual experiment started. Additionally, a short introduction to the SSVEP interaction technique was given.

Table 3. Tasks.

Number	Task
1	Internet radio: Switch the 5th channel
2	Internet radio: Increase volume by three steps
3	Fridge: Open and close door
4	Microwave: Choose cooking time 11:10 min and start
5	Microwave: Activate door release

Five subjects who used the eye tracker as a primary system were asked to select a target device by gazing at the marker. To evaluate the time needed for selection, users were required to gaze at each marker five times until the SSVEP stimuli started flickering. The time period between moving the head in the direction of the marker and triggering a selection event was measured. The ERD/ERS user imagined motions until the SSVEP stimuli started flickering and repeated this also five times. Each task starts with the selection of a device using either the eye tracker or an ERD/ERS-BCI and ends with turning off the SSVEP-LEDs.

6.2. Performance Measures

In this study the performance of the SSVEP-BCI was evaluated using the ITR (bit/min) [40] and accuracy (ACC) (%). The performance of the selection system (eye tracker or ERD/ERS) was obtained by measuring the time needed to select a specific device. The ITR was calculated based on the following formula:

$$B = \log_2 N + A\log_2 A + (1 - A)\log_2 \left(\frac{1 - A}{N - 1}\right) \tag{1}$$

In Equation (1), B represents the number of bits per trial, A represents the probability of correct classification and N is the number of choices. N is equal to 4, based on 4 LED for the control commands ("right", "select", "exit/standby", "left"). A is calculated as the number of correct classifications divided by the total number of classified commands. To obtain ITR in bits per minute, B is multiplied by the speed, which is the number of classified commands divided by time T. The time T was measured as the elapsed time from the moment of device selection until turning off the SSVEP-BCI.

6.3. Results

All six participants succeeded in performing the five requested tasks with good performance. None of them reported any pain or discomfort while wearing the sBCI-headset.

The performance of the SSVEP-BCI is presented in Table 4. All six subjects were able to use the hierarchical SSVEP-interface to complete the assigned tasks. For each task, two measures of performance were available: ACC_{SSVEP} and ITR_{SSVEP}. The information transfer rate was calculated according to Equation (1). The subjects achieved a mean ITR of 41.2 bit/min and a mean accuracy of 96.3%.

Table 5 shows the time that was needed to select a target device by subjects A–E who used the eye tracker.

Five participants who used eye tracker as a selection device gazed at each marker five times to activate the corresponding device. The values shown in Table 5a are the averaged values across these five selections. The best result was achieved by subject D who required only averaged 2.29 s to perform a selection. In contrast to this result, subject B needed 6.94 s on average. The most suitable reason for this difference was identified as a shortcoming of the image processing software. Excessive eye makeup of subject B seems to disturb the results of pupil detection.

Table 6 shows the results of the comparability test for the sBCI-ERD/ERS system.

Table 4. SSVEP-BCI: Information Transfer Rate (ITR)$_{SSVEP}$ (bit/min) and Accuracy (ACC)$_{SSVEP}$ (%) achieved during operation with the selected device.

Subject	Task 1		Task 2		Task 3		Task 4		Task 5		Mean	
Subject	ITR$_{SSVEP}$	ACC$_{SSVEP}$	ITR$_{SSVEP}$	ACC$_{SSVEP}$	ITR$_{SSVEP}$	ACC$_{SSVEP}$	ITR$_{SSVEP}$	ACC$_{SSVEP}$	ITR$_{SSVEP}$	ACC$_{SSVEP}$	ITR$_{SSVEP}$	ACC$_{SSVEP}$
A	55.4	100	52.1	100	58.4	100	25.8	89	73.9	100	53.1	98
B	34.4	100	33.1	88	41.5	100	25.2	93	52.2	100	37.3	96
C	69.7	100	64.0	100	64.0	100	40.9	100	64.9	100	60.7	100
D	9.4	82	17.4	100	34.3	100	7.5	91	9.2	88	15.6	92
E	57.9	100	33.3	88	54.2	100	32.4	93	64.9	100	48.6	96
F	29.5	100	23.6	100	43.9	100	28.9	90	34.2	89	32.0	96
Mean	42.7	97	37.3	96	49.4	100	26.8	92.7	49.9	96.2	41.2	96.3

Table 5. Selection time using eye tracker.

Subject	Time$_{microwave}$ (s)	Time$_{fridge}$ (s)	Time$_{radio}$ (s)	Time$_{mean}$ (s)
A	4.77	3.62	2.80	3.73
B	4.89	7.38	8.54	6.94
C	3.32	2.44	3.42	3.06
D	2.10	2.37	2.40	2.29
E	3.14	3.97	3.50	3.53
Mean	3.56	3.96	4.13	3.90

Table 6. Selection time using ERD/ERS-BCI.

Subject	Time$_{microwave}$ (s)	Time$_{fridge}$ (s)	Time$_{radio}$ (s)	Time$_{mean}$ (s)
F	36	10.5	8.75	20

Subject F who used ERD/ERS-BCI as selection system selected each device according to the five requested tasks: internet radio twice, fridge once, and microwave twice. Subject F spent an average of 20 s on the selection task. The average selection speed was 6.05 s per command. In the study published in [34], subject F achieved 8.81 s per command.

The results allow the assumption that the change to the new electrode cap induces no negative influence. The accuracy of the ERD/ERS-BCI for each assigned task achieved 100%.

The results suggest also that the combination of eye-tracker and SSVEP-BCI is the preferable solution. Readers should remember, however, that this can finally be judged only in relation to the capabilities of the disabled user. First trials with the eye tracker in an environment with varying illumination show a decrease in accuracy. The recognition time may increase and reliability may decrease further if the eye tracker is used with real objects and SIFT as recognition methods instead of ARToolkit markers. ERD/ERS-BCI as a self-paced BCI is much slower than the eye tracker but is important to switch the whole system or components on and off. In a forthcoming study with more participants who also will be trained for motion imagination, statistically sound results will be researched.

7. Discussion and Conclusions

A multimodal, hybrid BCI is designed which combines an eye tracker, an SSVEP-BCI, and a multiclass ERD/ERS-based BCI and offers the possibility for a detailed study in which different combinations of the system are researched and evaluated in relation to the disability of the user. For easy evaluation of this hybrid system, a multimodal sBCI-headset was designed. The proposed multimodal BCI system was used to control three devices that play an important role in future ADL application. The benefits of using two BCI modalities include the possibility to activate the eye- tracker and SSVEP-based control only on demand, i.e., both can independently be turned off during inactive periods. Thus, the hybrid setup of the system minimizes the number of involuntary selections and increases the convenience of the whole interface.

The multimodal sBCI-headset is a sensing system which integrates multi-channel EEG equipment, an eye tracking system and a visual stimulator for the SSVEP-BCI. During the designing phase of the headset, all effort has been made to optimize the long term wearing comfort, while maintaining the ergonomic and aesthetic appearance and also the quality of EEG-signals. Despite the substantial investment of time and resources, it was not possible to successfully develop a one-size helmet that fits onto any adult's head. All one-size prototypes of the sBCI-headset have failed the long term comfort tests. Consequently, the sBCI-headset is provided in three different sizes based on head circumference (Small: 56 cm, Medium: 58 cm, Large: 60 cm) in order to fit the head to the majority of adult users. Small distances up to 5 mm between the hard case cap and the skin were easily bridged by the soft springs of the electrode holders. Using such holders yields a double benefit. The wearing comfort is increased and the electrode-skin coupling enhanced.

A preliminary test with six able-bodied volunteers using the newly designed sBCI-headset was performed. Two fusion techniques were evaluated: Gaze-SSVEP and an ERD/ERS-SSVEP, called a physiological and pure interface [41], respectively. The performance measurements show that the sBCI system provides an effective environmental control method for all six subjects. The two fusion techniques are compared with a limited data set only. The eye tracker as the selection device is in the set up obviously much faster than an ERD/ERS-BCI (3.9 s *vs.* 20 s) and achieves a high accuracy. However, the accuracy of the device selection is until now only tested with ARToolKit markers. While there is a limited accuracy of the ERD/ERS-BCI the probability to select an undesired device with an ARToolKit marker was close to zero. But usage of markers requires preparation of the users' environment and limits the usage to such a prepared environment. To avoid markers, they will be replaced by object recognition based on SIFT features. That may decrease recognition speed and recognition accuracy especially if no constant illumination can be guaranteed. A statistically sound comparison of all features and possible combinations is in preparation.

The participants instantaneously accepted the multimodal BCI system based on eye gaze and SSVEP-BCI because both systems require little to no training. The ERD/ERS-based BCI system which requires training to learn operation via motor imagery could only be used by one trained user. The results can therefore not be compared with each other. However, as mentioned before, the ERD/ERS system is necessary to switch the eye-tracker and SSVEP-BCI on/off by disabled users without the need for additional support. The full integration of sBCI and the usage of the full potential of the new hardware requires a careful design and integration of the control strategy in an ambient assistant environment. The user should be able to operate all devices intuitively and without remembering specific sequences for each object.

One of our future pieces of work will focus on further improvement of the eye tracker hardware components, primarily the cameras. With the cameras used presently, the determination of the focus point is not precise enough to distinguish between two objects that are close to the line of sight but at different distances. The interaction technique based on dwell time can benefit from high resolution images, while reducing the selection time. We are also working on integration of fast SIFT object recognition method [37] and an easy to train object database in order to overcome the need of artificial markers.

Additionally, we will use sBCI to research optimal control modes for complex devices, to determine optimal blinking frequency and duty cycle, and research whether the measurement of error potential improves the robustness and how it can be used together with the eye tracker.

Interested research institutes may acquire the system (one cap, electrode holder, SSVEP diodes incl. control unit and eye tracker incl. cameras but without EEG electrodes, EEG amplifier and control PC) from a vendor for a budget price of approximately €8,000.

Acknowledgments

The authors especially thank Riad Hamadmad (evado design for business, Bremen, Germany) for designing and assembling the sBCI-headset. Our thanks go also to the Annastift Hospital (Department of Paediatric Orthopaedics and Neuro-Orthopaedics) in Hannover, Germany, especially to Hannelore Willenborg, for the opportunity to generate 3D-models of adult heads. We thank also the AiF (Arbeitsgemeinschaft industieller Forschungsvereinigungen) for providing the grant for research project (16136BG).

Author Contributions

Tobias Tetzel, Ulrich Krebs designed and developed sBCI hardware and developed image processing for eye tracker. Tatsiana Malechka, Diana Valbuena conceived, designed and performed the experiments and analyzed the data. Axel Graeser wrote the research proposal and supervised the research. Tatsiana Malechka, Tobias Tetzel and Axel Graeser wrote the paper.

Conflicts of Interest

The authors declare no conflict of interest.

References

1.   Gräser, A.; Heyer, T.; Fotoohi, L.; Lange, U.; Kampe, H.; Enjarini, B.; Heyer, S.; Fragkopoulos, C.; Ristić-Durrant, D. A Supportive FRIEND at work: Robotic workplace assistance for the disabled. *Robot. Autom. Mag.* **2013**, *20*, 148–159.

2. Gräser, A. *Assistenzsysteme zur Unterstützung Behinderter Personen, Automatisierungstechnische Praxis (atp)*; Oldenbourg Verlag: Oldenbourg, Germany, 2000; Volume 42, pp. 43–49. (In German)
3. Grigorescu, S.M.; Lüth, T.; Fragkopoulos, C.; Cyriacks, M.; Gräser, A. A BCI-controlled robotic assistant for quadriplegic people in domestic and professional life. *Robotica* **2011**, *30*, 419–431.
4. Friman, O.; Volosyak, I.; Gräser, A. Multiple channel detection of steady-state visual evoked potentials for brain-computer interfaces. *IEEE Trans. Biomed. Eng.* **2007**, *54*, 742–750.
5. Volosyak, I.; Cecotti, H.; Valbuena, D.; Gräser, A. Evaluation of the Bremen SSVEP based BCI in real world conditions. In Proceedings of the 11th International Conference on Rehabilitation Robotics, Kyoto, Japan, 23–26 June 2009; pp. 322–331.
6. Volosyak, I.; Valbuena, D.; Luth, T.; Graeser, A. Towards an SSVEP Based BCI With High ITR. Available online: http://elib.suub.uni-bremen.de/edocs/00102056-1.pdf (accessed on 25 February 2015).
7. Kus, R.; Valbuena, D.; Zygierewicz, J.; Malechka, T.; Gräser, A.; Durka, P. Asynchronous BCI based on motor imagery with automated calibration and neurofeedback training. *IEEE Trans. Neural Syst. Rehabil. Eng.* **2012**, *1*, 1–13.
8. Wang, Y.; Wang, R.; Gao, X.; Hong, B.; Gao, S. A practical vepbased brain-computer interface. *IEEE Trans. Neural Syst. Rehabil. Eng.* **2006**, *14*, 234–239.
9. Volosyak, I.; Valbuena, D.; Lüth, T.; Malechka, T.; Gräser, A. BCI Demographics II: How many (and what kinds of) people can use an SSVEP BCI? *IEEE Trans. Neural Syst. Rehabil. Eng.* **2011**, *19*, 232–239.
10. Sellers, E.W.; Krusienski, D.J.; McFarland, D.J.; Vaughan, T.M.; Wolpaw, J.R. A P300 event-related potential brain-computer interface (BCI): The effects of matrix size and inter stimulus interval on performance. *Biol. Psychol.* **2006**, *73*, 242–252.
11. Pfurtscheller, G.; da Silva, F.H.L. Event-related EEG/MEG synchronization and desynchronization: Basic principles. *Clin. Neurophysiol.* **1999**, *110*, 1842–1857.
12. Neuper, C.; Pfurtscheller, G. Event-related dynamics of cortical rhythms: Frequency-specific features and functional correlates. *Int. J. Psychophysiol.* **2001**, *43*, 41–58.
13. Shenoy, P.; Krauledat, M.; Blankertz, B.; Rao, R.; Müller, K. Towards adaptive classification for BCI. *J. Neural Eng.* **2006**, *3*, 13–23.
14. Vidaurre, C.; Scherer, R.; Cabeza, R.; Schlögl, A.; Pfurtscheller, G. Study of discriminant analysis applied to motor imagery bipolar data. *Med. Biol. Eng. Comput.* **2007**, *45*, 61–68.
15. Pastor, M.A.; Artieda, J.; Arbizu, J.; Valencia, M.; Masdeu, J.C. Human cerebral activation during steady-state visual-evoked responses. *J. Neurosci.* **2003**, *23*, 11621–11627.
16. Pfurtscheller, G.; Allison, B.Z.; Brunner, C.; Bauernfeind, G.; Escalante, T.S.; Scherer, R.; Zander, T.O.; Mueller-Putz, G.; Neuper, C.; Birbaumer, N. The hybrid BCI. *Front. Neurosci.* **2010**, *2*, 1–11.
17. Brunner, C.; Allison, B.Z.; Krusienski, D.J.; Kaiser, V.; Putz, G.R.M.; Pfurtscheller, G.; Neuper, C. Improved signal processing approaches in an offline simulation of a hybrid brain-computer interface. *J. Neurosci. Methods* **2010**, *188*, 165–173.
18. Allison, B.Z.; Brunner, C.; Kaiser, V.; Müller-Putz, G.R.; Neuper, C.; Pfurtscheller, G. Toward a hybrid brain-computer interface based on imagined movement and visual attention. *J. Neural Eng.* **2010**, *7*, 026007.

19. Ortner, R.; Allison, B.; Korisek, G.; Gaggl, H.; Pfurtscheller, G. An SSVEP BCI to control a hand orthosis for persons with tetraplegia. *IEEE Trans. Neural Syst. Rehabil. Eng.* **2010**, *19*, 1–5.

20. Pfurtscheller, G.; Solis-Escalante, T.; Ortner, R.; Linortner, P. Selfpaced operation of an SSVEP-based orthosis with and without an imagery-based "brain switch": A feasibility study towards a hybrid BCI. *IEEE Trans. Neural Syst. Rehabil. Eng.* **2010**, *18*, 409–414.

21. Ferrez, P.W.; del Millan, J. Error-related EEG potentials generated during simulated brain-computer interaction. *IEEE Trans. Biomed. Eng.* **2008**, *55*, 923–929.

22. Scherer, R.; Müller-Putz, G.R.; Pfurtscheller, G. Self-initiation of EEG-based brain-computer communication using the heart rate response. *J. Neural Eng.* **2007**, *4*, L23–L29.

23. Leeb, R.; Sagha, H.; Chavarriaga, R.; Millan, J.D.R. A hybrid brain-computer interface based on the fusion of electroencephalographic and electromyographic activities. *J. Neural Eng.* **2011**, *8*, 025011.

24. Vilimek, R.; Zander, T.O. BC(eye): Combining Eye-Gaze Input with Brain-Computer Interaction. In *Universal Access in Human-Computer Interaction. Intelligent and Ubiquitous Interaction Environments*; Lecture Notes in Computer Science Volume 5615; Springer: Berlin, Germany, 2009; pp. 593–602.

25. Zander, T.; Gaertner, M.; Kothe, C.; Vilimek, R. Combining eye gaze input with a brain-computer interface for touchless human-computer interaction. *Int. J. Hum. Comput. Interact.* **2011**, 27, 38–51.

26. Liao, L.; Chen, C.; Wang, I.; Chen, S.; Li, S.; Chen, B.; Chang, J.; Lin, C. Gaming control using a wearable and wireless EEG-based brain-computer interface device with novel dry foam-based sensors. *J. NeuroEng. Rehabil.* **2012**, *9*, 1–5.

27. Dias, N.; Carmo, J.; da Silva, A.F.; Mendes, P.; Correia, J. New dry electrodes based on iridium oxide (IrO) for non-invasive biopotential recordings and stimulation. *Sens. Actuators A Phys.* **2010**, *164*, 28–34.

28. Ruffini, G.; Dunne, S.; Farr'es, E.; Cester, I.; Watts, P.; Ravi, S.; Silva, P.; Grau, C.; Fuentemilla, L.; Marco-Pallares, J.; *et al.* Enobio dry electrophysiology electrode; first human trial plus wireless electrode system. In Proceeding of 29th Annual International Conference of the IEEE Engineering in Medicine and Biology Society, Lyon, France, 22–26 August 2007; pp. 6689–6693.

29. Volosyak, I.; Valbuena, D.; Malechka, T.; Peuscher, J.; Gräser, A. Brain-computer interface using water-based electrodes. *J. Neural Eng.* **2010**, *7*, 066007.

30. Cester, I.; Dunne, S.; Riera, A.; Ruffini, G. ENOBIO: Wearable, wireless, 4-channel electrophysiology recording system optimized for dry electrodes. In Proceedings of the Health Conference, Valencia, Spain, 21–23 May 2008.

31. Ranky, G.; Adamovich, S. Analysis of a commercial EEG device for the control of a robot arm. In Proceedings of the IEEE 36th Annual Northeast Bioengineering Conference, New York, NY, USA, 26–28 March 2010.

32. Singh, H.; Singh, J. Human eye tracking and related issues: A review. *Int. J. Sci. Res. Publ.* **2012**, *2*, 1–9.

33. ARToolKit. Available online: http://www.hitl.washington.edu/artoolkit/ (accessed on 25 February 2015).

34. Alhwarin, F. *Fast and Robust Image Feature Matching Methods for Computer Vision Applications*; Shaker Verlag: Aachen, Germany, 2011.

35. Alhwarin, F.; Ristic-Durrant, D.; Graeser, A. VF-SIFT: very fast sift feature matching. *Lect. Notes Comput. Sci.* **2010**, *6376*, 222–231.

36. Kato, H.; Billinghurst, M. Marker tracking and HMD calibration for a video-based augmented reality conferencing system. In Proceedings of 2nd IEEE and ACM International Workshop on the Augmented Reality, San Francisco, CA, USA, 20–21 October **1999**; pp. 85–94.

37. Dornhege, G.; Blankertz, B.; Curio, G.; Muller, K.-R. Boosting bit rates in noninvasive EEG single-trial classifications by feature combination and multiclass paradigms. *IEEE Trans. Biomed. Eng.* **2004**, *51*, 993–1002.

38. Fotoohi, L.A. Graeser: Building a safe care-providing robot. In Proceedings of the 12th IEEE International Conference on Rehabilitation Robotics-ICORR'11, Zurich, Switzerland, 29 June–1 July 2011.

39. Schalk, G.; McFarland, D.; Hinterberger, T.; Birbaumer, N.; Wolpaw, J. BCI2000: A general-purpose brain-computer interface (BCI) system. *IEEE Trans. Biomed. Eng.* **2004**, *51*, 1034–1043.

40. Wolpaw, J.; Birbaumer, N.; McFarland, D.; Pfurtscheller, G.; Vaughan, T. Brain-computer interfaces for communication and control. *Clin. Neurophysiol.* **2002**, *113*, 767–791.

41. Allison, B.; Leeb, R.; Brunner, C.; Müller-Putz, G.; Bauernfeind, G.; Kelly, J.; Neuper, C. Toward smarter BCIs: Extending BCIs through hybridization and intelligent control. *J. Neural Eng.* **2012**, *9*, 013001.

13

Performance of SU-8 Membrane Suitable for Deep X-Ray Grayscale Lithography

Harutaka Mekaru

Research Center for Ubiquitous MEMS and Micro Engineering (UMEMSME), National Institute of Advanced Industrial Science and Technology (AIST), 1-2-1 Namiki, Tsukuba, Ibaraki 305-8564, Japan; E-Mail: h-mekaru@aist.go.jp

Academic Editor: Arnaud Bertsch

Abstract: In combination with tapered-trench-etching of Si and SU-8 photoresist, a grayscale mask for deep X-ray lithography was fabricated and passed a 10-times-exposure test. The performance of the X-ray grayscale mask was evaluated using the TERAS synchrotron radiation facility at the National Institute of Advanced Industrial Science and Technology (AIST). Although the SU-8 before photo-curing has been evaluated as a negative-tone photoresist for ultraviolet (UV) and X-ray lithographies, the characteristic of the SU-8 after photo-curing has not been investigated. A polymethyl methacrylate (PMMA) sheet was irradiated by a synchrotron radiation through an X-ray mask, and relationships between the dose energy and exposure depth, and between the dose energy and dimensional transition, were investigated. Using such a technique, the shape of a 26-μm-high Si absorber was transformed into the shape of a PMMA microneedle with a height of 76 μm, and done with a high contrast. Although during the fabrication process of the X-ray mask a 100-μm-pattern-pitch (by design) was enlarged to 120 μm. However, with an increase in an integrated dose energy this number decreased to 99 μm. These results show that the X-ray grayscale mask has many practical applications. In this paper, the author reports on the evaluation results of SU-8 when used as a membrane material for an X-ray mask.

Keywords: SU-8; X-ray lithography; X-ray mask; grayscale; tapered trench etching; PMMA; synchrotron radiation

1. Introduction

In a diffraction phenomenon where a light ray can bend around the edges of its target and reach to its backside, a bokeh of the size near a wavelength is often observed. Therefore, in optical transfer of a fine pattern, it is necessary to use lights with short wavelengths that relate to the pattern size so that this bending of light by diffraction does not occur. In semiconductor device manufacturing, the light source for lithography has been reducing with time from visible lights to ultraviolet (UV) lights, and down to soft X-rays [1,2]. Also, when exposing deeper into a material to produce a thick structure, such as in microelectro-mechanical-systems (MEMS), X-rays with short wavelengths have been successfully employed [3,4]. For a powerful light that can penetrate straight into a photoresist film, the X-rays from a synchrotron radiation (SR) have been used. A LIGA (lithographie, galvanoformung and abformung) [5] process was proposed as a mass-producing technology that can be a cost-effective way of combining the high-cost X-ray lithography with a relatively less-expensive electroforming and replication technologies. As for the photoresists to be used for the X-ray lithography, there is polymethyl methacrylate (PMMA) as a positive-tone photoresist [5] and there is SU-8 (MicroChem, Newton, MA, USA) as a negative-tone photoresist [6]. An X-ray mask wields significant influence on the transfer accuracy of X-ray lithography. The process comprises a membrane through which the X-rays can pass efficiently, and an absorber, which can serve as an opaque barrier to the X-rays. To serve as a membrane of an X-ray mask, various materials have been used in many cases. Those membrane materials are required to have the following characteristics [7]:

(1) High transmittance in the X-ray energy region;
(2) Dimensional stability during the X-ray exposure;
(3) Durability for an extended X-ray exposure time;
(4) Sufficient mechanical strength serving as a self-supporting membrane;
(5) Simple film-forming method and high compatibility with other processes.

Table 1 shows several kinds of conventional X-ray masks. Stainless steel [8] and Si [9] stencil X-ray masks being independent of any film size cannot form isolated patterns because these masks have no supporting membranes. In a large-size X-ray mask, that comprises polyimide (Kapton) [10] or polyester (Mylar) [11] membranes, the adhesion forces between the membrane and absorbers happen to be weak because only the frame of the mask holds these films. Moreover, X-ray mask has been developed by an electroplating of Au on a SU-8 film previously spin-coated on the surface of a PMMA photoresist [12]. However, there exists an unsubstantiated notion that X-ray masks are not suitable for multiple uses. Although there are inorganic materials, such as Si [13], SiC [14], and SiN [15], that are used on masks, and they are excellent in their dimensional stability and chemical resistance, but forming a thin layer on them is relatively a difficult task. In addition, an X-ray mask with a graphite membrane, of which the thermal resistance is same as that of an inorganic material, has also been made [16]. In the case of absorber materials, an Au [17] absorber, which can easily be manufactured by electroplating, has been employed by many researchers. Materials like Ta [18] and W [19], on which reactive ion etching is possible, have also been examined. Moreover, TaGeN [20], TaBN [20], *etc.*, have been used to prevent the worsening the transfer accuracy caused by the oxidization of absorbers. However, the X-ray masks mentioned above can be used to form photoresist structures that can be only rectangular in shape with vertical sidewalls. We fabricated a three-dimensional Si absorber employing a tapered-trench-etching technique using a mixed gas of SF_6, C_4F_8, and O_2; and we had also developed a stencil X-ray mask [21] and a SU-8 membrane X-ray mask [22]. Furthermore, by combining these X-ray masks with a LIGA process, a micro-needle array was formed [23]. SU-8 was designed as a negative-tone photoresist for a permanent film based on an epoxy resin so that photoresist structures are stable chemically and thermally. Therefore, the author believed such special quality could be suited for a membrane material of X-ray masks used under a harsh environment caused by X-ray's irradiation. Although 3-dimensional Si absorbers have been fully described in these reports, there has not been any report published on the evaluation of SU-8 when used as a membrane material. And moreover, the performance of the SU-8 as a photoresist (before photo-curing) has been reported as regard to its exposure to electron beam [24] and proton beam [25], and not to forget UV lights and X-rays. However, any evaluation of the usefulness of photo-cured SU-8 as a component of any incorporated device has hardly been reported. This paper reports on the evaluation results of SU-8 as regard to three characteristics namely transmittance, dimensional stability, and durability that are required of a membrane of any X-ray mask. This information, that shows excellent characteristics of SU-8 in the X-ray energy region, is expected to be quite valuable to the researchers in this field.

Table 1. Materials of X-ray masks.

Category	Absorber	Membrane	Coefficient of Thermal Expansion on Membrane
Stencil	Stainless steel	None	-
	Si	None	-
Polymer membrane	Au	Kapton (Polyimide)	20×10^{-6}/K [26]
		Mylar (Polyester)	17×10^{-6}/K [27]
Built-on	Au	SU-8	52×10^{-6}/K [28,29]
SiX membrane	Au, Ta, W	Si	2.6×10^{-6}/K [30]
		SiNx	3.3×10^{-6}/K [31]
		SiC	3.8×10^{-6}/K [32]
Oxidation inhibiting	TaGeN, TaBN	SiC	3.8×10^{-6}/K [32]
Others	Au	Graphite	3.8×10^{-6}/K [33]
Grayscale	Si	None	-
(Our product)	Si	SU-8	52×10^{-6}/K [28,29]

2. Experimental Section

2.1. SU-8 Built-In X-Ray Mask

When designing an X-ray mask, it is necessary to understand the environment in which it is going to be used. An X-ray lithography experiment was carried out at the TERAS [34] SR facility of the National Institute of Advanced Industrial Science and Technology (AIST). Figure 1 shows a calculation result of the output beam spectrum of a beamline BL-4 [35], and of the transmittance of a membrane and of an absorber in a designed X-ray mask. The membrane is made of a 30-μm-thick SU-8, and the absorber consists of a Si film with a maximum thickness of 25 μm and an additional 5-μm-thick SU-8. The spectrum was read of the output beam that was transmitted through a 50-μm-thick Be window and then allowing the beam to pass through a 70-mm long passage of a flowing stream of the He gas (pressure: 1 atm). The beam was generated by an SR that comprised a 10-m-curvature-radius-bending magnet inserted into an electron storage ring of TERAS facility. The optical arrangement of the beamline BL-4 is mentioned later in this paper. In the output beam from the beamline BL-4, when photon energy nears 2 keV, a photon flux appears and serves as a peak. The transmittance of the SU-8 membrane at 2 keV is approximately 30%, and the X-ray hardly penetrates the Si absorber (SU-8 of 5-μm thickness being included). Moreover, at near 4 keV, the transmittance of SU-8 membrane is 75%, and the value of the absorber is only 5%; and it thus amounts to one-fifteenth of the membrane's value. At the photon energy of 6 keV, although the difference between membrane and absorber is decreased, the contrast ratio of "membrane:absorber" remains 2:1.

Figure 2 shows a process flow of an X-ray grayscale mask's manufacturing. A silicon on insulator (SOI) wafer that consists of a 30-μm-thick active Si layer, a 1-μm-thick SiO_2 layer, and a 525-μm-thick Si substrate, was prepared, and a 3-μm-thick positive-tone photoresist S1830 (Shipley, Kaysville, UT, USA) was spin-coated on the active Si layer (Figure 2a). Cr patterns on a quartz reticle were then transferred to the S1830 photoresist by a UV stepper 1500 MVS R-PC system (Ultratech, San Jose, CA, USA). And the resist was then developed (Figure 2b,c). After the development, there appeared a circular dotted pattern 50 μm in diameter and a square dotted pattern 50 μm in width. Those patterns then served as a masking layer for the next step where the active Si was tapered-trench-etched using a reactive-ion-etching (RIE) system Alcatel 601E system (Alcatel Vacuum Technology, Annecy, France) (Figure 2d). In a tapered-trench-etching technique, an inductively coupled plasma (ICP) source has been used, and the shape of the Si absorbers was controlled by a mixed gas pressure with SF_6, C_4F_8, and O_2 gas, and by the etching time. The optimization conditions have been described in References [21,22]. In this experiment, the mixed gas pressure and the etching time were more than 10 Pa and 11 min, respectively.

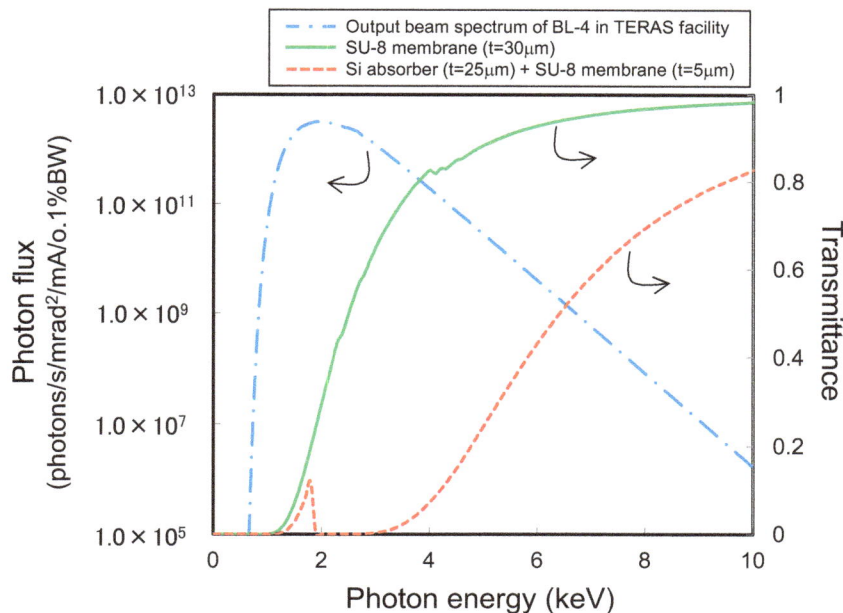

Figure 1. Calculated output beam spectrum of beamline BL-4 at TERAS synchrotron radiation facility and transmittance of SU-8 membrane (30 μm thickness) and Si (25 μm thickness) absorber on SU-8 film (5 μm thickness) depended on photon energy. "t" in graph legend in the figure means thickness.

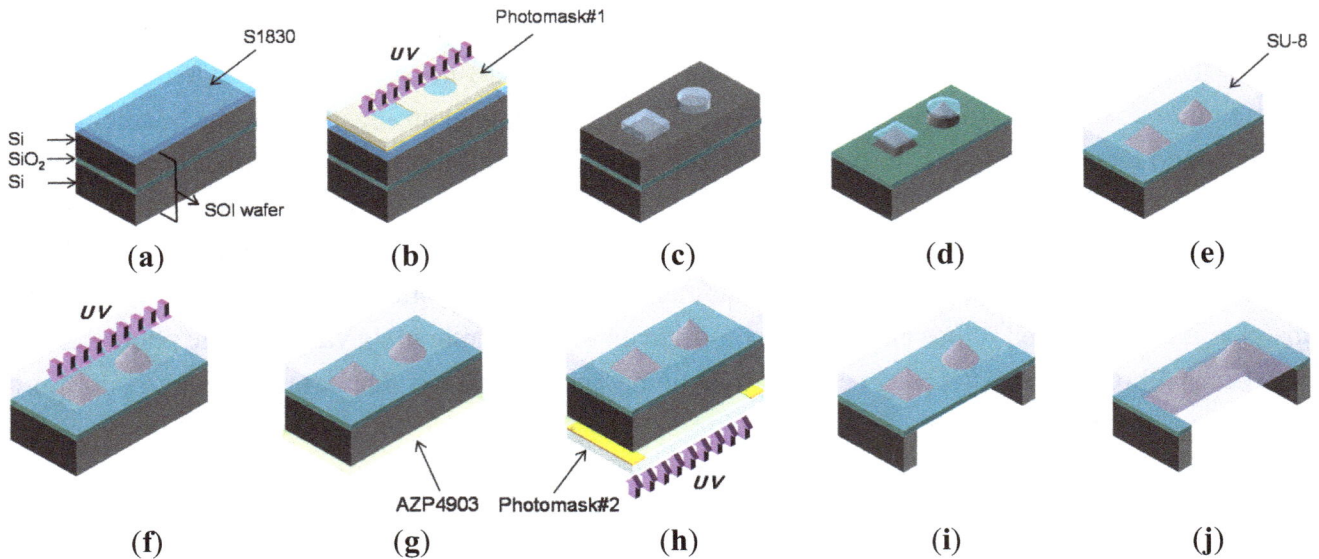

Figure 2. Process flow to fabricate X-ray mask composed with SU-8 membrane and Si absorber for grayscale lithography. Illustrations show the cross-sectional structure of the X-ray grayscale mask. (**a**) 1st spin-coating, (**b**) 1st UV exposure, (**c**) 1st development, (**d**) taper-RIE of Si, (**e**) removal of photoresist and 2nd spin-coating, (**f**) 2nd UV exposure and post-baking, (**g**) 3rd spin-coating, (**h**) 3rd UV exposure, (**i**) 2nd development and deep RIE of Si, (**j**) RIE of SiO_2.

Figure 3a,b show the scanning-electron-microscope (SEM) images of cone- and pyramid-shaped Si structures before the removal of the photoresist layer. The sidewall of the Si structures under the masking layer was inclined, and it can be confirmed that the tapered-trench-etching was successful. Figure 3c,d show Si absorbers after the masking layer being removed. The tips of cone- and pyramid-shaped Si structures appeared sharp and firm, and it was checked that shape of the fabricated tip was sharp. After the remaining S1830 masking layer was removed by acetone, the Si structures on the SiO_2 layer were covered by a spin-coated 30-μm-thick SU-8 25 photoresist (MicroChem) (Figure 2e). While keeping a sufficient relaxation time to ease an internal stress, a pre-baking at 95 °C for 10 min, UV irradiation for 1 min using a contact aligner PEM-800 (Union Optical, Tokyo, Japan), a post-baking at 95 °C for 10 min, and another relaxation over night were performed (Figure 2f). After the spin-coating of a 16-μm-thick photoresist AZ4903 (AZ Electronic Materials, Luxembourg, Luxembourg) on the backside of the SOI wafer (Figure 2g), a frame pattern was transferred to AZP4903 photoresist by the same contact aligner (Figure 2h). By a deep RIE using the AZP4903 masking layer and an RIE system MUC-2 (Sumitomo Precision Products, Amagasaki, Japan), the 525-μm-thick Si substrate was etched completely (Figure 2i), and the SiO_2 layer was also successfully etched using another RIE system Model RIE-10NRS (Samco, Kyoto, Japan) without breaking the SU-8 membrane (Figure 2j). In the final step, the masking layer that remained on the Si frame was removed by O_2 plasma ashing, and an X-ray grayscale mask was thus completed. The details on the process conditions are summarized in Table 2. Figure 4a shows a photograph of the completed X-ray grayscale mask. The square-shaped X-ray mask of 25 mm^2 area is surrounded by a 3-μm-wide Si frame. From this photograph, it was checked that the SU-8 membrane has good transmittance in the visible-light region.

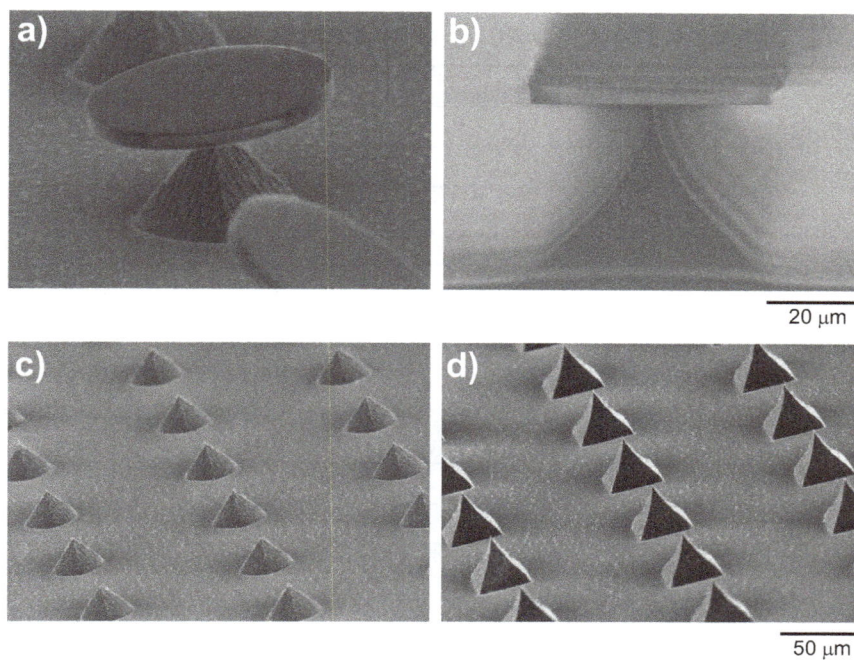

Figure 3. Scanning-electron-microscope (SEM) images of 3-dimensional Si absorbers before (**a**) cone-shaped and (**b**) pyramid-shaped; and after (**c**) cone-shaped and (**d**) pyramid-shaped the removal of S1830 photoresist.

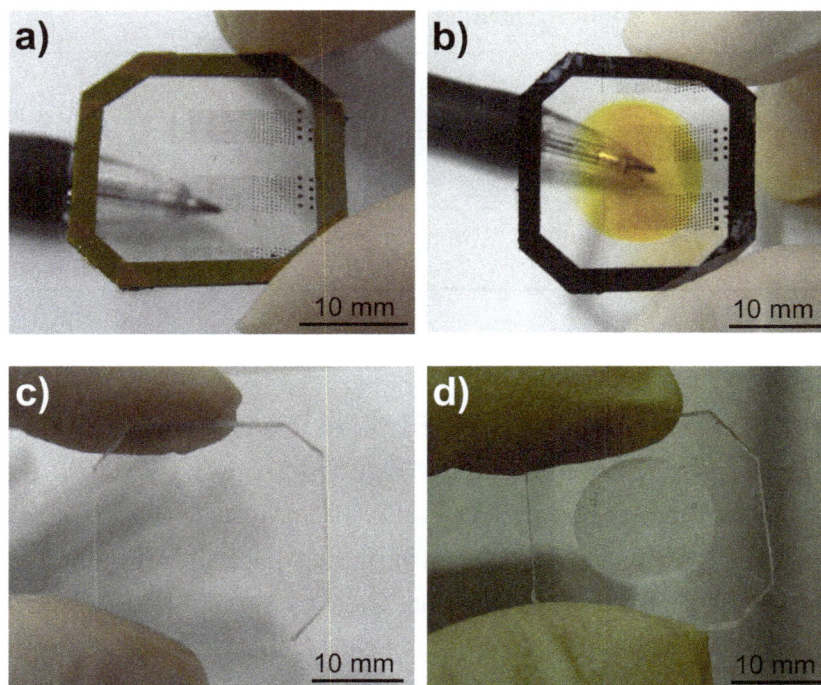

Figure 4. Photographs of X-ray mask (**a**) before; and (**b**) after X-ray irradiation; and of PMMA sheet (**c**) before; and (**d**) after X-ray exposure and development.

Table 2. Process conditions to fabricate X-ray mask.

Process	Material	Parameter	Condition
1st Spin-coating	S1830	thickness	3 µm
1st UV exposure		time	15 s
Pre-baking		temperature	120 °C
		time	20 min
1st Development	NF-319	temperature	Room Temperature
		time	5 min
Tapered trench etching		gas	$SF_6 + C_4F_8 + O_2$
		pressure	3.7–9.5 Pa
		time	11 s
Removal (+Ultrasonic)	Acetone	time	10 min
2nd Spin-coating	SU-8 25	thickness	30 µm
Pre-baking		temperature	95 °C
		time	10 min
2nd UV exposure		time	1 min
Post-baking		temperature	95 °C
		time	10 min
3rd Spin-coating	AZP4903	thickness	16 µm
3rd UV exposure		time	35 s
2nd Development	AZ400k + Distilled water (1:3)	temperature	Room Temperature
		time	5 min
Deep RIE of Si		gas	$SF_6 + C_4F_8$
		time	150 min
RIE of SiO_2		gas	CHF_3
		time	50 min
Ashing		gas	O_2
		time	10 min

2.2. X-Ray Exposure and Development

Figure 5 shows an end station of the beamline BL-4. In this experiment, 1-mm-thick PMMA sheet was chosen as a photoresist for X-ray lithography. Figure 4c shows a PMMA sheet before the X-ray exposure. In keeping up with the requirements of an X-ray grayscale mask's size, the four corners of a 25-µm-square PMMA sheet were cut out. As shown in the right half of Figure 5a, the X-ray grayscale mask and the PMMA sheet were fixed on an exposure stage by two spring-loaded plates so that the SU-8 membrane could make contact with the surface of the PMMA sheet. The exposure stage with a diameter of 70 mm was inserted at the end of the beamline BL-4.

Figure 5b shows an optical arrangement of the beamline BL-4. The synchrotron radiation, which was generated from the bending magnet is led to an exposure chamber through a Be window with a thickness of 50 µm arranged at a distance of 10 m from the bending magnet. In an exposure chamber, the air was replaced by a flowing stream of He gas through the chamber, and the exposure stage was set from the Be window at a distance of 70 mm. The X-ray exposure energy was estimated and controlled as a dose energy (mA·h), which is a product of a storage ring current and X-ray irradiation time. At the TERAS

SR facility, after an electron is accelerated to 750 MeV and accumulated until the storage ring current reaches to 270 mA, it is then reduced to 50 mA during an operation for a day. The X-ray irradiation experiment was conducted over several days using the system to its full capacity during its operation time of 5–6 h, a day. The PMMA sheet used after the X-ray exposure, was developed by a GG developer consisting of diethyleneglycol-monobutylether (60 vol %), morpholin (20 vol %), ethanol-amine (5 vol %) at 25 °C for 18 h, and distilled water (15 vol %); and then was washed by a distilled water. The PMMA sheet after the X-ray exposure and development is shown in Figure 4d. The irradiated area by the X-rays can be clearly observed as a 15-mm-diameter circle. In the next step, the completed PMMA structure and the X-ray mask, after the X-ray irradiation was evaluated using SEM and an optical microscope.

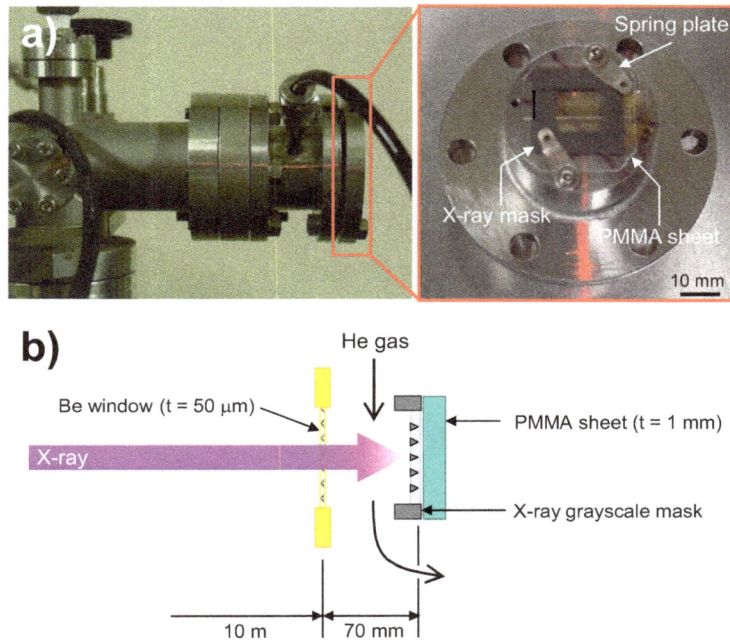

Figure 5. (a) Photographs of the end station of the beamline BL-4 (left) and exposure stage (right); **(b)** Cross-sectional schematic view of the end station of the beamline BL-4 during the X-ray exposure.

3. Results and Discussion

3.1. Transmission Property of SU-8 in the X-Ray Energy Region

The exposure depths on the PMMA by the output beam from the beamline BL-4 in both cases of the absence and presence of the 30-μm-thick SU-8 membrane are compared in Figure 6. A curved line in the figure indicates an approximation. The difference between both cases lies in their maximum values when the dose energy is 100 mA·h. The exposure depth in the absence and presence of the SU-8 membrane was measured as 70 and 35 μm, respectively; and specifically the ratio was estimated as 2:1. As shown in Figure 1, in the photon energy range of 2–6 keV, the transmittance of SU-8 of 30-μm thickness increases gradually from 20% to 95%. Because a peak of the photon flux of the output beam from the beamline BL-4 shows up at near 2 keV and the transmittance of SU-8 at near 2 keV is 40% and more, the result, that the exposure depth in the presence of the SU-8 membrane is roughly half compared to the value in the absence of the SU-8 membrane, is quite understandable.

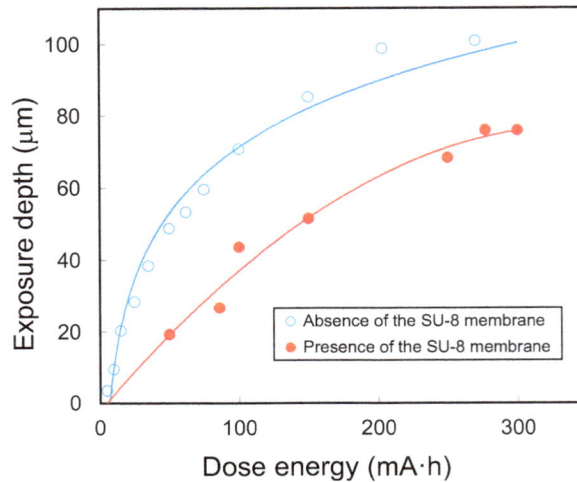

Figure 6. Exposure depth of PMMA sheet in cases of the absence and presence of the SU-8 membrane depended on the dose energy.

3.2. Transition of Pattern Width with SU-8 Thermal Expansion

Since the 1-mm-thick PMMA sheet was hard, the PMMA sheet after X-ray lithography was put into liquid nitrogen, and then cleaved by hand. And then in order to prevent any charge built-up, a 10-nm-thick Pt was sputter-deposited on the sample surface before an SEM observation was made. The left and right halves of Figure 7 are the results of pattern transferring from cone- and pyramid-shaped Si absorbers to a PMMA structure, respectively. As for the exposure conditions, in the upper figures, the dose energy was set at 50 mA (Figure 7a,b); another case of a 300-mA dose energy is shown in the lower figures (Figure 7c,d).

From these SEM images, it is confirmed that the X-ray grayscale mask has a sufficient contrast between the absorber and membrane. As shown in Figure 3, a cone-shaped Si absorber was 46 μm in diameter and 27 μm in height; and a pyramid-shaped Si absorber was 42 μm in the bottom width and 27 μm in height. Figure 7 shows a transition of pattern widths on the PMMA sheets exposed by the various dose energies. Although the 50-μm-diameter circular and 50-μm-wide square dotted patterns were accurately transferred from the photomask to the photoresist, at the early step of fabricating the X-ray grayscale mask, a pattern size reduction of 8%–16% was made by the tapered-trench-etching. After that, cone- and pyramid-shaped Si absorbers were transferred by deep X-ray lithography to PMMA structures of 34.5 μm in diameter, and 36.2 μm in width on average. In this process, the pattern width decreased further down to 75%–86% of Si absorbers.

This pattern width reduction originates from the diffraction effect in which case the X-rays bending around the edges and reaching to the backside of Si absorbers occurs. Usually, in the area where the X-rays were irradiated, a certain degree of heating took place and the absorber, and the membrane expanded depending on their coefficients of thermal expansion. In the beamline BL-4, infrared rays are significantly cut by the Be window; and a device was equipped to cool the exposure stage continuously by a flowing stream of the He gas. However, a certain degree of thermal expansion is unavoidable, and a degradation of the pattern transfer accuracy does occur, and remains as a matter of concern. While the X-ray grayscale mask is irradiated with X-rays, the irradiated area gradually reaches to a high temperature, and the membrane also expands. Therefore, according to the expansion of the membrane, the pattern's position is shifted because the isolated Si absorbers were embedded within the SU-8 membrane. As shown in Figure 8, the bottom width of the PMMA structures does not exceed the width of the Si absorbers; it rather shrinks to a smaller dimension as dictated by the diffraction effect of the X-rays. This result is the proof, which was about maintaining a very stable accuracy of position, without the expanding of the SU-8 membrane during the X-ray irradiation.

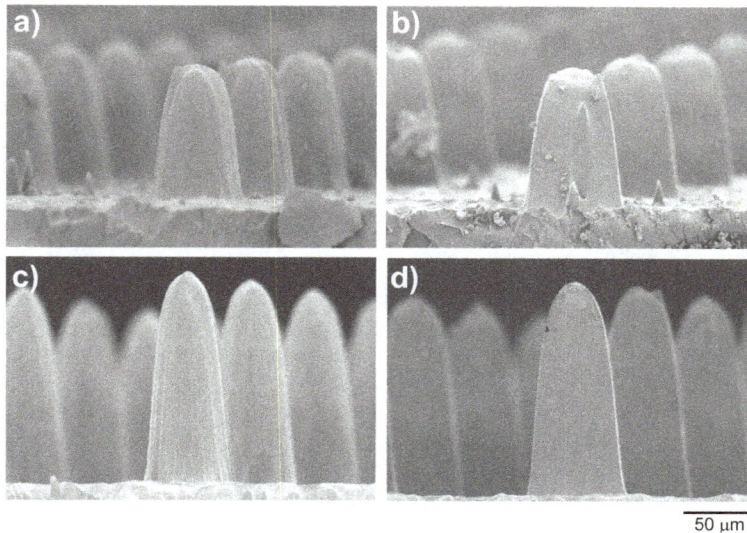

Figure 7. Cross-sectional SEM images of the PMMA structures, which (**a,c**) cone and (**b,d**) pyramid shapes of Si absorber were transferred by the X-rays at dose energies of 50 and 300 mA, respectively.

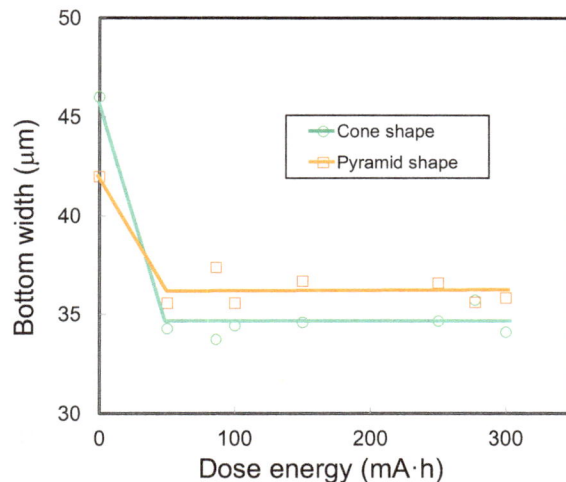

Figure 8. Bottom width of the PMMA structures fabricated by X-rays which penetrated through the cone and pyramid shaped structures of the Si absorbers depended on the dose energy.

3.3. Tolerance of SU-8 to Synchrotron Radiation

Figure 9 shows the SEM images of the Si absorbers on the X-ray grayscale mask and the PMMA structures after X-ray lithography. Using these top view images, a pitch between the structures was measured. The investigation result of the durability of the grayscale mask to X-rays is shown in Figure 10. In the figure, the average values of both patterns' pitches in X and Y directions (P_x and P_y) are plotted, and the data has been approximated and represented in form of two curves. In the X-ray mask manufacturing step, the SU-8 spin-coated on the SiO_2 layer was baked twice, and an internal stress was accumulated in a relatively soft SU-8. The coefficients of the thermal expansion of SU-8 (52×10^{-6}/K) [28,29] and SiO_2 (0.7×10^{-6}/K) [36] are quite different from each other. When the SiO_2 film was removed by the RIE as a post process, the internal stress of the SU-8 was eased, resulting in them being expanded, and in the bending of the SU-8 membrane. As a result, the circular and square dotted patterns, which were of 100-μm pitches on the photomask, are spread to 120 and 117 μm, respectively.

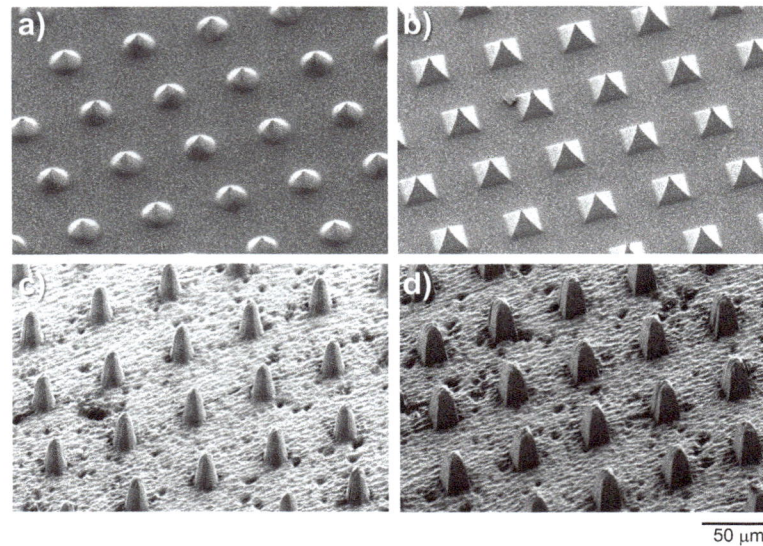

Figure 9. SEM images of the Si absorbers on the X-ray grayscale mask before spin-coating of SU-8: (**a**) cone and (**b**) pyramid shape. SEM images of the PMMA microstructures fabricated by X-ray lithography at the dose energy of 300 mA·h; (**c**) cone and (**d**) pyramid shape. Reprinted with permission from [22].

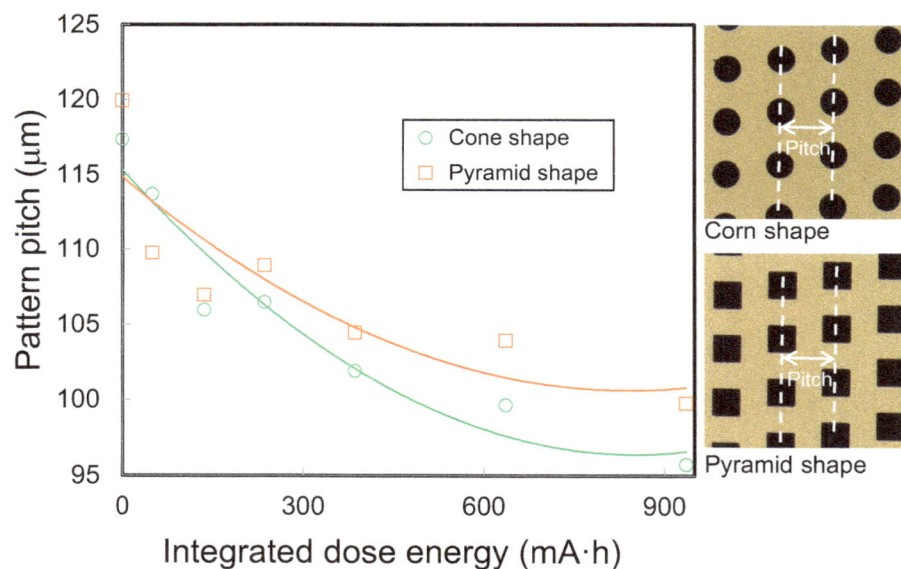

Figure 10. Pattern pitch of PMMA structures which cone, and pyramid shapes of Si absorbers was transferred by X-rays depended on the integrated dose energy. Pitches in X and Y directions indicated as P_x and P_y in the figure, respectively.

4. Conclusions

The performance of an X-ray grayscale mask that consisted of Si absorbers processed by tapered-trench-etching, and of a SU-8 membrane was evaluated using a beamline BL-4 of TERAS SR facility at AIST. Three kinds of characteristic are required of a membrane material for X-ray lithography. The first is a transmission property of X-rays. When a 30-μm-thick SU-8 membrane was absent or present, an exposure depth of a PMMA photoresist was measured and was compared with a calculated transmittance of SU-8 in an X-ray energy region. In fact, the measured value and the calculated value were well in agreement. Next, it is important that a pattern width does not change with a membrane's thermal expansion during the X-ray irradiation. This transition leads to a fatal defect in grayscale lithography. On a positive note, it was learned that a displacement of the pattern width at different dose energies was controllable within 2 μm. Finally, it remains to be addressed if SU-8 has plenty of tolerance to synchrotron radiation. If an operating time of the X-ray mask is increased, the polymerization by cross-linking in the SU-8 film was moved forward and a solvent that remains inside the SU-8 membrane can evaporate and the membrane could become brittle. Although the membrane's color would also be changed, a reduction of a pattern pitch was found to be saturated when the integrated dose energy reached to 900 mA·h. Moreover, the measured pitch became almost equal to the designed value. These results proved that the X-ray grayscale mask with the SU-8 membrane does perform well for all practicality and can be used in an X-ray mask after a suitable baking process by irradiation of X-rays. It is also believed that with this method, the pattern transfer quality can be achieved to be very high.

Acknowledgments

X-ray exposure experiments were executed using the SR Facility TERAS of the National Institute of Advanced Industrial Science and Technology (AIST). The authors would like to thank Koichi Awazu of Electronics and Photonics Research Institute, and to the Quantum Radiation Research group at the Research Institute of Instrumentation Frontier in AIST. The author accepted technical supports in Si taper-trench-etching from Takayuki Takano of the Advanced Manufacturing Research Institute in AIST (at the time).

Conflicts of Interest

The author declares no conflict of interest.

References and Notes

1. Wu, B.; Kumar, A. Extreme ultraviolet lithography: A review. *J. Vac. Sci. Technol. B* **2007**, *25*, 1743–1761.
2. Kinoshita, H.; Wood, O. An Historical Perspective. In *EUV Lithography*; Bakshi, V., Ed.; SPIE Press: Bellingham, WA, USA, 2007; pp. 1–54.
3. Malek, C.K.; Saile, V. Applications of LIGA technology to precision manufacturing of high-aspect-ratio micro-components and systems: A review. *Microelectron. J.* **2004**, *35*, 131–143.
4. Salker, V. Introduction: LIGA and Its Applications. In *LIGA and Its Applications*; Saile, V., Wallrabe, U., Tabata, O., Korvink, J.G., Eds.; Wiley-Vch: Weinheim, Germany, 2009; pp. 1–10.

5. Becker, E.W.; Ehrfeld, W.; Hagmann, P.; Manwe, A.; Münchmeyer, D. Fabrication of microstructures with high aspect ratios and great structural heights by synchrotron radiation lithography, galvanoforming, and plastic moulding (LIGA process). *Microelectron. Eng.* **1986**, *4*, 35–56.

6. Bogdanov, A.L.; Peredlkov, S.S. Use of SU-8 photoresist for very high aspect ratio X-ray lithography. *Microelectron. Eng.* **2000**, *53*, 493–496.

7. Desta, Y.; Goettert, J. X-Ray Masks for LIGA Microfabrication. In *LIGA and Its Applications*; Saile, V., Wallrabe, U., Tabata, O., Korvink, J.G., Eds.; Wiley-Vch: Weinheim, Germany, 2009; pp. 11–50.

8. Utsumi, Y.; Kishimoto, T.; Hattori, T.; Hara, H. Large-area x-ray lithography system for liga process operating in wide energy range of synchrotron radiation. *Jpn. J. Appl. Phys.* **2005**, *44*, 5500–5504.

9. Mekaru, H.; Takano, T.; Ukita, Y.; Utsumi, Y.; Takahashi, M. A Si stencil mask for deep X-ray lithography fabricated by MEMS technology. *Microsyst. Technol.* **2008**, *14*, 1335–1342.

10. Flanders, D.C.; Smith, I. Polyimide membrane X-ray lithography masks—Fabrication and distortion measurements. *J. Vac. Sci. Technol.* **1978**, *15*, 995–997.

11. Spiller, E.; Feder, R.; Topalian, J.; Castellani, E.; Romankiw, L.; Heritage, M. X-ray lithography for bubble devices. *Solid State Technol.* **1979**, *19*, 62–68.

12. Lee, S.; Kim, D.; Jin, Y.; Han, Y.; Desta, Y.; Bryant, M.D.; Goettert, J. A Micro corona motor fabricated by a SU-8 built-on X-ray mask. *Microsyst. Technol.* **2004**, *10*, 522–526.

13. Schmidt, C.J.; Lenzo, P.V.; Spencer, E.G. Preparation of thin windows in silicon masks for X-ray lithography. *J. Appl. Phys.* **1975**, *46*, 4080–4082.

14. Watts, R.K. X-ray lithography. *Solid State Technol.* **1979**, *22*, 68–82.

15. Bassous, E.; Feder, E.; Spiller, E.; Topalian, J. High transmission X-ray masks for lithographic applications. *Proc. Int. Electron Devices Meet.* **1975**, *21*, 17–19.

16. Desta, Y.M.; Aigeldinger, G.; Zanca, K.J.; Coane, P.J.; Goettert, J.; Murphy, M.C. Fabrication of graphite masks for deep and ultradeep X-ray lithography. *Proc. Mater. Device Charact. Micromach. III* **2000**, *4175*, 122–130.

17. Hofer, D.; Powers, J.; Grobman, W.D. X-ray lithographic patterning of magnetic bubble circuits with submicron dimensions. *J. Vac. Sci. Technol.* **1979**, *16*, 1968–1972.

18. Yamada, M.; Nakaishi, M.; Kudou, J.; Eshita, T.; Furumura, Y. An X-ray mask using Ta and heteroepitaxially grown SiC. *Microelectron. Eng.* **1989**, *9*, 135–138.

19. Lüthje, H.; Harms, H.; Matthiessen, B.; Bruns, A. X-ray lithography: Novel fabrication process for SiC/W steppermasks. *Jpn. J. Appl. Phys.* **1989**, *28*, 2342–2347.

20. Iba, Y.; Kumasaka, F.; Iizuka, T.; Yamabe, M. Amorphous refractory compound film material for X-ray mask absorbers. *Jpn. J. Appl. Phys.* **2000**, *39*, 5329–5333.

21. Mekaru, H.; Takano, T.; Awazu, K.; Takahashi, M.; Maeda, R. Fabrication and evaluation of a grayscale mask for X-ray lithography using MEMS technology. *J. Micro/Nanolithogr. MEMS MOEMS* **2008**, *7*, 013009.

22. Mekaru, H.; Takano, T.; Awazu, K.; Takahashi, M.; Maeda, R. Fabrication of a needle array using a Si gray mask for X-ray lithography. *J. Vac. Sci. Technol. B* **2007**, *25*, 2196–2201.

23. Mekaru, H.; Takano, T.; Awazu, K.; Takahashi, M.; Maeda, R. Demonstration of fabricating a needle array by the combination of X-ray grayscale mask with the lithografie, galvanoformung, abformung process. *J. Micro/Nanolithogr. MEMS MOEMS* **2009**, *8*, 033010.

24. Kudryashov, V.; Yuan, X.; Cheong, W.; Radhakrishnan, K. Grey scale structures formation in SU-8 with e-beam and UV. *Microelectron. Eng.* **2003**, *67–68*, 306–311.

25. Watt, F. Focused high energy proton beam micromachining: A perspective view. *Nucl. Instrum. Methods Phys. Res. B* **1999**, *158*, 165–172.

26. DuPont™ Kapton® HN, Polyimide Film. Available online: http://www.dupont.com/content/dam/ assets/products-and-services/membranes-films/assets/DEC-Kapton-HN-datasheet.pdf (accessed on 5 February 2015).

27. Product Information, Mular®, Polyester Film. Available online: http://usa.dupontteijinfilms.com/ informationcenter/downloads/Physical_And_Thermal_Properties.pdf (accessed on 5 February 2015).

28. Del Campo, A.; Greiner, C. SU-8: Aphotoresist for high-aspect-ratio and 3D submicron lithography. *J. Micromech. Microeng.* **2007**, *17*, R81–R95.

29. Lorenz, H.; Laudon, M.; Renaud, P. Mechanical characterization of a new high-aspect-ratio near UV-photoresist. *Microelectron. Eng.* **1998**, *41*, 371–374.

30. Okada, Y.; Tokumura, Y. Precise determination of lattice parameter and thermal expansion coefficient of silicon between 300 and 1500 K. *J. Appl. Phys.* **1984**, *56*, 314–320.

31. Tien, C.; Lin, T. Thermal expansion coefficient and thermomechanical properties of SiN_x thin films prepared by plasma-enhanced chemical vapor deposition. *Appl. Opt.* **2012**, *51*, 7229–7235.

32. Goldberg, Y.; Levinshtein, M.E.; Rumyantsev, S.L. Chapter 5: Silicon Carbide (SiC). In *Properties of Advanced Semiconductor Materials: GaN, AlN, InN, BN, SiC, SiGe*; Levinshtein, M.E., Rumyantsev, S.L., Shur, M.S., Eds.; John Wiley & Sons: Hoboken, NJ, USA, 2001; pp. 93–148.

33. The coefficient of thermal expansion on graphite was applied by Tokai Carbon Co., Ltd. (Tokyo, Japan).

34. Toyokawa, H.; Awazu, K.; Hohara, S.; Koike, M.; Kuroda, R.; Morishita, Y.; Saito, N.; Saito, T.; Tanaka, M.; Watanabe, K.; *et al.* Present status of the electron storage ring TERAS of AIST. In Proceedings of the First Annual Meeting of Particle Accelerator Society of Japan and the 29th Linear Accelerator Meeting in Japan, Funabashi, Japan, 4–6 August 2004; pp. 203–205.

35. Awazu, K.; Wang, X.; Fujimaki, M.; Kuriyama, T.; Sai, A.; Ohki, Y.; Imai, H. Fabrication of two- and three-dimensional photonic cystals of titania with submicrometer resolution by deep X-ray lithography. *J. Vac. Sci. Technol. B* **2005**, *23*, 934–939.

36. Tada, H.; Kumpel, A.E.; Lathrop, R.E.; Stanina, J.B.; Nieva, P.; Zavracky, P.; Miaoulis, I.N.; Wong, P.Y. Thermal expansion coefficient of polysrystalline silicon and silicon dioxide thin films at high temperatures. *J. Appl. Phys.* **2000**, *87*, 4189–4193.

14

Micromachining of AlN and Al$_2$O$_3$ Using Fiber Laser

Florian Preusch, Benedikt Adelmann * and Ralf Hellmann

Applied Laser and Phonics Group, University of Applied Sciences Aschaffenburg, Wuerzburger Strasse 45, D-63743 Aschaffenburg, Germany; E-Mails: florian.preusch@h-ab.de (F.P.); ralf.hellmann@h-ab.de (R.H.)

* Author to whom correspondence should be addressed; E-Mail: benedikt.adelmann@h-ab.de

External Editors: Maria Farsari and Costas Fotakis

Abstract: We report on high precision high speed micromachining of Al$_2$O$_3$ and AlN using pulsed near infrared fiber laser. Ablation thresholds are determined to be 30 J/cm^2 for alumina and 18 J/cm^2 for aluminum nitride. The factors influencing the efficiency and quality of 3D micromachining, namely the surface roughness, the material removal rate and the ablation depth accuracy are determined as a function of laser repetition rate and pulse overlap. Using a fluence of 64 J/cm^2, we achieve a material removal rate of up to 94 mm^3/h in Al$_2$O$_3$ and 135 mm^3/h in AlN for high pulse overlaps (89% and 84%). A minimum roughness of 1.5 μm for alumina and 1.65 μm for aluminum nitride can be accomplished for medium pulse overlaps (42% to 56%). In addition, ablation depth deviation of the micromachining process of smaller than 8% for alumina and 2% for aluminum nitride are achieved. Based on these results, by structuring exemplarily 3D structures we demonstrate the potential of high quality and efficient 3D micromachining using pulsed fiber laser.

Keywords: ceramics; Al$_2$O$_3$; AlN; micromachining; fiber laser

1. Introduction

Micro structuring of materials with limited machinability by conventional machining methods, such as ceramics, can be executed by laser ablation employing pulsed lasers [1–3]. While, e.g., mechanical removal, electrical discharge machining, chemical or electrochemical treatment are less appropriate for ceramics due to their high hardness, brittleness and chemical inertness, pulsed laser structuring evades these difficulties and combines the advantages of low mechanical stress, precision, process controllability and low thermal load. The latter being associated with a reduced heat affected zone surrounding the interaction zone of the laser impinging the processed specimen [4,5].

Technical ceramics such as Al_2O_3 and AlN are indispensable substrate materials for, e.g., microelectronics, electronic packaging and micro electro mechanical systems (MEMS) [5–7]. Their salient properties are a high electrical resistance, a high thermal conductivity and a great chemical resistance. With respect to their application in micro tooling, hot embossing dies for microfluidics, electronics and sensing both the ceramic micro structuring of the surface as well as the volume (3D structures) is of upmost importance [8]. Decreasing feature size and increasing requirements on quality and cost efficiency, expedite a further development of precise and efficient ablation processes with high removal rates and attainable surface roughness allowing the fabrication of complex 3D structures for, e.g., tooling inserts, to be fabricated [5]. Further quality aspects of the micro-structuring are spatial accuracy and steep ablated edges.

In case of ceramics, CO_2 lasers are primarily used for processes such as cutting, drilling and perforating [6,9]. For micro structuring, however, primarily pulsed solid state [4,10], and Excimer-lasers [11–13] have been applied with pulse durations in the nanosecond, picosecond and femtosecond regime. However, due to the substantial progress, also fiber lasers have recently been employed to efficiently process ceramics [4,14]. With regards to micromachining, fiber lasers offer a competitive edge by a high beam quality over their complete power range, a high reliability and long lifetime, as well as low operating costs. In addition, their compact design and inherent fiber delivery simplify the setup of workstations. Moreover, compared to CO_2 lasers, fiber lasers are capable to realize structures significantly smaller than 100 μm in consequence of to a lower beam parameter product associated with the emission wavelength around 1 μm [10].

With the focus on attainable surface roughness and processing speed, Sciti *et al.* utilized a KrF Excimer laser (248 nm) with pulse duration of 25 ns and fluences between 1.8 and 7.5 J/cm² to micro-structure unpolished Al_2O_3 [11]. A surface roughness R_a between 0.8 and 1.0 μm was accomplished at a process rate between 1.0 and 2.0 mm/s. In addition, evidence was found that polished samples have a lower surface roughness R_a after laser treatment being on the order of 0.25 to 0.4 μm. Also using a KrF laser, Oliveira *et al.* worked on Al_2O_3-TiC ceramics (fluence between 2 and 8 J/cm²) and report on an increase of the surface roughness with increasing pulse number up to 200 pulses and an increase with fluence from 0.2 μm at 2 J/cm² and 0.7 μm at 8 J/cm³, respectively [12].

Weichenhain *et al.* [10] deployed a frequency tripled pulsed diode-pumped solid-state laser (DPSSL) at 355 nm with fluences between 1.9 and 16.2 J/cm² (pulse duration 10 ns, repetition rate 1 kHz) to process the non-oxide ceramics SiC and Si_3N_4 as a function of pulse overlap and fluence. For SiC the surface roughness remains almost constant at about 0.4 μm for pulse overlaps up to 80%, being almost independent on fluence up to 16.2 J/cm². Ablating the material in different atmospheres (e.g., vacuum, ambient air, oxygen or nitrogen) does not significantly alter this behavior. Yet, above a pulse overlap of 80% the surface roughness raises sharply with an ambiguous dependence on fluence. For Si_3N_4, however, the surface roughness reveals a more pronounced dependence on fluence also for low pulse overlaps with a varying behavior *versus* pulse overlap at different fluence levels. Moreover, the ambient gas atmosphere appears to have a stronger influence in case of Si_3N_4 as compared to SiC. A further discussion on the chemical composition of the treated surface upon laser irradiation in different atmospheres is, however, not presented. Nonetheless, 3D micro geometries with feature sizes between 100 and 500 μm are generated to prove the concept of three dimensional micro structuring. To achieve high precision with low roughness and little residues, the process speeds had to be chosen below 10 mm/s with the fluence below 15 J/cm² [10].

A study on the near surface chemical composition after laser ablation of AlN by a KrF Excimer laser is presented by Hirayama *et al.* [9] using X-ray photoelectron spectroscopy. Examining the concentration depth profile within about 1.6 μm beneath the treated surface, it is shown that due to the nanosecond laser induced thermal decomposition process an aluminum layer is formed on the surface. Similar results are obtained using CO_2 lasers [6,9].

This contribution encompasses a study and optimization of the laser micro machining of Al_2O_3 and AlN using nanosecond fiber laser. The optimization targets high surface quality evaluated by the surface roughness and the accuracy of ablation depth for 3D volume ablation. As processing parameters the fluence, pulse overlap and laser repetition rate, respectively the process velocity, and the number of passes in a multi-pass process are varied. In addition, ablation thresholds and removal rates are determined and, finally, the distinguished quality of 3D micro structuring achievable with fiber laser is exemplified by realizing selected 3D structures.

2. Experimental Section

2.1. The Laser-System

For the experiments a pulsed 20 W fiber laser is used (YLP-1/100/20, IPG Photonics Corporation, Oxford, MA, USA), which is characterized by a pulse length of 100 ns, a beam quality of $M^2 = 1.6$ and an emission wavelength of 1064 nm. The repetition rate can be varied between 2 and 80 kHz with a nominal average power of 20 W at 20 kHz. The maximum pulse energy is 1 mJ and the maximum fluence is 64 J/cm³ at a 44 μm spot diameter, respectively. While above 20 kHz the average power is constant with decreasing pulse energy at increasing repetition rate, below 20 kHz the average power decreases at constant pulse energy. At fixed repetition rate, the output power can get adjusted between 10% and 100% of the nominal power.

The laser beam is positioned over the specimen by a galvo scanner (Raylase SuperScan) with a focal length of 163 mm. The scanner deflects the laser spot over a square of 110×110 mm² with a maximum positioning speed of larger than 7 m/s.

2.2. Materials

In this study, alumina and aluminum nitride with a purity of 96% and grain size of 3 to 5 μm (Ceramtec) are laser structured. The substrate platelets have a thickness of 0.33, 0.63 and 1 mm, respectively. Aluminum nitride has a specified thermal conductivity of 140 W/mK, alumina of 20 W/mK. For a spectral analysis of the absorption properties, we applied a spectroradiometer with equipped goniometer and integrating sphere in the detection arm to collect scattered light. For Al_2O_3, our measurements at room temperature reveal a reflection of 81% and a transmission of 16% at 1064 nm (substrate thickness 0.63 mm). This implies a calculated absorption of 3%. For AlN, the refection is 44% and the transmission 12%, which yields an absorption of 44% (substrate thickness 1 mm).

3. Results and Discussion

3.1. Ablation Threshold

In the conducted study, ablation of Al_2O_3 and AlN is initiated only for number of pulses. For instance, at a pulse repetition rate of 20 kHz ablation of Al_2O_3 requires about 1000 pulses. For comparative reasons, we therefore determine the ablation threshold fluence (F_{th}) for both materials for a sequence of 1000 pulses. For lasers having a Gaussian beam profile, F_{th} is typically determined by measuring the diameter of the generated ablation crater for varying fluences [15,16].

Figure 1 depicts the squared diameter of the ablation crater as measured by a confocal microscope *versus* the single pulse fluence in the center of the laser beam between 35 J/cm² and 64 J/cm² (fluence adjusted by changing the laser pulse energy, respectively the average power). According to [15], the intersection of a linear regression with the axis of abscissae (fluence), *i.e.*, a hypothetical vanishing crater diameter, yields the ablation threshold.

Figure 1. Squared diameters of Al_2O_3 and AlN as function of laser fluence.

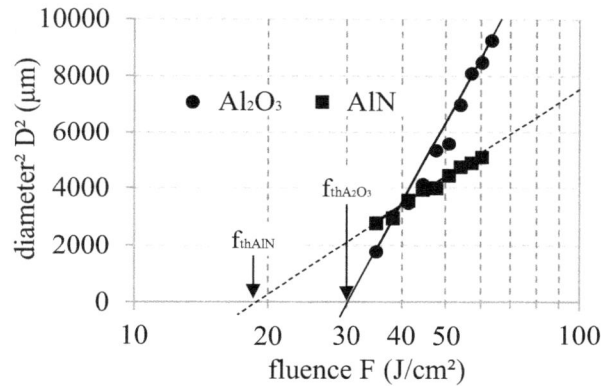

For Al_2O_3 and AlN we determine an ablation threshold F_{th} of 30 and 18 J/cm², respectively. The difference of F_{th} between Al_2O_3 and AlN in this study can be assigned to the different absorption of the materials at 1064 nm. Since ablation thresholds depend on, e.g., the optical absorption that varies with wavelength [17,18] and on the number of pulses [19,20], F_{th} cannot be directly compared between different studies. In addition also the pulse length has an influence on the ablation threshold with by trend shorter pulses result in smaller thresholds [13]. Nonetheless, for a simplified assessment, the values found in this study correspond to those reported for CO_2 laser ablation [9]. In contrast, for excimer lasers, significantly lower ablation thresholds are determined in [9], which can be attributed to the higher absorption in the UV. Moreover, for femtosecond pulsed laser also lower threshold have been determined [21].

3.2. Correlation between Pulse Overlap and Roughness

For the application of micro structured ceramics, the surface roughness after laser ablation is of upmost importance. As previous studies have shown, it is strongly influenced by the pulse overlap [7,10]. Applying a meander type scanning strategy, we have investigated the surface roughness as a function of pulse repetition rate and pulse overlap of the laser scanning across the workpiece. The distance between two adjacent lines of the meander path is set equal to the distance between two following pulses within one line. In addition, ablation is performed using the highest possible fluence of 64 J/cm² for high ablation rate and scanning the laser in five passes over an area of 1 × 1 mm². In addition, the pulse repetition rate is varied between 5 and 20 kHz.

Figure 2 shows the mean surface roughness R_a, measured by a laser scanning microscope, for alumina for different pulse overlaps between 42% and 89%. Below 42% pulse overlap, no material removal was observed for Al_2O_3. Please note, that the untreated alumina has a roughness R_a of 3.4 µm. After laser ablation, R_a is reduced to 1.50 µm at a pulse overlap of 42% and remains below the roughness of the untreated surface up to a pulse overlap of 56%. Between an overlap of 41% and about 80%, R_a increases independently of the repetition rate. Please note, that at constant pulse overlap an increase of the pulse repetition rate results in an equivalent increase of the velocity. For a pulse overlap exceeding 84%, the surface roughness increases sharply which is accompanied by an extended melt formation. A similar correlation between melt formation and roughness at high pulse overlap has been reported for laser structuring Y-TZP ceramics, being attributed to a hampered material ejection due to melt coverage [7].

In general, for AlN similar results are obtained as shown on Figure 3. However, due to the higher absorption material removal is already initiated at a pulse overlaps of 36%. Between 36% and 65% the surface roughness remains almost constant (R_a about 1.65 µm) and rises sharply for an overlap above of 80%. Yet, since AlN does not form a liquid phase [22], we associate this sharp increase with melt formation of liquid aluminum [9]. A comparable behavior, being explained by the presence of oxygen, has been observed by Weichenhain et al. when processing Si_3N_4 [10].

Figure 2. Mean roughness R_a as a function of pulse overlap at Al_2O_3.

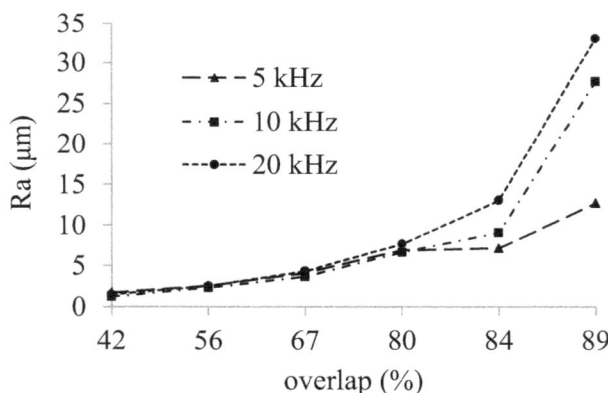

Figure 3. Mean roughness R_a as a function of pulse overlap at AlN.

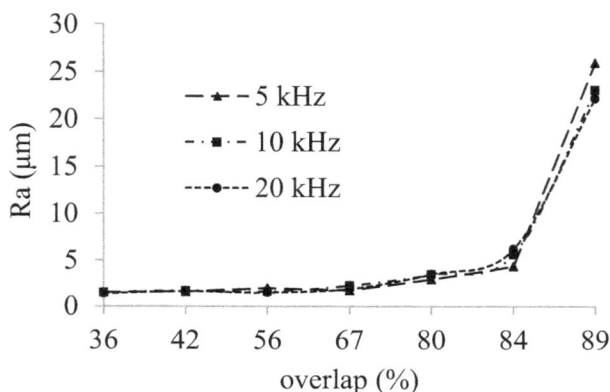

3.3. Material Removal Rate

The removal rate is defined by the quotient of the ablated volume and the processing time. Figure 4 summarizes the removal rate for Al_2O_3 and AlN as a function of pulse overlap and for varying laser repetition rate. As a consequence of the lower ablation threshold for aluminum nitride, the removal rate of AlN is, in general, higher as compared to alumina. In particular, for alumina a maximum ablation rate of 94 mm³/h (84% pulse overlap, 20 kHz) and for aluminum nitride 135 mm³/h (89% pulse overlap, 20 kHz) is achieved.

Figure 4. Removal rate as a function of pulse overlap and for varying repetition rate for Al_2O_3 (grey line) and AlN (black line).

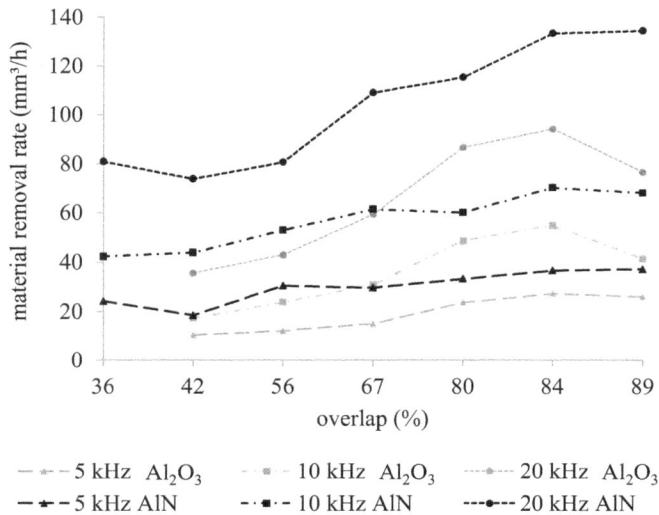

In addition, the removal rate increases for both pulse overlap and repetition rate. For example, in case of Al_2O_3 a maximum of 94 mm³/h is achieved for a pulse overlap of 84% at 20 kHz. This is significantly higher as compared to 10 mm³/h reported for laser ablation of alumina using a diode pumped solid state laser at a wavelength of 355 nm [23]. In addition, it is worthwhile to note that contrary to our findings in [23] the removal rate decreases with increasing pulse overlap. This can be attributed to the reduced absorption of the infrared fiber laser in the plasma that is generated during laser ablation. A higher plasma absorption in the UV reduces the energy input and thus the removal rate.

Since the laser pulse energy remains constant up 20 kHz, an increase of pulse repetition rate corresponds to an increase of average laser power and consequently to a higher ablation rate (Figure 4).

3.4. Ablation Depth Accuracy

For micromachining a 3D structure, the laser is scanned in a meander type pattern in multiple passes across the specimen until the targeted depth is achieved. For high quality 3D micromachining it is imperative to guarantee a high accuracy of the depth of ablation for multiple scan passes. In addition, a high material removal rate and a low surface roughness are desirable to ensure an efficient micromachining process and a high surface quality. To facilitate a good compromise between material removal rate and surface roughness, a fluence of 64 J/cm², a pulse overlap of 56% (Al_2O_3) and 77% (AlN) at a pulse repetition rate of 10 kHz are chosen to determine the depth of ablation accuracy. As each scan cycle removes an equal amount of material with a depth of about 6 μm for Al_2O_3 and 25 μm for AlN, respectively, the total ablation depth can be varied with the number of passes.

Figure 5 reveals a linear dependence of the ablation depth as a function of the number of scan passes (meander type) for both alumina and aluminum nitride. The scanning velocities for these experiments are 200 mm/s for Al_2O_3 and 100 mm/s for AlN, respectively. Please note, that the linear regression does not intercept the origin of coordinates as for the first scan the ablation depth is smaller due to reduced absorption.

Figure 5. Removal rate as a function of pulse overlap and for varying repetition rate for Al_2O_3 (solid line) and AlN (dashed line).

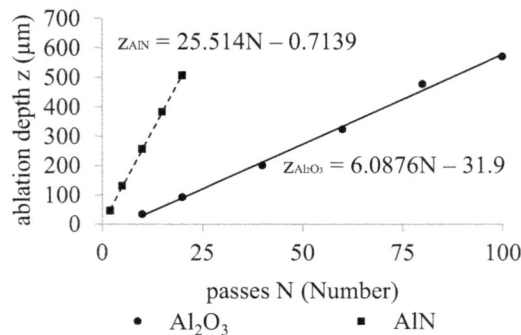

Using the observed linear dependence, the number of required scan passes for a targeted ablation depth can be calculated. However, as this does not necessarily lead to an integer number, the number of scans has to be rounded up or down. As a result, the ablated depth deviates from the targeted value. Tables 1 and 2 compare the targeted, calculated and measured for a varying number of scans. Though the deviation between calculated and realized depth depends on the overall targeted depth and material, Table 1 highlights that the deviation is rather small, e.g., for AlN 1.5% for a targeted depth of 50 μm and <0.01% for a targeted depth of 500 μm. The reason for the higher accuracy of AlN ablation is assumed in the higher optical absorption compared to alumina which allows a more defined ablation. Based on this accuracy, high quality 3D structures can be realized in alumina and aluminum nitride.

Table 1. Comparison of targeted, calculated and measured ablation depth for Al_2O_3.

Characteristic	TARGET (μm)			
	50	**100**	**300**	**500**
number of passes (N)	14	22	55	87
calculated (μm)	53.3	102.0	302.9	497.7
measured (μm)	49.1	95.3	287.6	482.4
Deviation in %	7.9	6.6	5.0	3.0

Table 2. Comparison of targeted, calculated and measured ablation depth for AlN.

Characteristic	TARGET (μm)			
	50	**100**	**300**	**500**
number of passes (N)	2	4	12	20
calculated (μm)	50.3	101.3	305.5	509.6
measured (μm)	51.1	103.4	305.3	508.9
Deviation in %	1.5	2.0	<0.1	<0.1

4. Example of 3D Microstructures

Based on the experimental results presented above, defined 3D microstructures can be realized with high geometrical accuracy as well as high surface and edge quality. To prove the potential of fiber laser micromachining, Figure 6 depicts exemplarily an inversed pyramid realized in AlN that has been generated by sequentially ablating three squares having side lengths of 0.9, 0.6 and 0.3 mm and a depth of 300 µm per step. In contrast, Figure 7 illustrates a submerged pyramid in AlN, which has been realized by ablating seven steps with a depth of 100 µm each. For both microstructures an edge steepness between 70° and 75° is achieved.

Figure 6. Laser scanning microscope image of a fiber laser micro machined inverse pyramid in AlN.

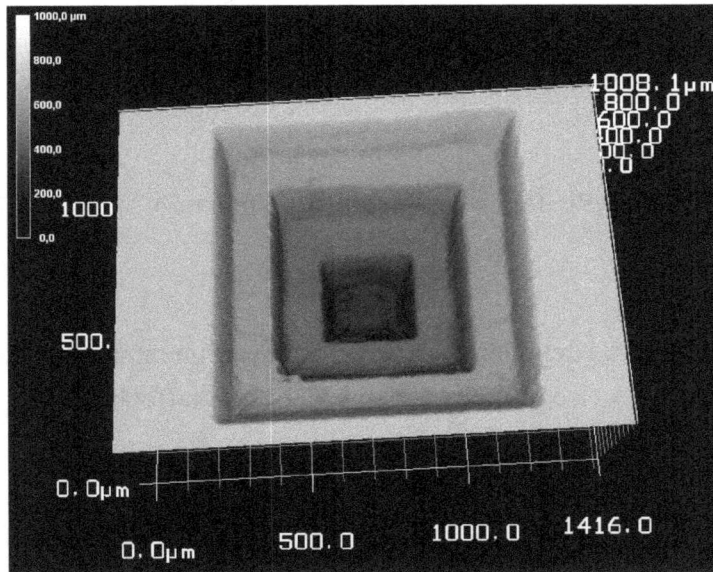

Figure 7. Laser scanning microscope image of a fiber laser micro machined submerged pyramid in AlN.

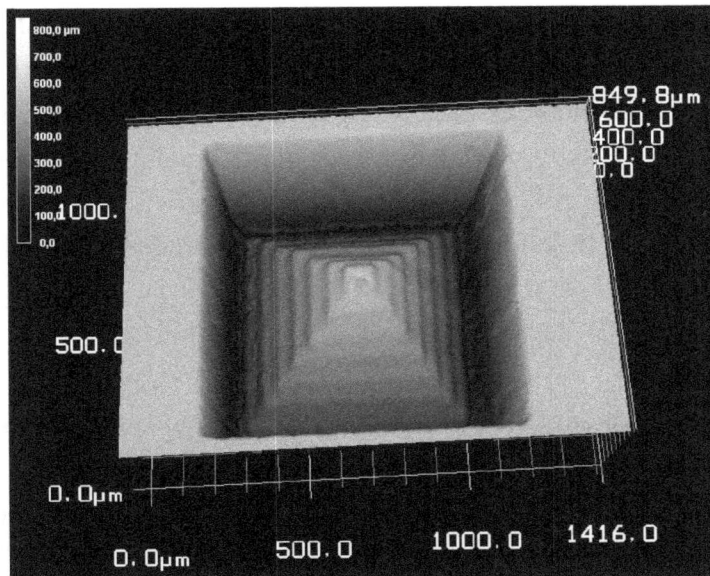

5. Conclusions

We have demonstrated that near infrared pulsed fiber lasers are eligible tools for efficient and high quality 3D micromachining. The pulse overlap and repetition rate are shown to be the most influencing parameters determining the achievable surface roughness and the material removal rate. The latter can be maximized for high repetition rates and overlaps, reaching a maximum of 94 mm³/h in Al₂O₃ and 135 mm³/h in AlN at a repetition rate of 20 kHz and pulse overlaps exceeding 84% and 89% for alumina and aluminum nitride, respectively. The surface roughness R_a can be minimized, however, for lower pulse overlaps being on the order of 50% reaching a level of 1.5 µm for Al₂O₃ and 1.65 µm for AlN. The ablation depth accuracy for multi-pass laser ablation is better than 8% for alumina and 2% for aluminum nitride, enabling high precision micro structuring.

Using a compromised parameter combination of maximum laser fluence (64 J/cm²), 10 kHz pulse repetition rate and medium pulse overlap (56% for Al₂O₃ and 77% for AlN) complex 3D structures in alumina and aluminum nitride are fabricated with high geometrical accuracy and high surface quality.

Acknowledgments

This contribution is funded by the Bavarian Research Foundation.

Author Contributions

Florian Preusch and Benedikt Adelmann performed the experimental work, Ralf Hellmann is head of the Applied Laser and Photonics Group at the University of Aschaffenburg.

Conflicts of Interest

The authors declare no conflict of interest.

References

1. Kibria, G.; Doloi, B.; Bhattacharyya, B. Predictive model and process parameters optimization of Nd: YAG laser micro-turning of ceramics. *Int. J. Adv. Manuf. Technol.* **2013**, *65*, 213–229.
2. Knowles, M.R.H.; Rutterford, G.; Karnakis, D.; Ferguson, A. Micro-machining of metals, ceramics and polymers using nanosecond lasers. *Int. J. Adv. Manuf. Technol.* **2007**, *33*, 95–102.
3. Pham, D.T.; Dimov, S.S.; Petkov, P.V. Laser milling of ceramic components. *Int. J. Mach. Tools Manuf.* **2007**, *47*, 618–626.
4. Adelmann, B.; Hellmann, R. Investigation on flexural strength during fiber laser cutting of alumina. *Phys. Procedia* **2013**, *41*, 398–400.
5. Samant, A.N.; Narendra, B.D. Laser machining of structural ceramics—A review. *J. Eur. Ceram. Soc.* **2009**, *29*, 969–993.
6. Molian, R.; Shrotriya, P.; Molian, P. Thermal stress fracture mode of CO_2 laser cutting of aluminum nitride. *Int. J. Adv. Manuf. Technol.* **2008**, *39*, 725–733.
7. Wang, X.; Shephard, J.D.; Dear, F.C.; Hand, D.P. Optimized Nanosecond Pulsed Laser Micromachining of Y-TZP Ceramics. *J. Am. Ceram. Soc.* **2008**, *91*, 391–397.

8. Vanko, G.; Hudek, P.; Zehetner, J.; Dzuba, J.; Choleva, P.; Kutiš, V.; Vallo, M. Rýger, I., Lalinský, T. Bulk micromachining of SiC substrate for MEMS sensor applications. *Microelectron. Eng.* **2013**, *110*, 260–264.

9. Hirayama, Y.; Yabe, H.; Obara, M. Selective ablation of ALN ceramic using femtosecond, nanosecond, and microsecond pulsed laser. *J. Appl. Phys.* **2001**, *89*, 2943–2949.

10. Weichenhain, R.; Horn, A.; Kreutz, E.W. Three dimensional microfabrication in ceramics by solid state lasers. *Appl. Phys. A* **1999**, *69*, 855–858.

11. Sciti, D.; Melandri, A.; Bellosi, A. Excimer laser-induced microstructural changes of alumina and silicon carbide. *J. Mater. Sci.* **2000**, *35*, 3799–3810.

12. Oliveira, V.; Vilar, R.; Conde, O. Excimer laser ablation of Al2O3—TiC ceramics: Laser induced modifications of surface topography and structure. *Appl. Surf. Sci.* **1998**, *127*, 831–836.

13. Ihlemann, J.; Wolff-Rottke, B. Excimer laser micro machining of inorganic dielectrics. *Appl. Surf. Sci.* **1996**, *106*, 282–286.

14. Shah, L.; Fermann, M.E.; Dawson, J.W.; Barty, C.P.J. Micromachining with a 50 W, 50 μJ, subpicosecond fiber laser system. *Opt. Express* **2006**, *14*, 12546–12551.

15. Liu, J.M. Simple technique for measurements of pulsed Gaussian-beam spot sizes. *Opt. Lett.* **1982**, *7*, 196–198.

16. Martin, S.; Hertwig, A.; Lenzner, M.; Krüger, J.; Kautek, W. Spot-size dependence of the ablation threshold in dielectrics for femtosecond laser pulses. *Appl. Phys. A* **2003**, *77*, 883–884.

17. Torrisi, L.; Borrielli, A.; Margarone, D. Study on the ablation threshold induced by pulsed lasers at different wavelengths. *Nucl. Instrum. Methods Phys. Res. Sect. B Beam Interact. Mater. At.* **2007**, *255*, 373–379.

18. Borowiec, A.; Tiedje, H.F.; Haugen, H.K. Wavelength dependence of the single pulse femtosecond laser ablation threshold of indium phosphide in the 400–2050 nm range. *Appl. Surf. Sci.* **2005**, *243*, 129–137.

19. Kautek, W.; Krüger, J.; Lenzner, M.; Sartania, S.; Spielmann, C.; Krausz, F. Laser ablation of dielectrics with pulse durations between 20 fs and 3 ps. *Appl. Phys. Lett.* **1996**, *69*, 3146–3148.

20. Stuart, B.C.; Feit, M.D.; Rubenchik, A.M.; Shore, B.W.; Perry, M.D. Laser-induced damage in dielectrics with nanosecond to subpicosecond pulses. *Phys. Rev. Lett.* **1995**, *74*, 2248.

21. Kim, S.H.; Sohn, I.k.; Jeong, S. Ablation characteristics of aluminum oxide and nitride ceramics during femtosecond laser micromachining. *Appl. Surf. Sci.* **2009**, *255*, 9717–9720.

22. Bengisu, M. *Engineering Ceramics*; Springer: Heidelberg, Germany, 2001.

23. Hellrung, D. Material Removal by Laser Radiation for the Fabrication of Three Dimensional Micro Molding Tools Made of Hard Materials. Ph.D. Thesis, RWTH Aachen University, Aachen, Germany, 2000. (In German)

Permissions

List of Contributors

Jenelle Armstrong Piepmeier
Systems Engineering, United States Naval Academy, 105 Maryland Avenue, Annapolis, MD 21402, USA

Samara Firebaugh
Electrical and Computer Engineering, United States Naval Academy, 105 Maryland Avenue, Annapolis, MD 21402, USA

Caitlin S. Olsen
Systems Engineering, United States Naval Academy, 105 Maryland Avenue, Annapolis, MD 21402, USA

Anton A. Smirnov
Institute of Applied Physics of Russian Academy of Sciences, 46 Ul'yanov Street, Nizhny Novgorod 603950, Russia

Alexander Pikulin
Institute of Applied Physics of Russian Academy of Sciences, 46 Ul'yanov Street, Nizhny Novgorod 603950, Russia

Natalia Sapogova
Institute of Applied Physics of Russian Academy of Sciences, 46 Ul'yanov Street, Nizhny Novgorod 603950, Russia

Nikita Bityurin
Institute of Applied Physics of Russian Academy of Sciences, 46 Ul'yanov Street, Nizhny Novgorod 603950, Russia

Chung-Liang Chang
Department of Biomechatronics Engineering, National Pingtung University of Science and Technology, No. 1 Shuefu Road, Neipu, Pingtung County 91201, Taiwan

Jin-Long Shie
Department of Biomechatronics Engineering, National Pingtung University of Science and Technology, No. 1 Shuefu Road, Neipu, Pingtung County 91201, Taiwan

Shinya Sakuma
Department of Mechanical Engineering, Osaka University, 2-1 Yamadaoka, Suita, Osaka 565-0871, Japan

Keisuke Kuroda
Department of Mechanical Engineering, Osaka University, 2-1 Yamadaoka, Suita, Osaka 565-0871, Japan

Fumihito Arai
Department of Micro-Nano Systems Engineering, Nagoya University, Furo-Cho, Chikusa-Ku, Nagoya 464-8601, Japan

Tatsunori Taniguchi
Department of Cardiovascular Medicine, Osaka University, 2-1 Yamadaoka, Suita, Osaka 565-0871, Japan

Tomohito Ohtani
Department of Cardiovascular Medicine, Osaka University, 2-1 Yamadaoka, Suita, Osaka 565-0871, Japan

Yasushi Sakata
Department of Cardiovascular Medicine, Osaka University, 2-1 Yamadaoka, Suita, Osaka 565-0871, Japan

Makoto Kaneko
Department of Mechanical Engineering, Osaka University, 2-1 Yamadaoka, Suita, Osaka 565-0871, Japan

Song-Bin Huang
Graduate Institute of Biochemical and Biomedical Engineering, Chang Gung University, Taoyuan 333, Taiwan

Yang Zhao
State Key Laboratory of Transducer Technology, Institute of Electronics, Chinese Academy of Sciences, Beijing 100190, China

Deyong Chen
State Key Laboratory of Transducer Technology, Institute of Electronics, Chinese Academy of Sciences, Beijing 100190, China

Shing-Lun Liu
Graduate Institute of Biochemical and Biomedical Engineering, Chang Gung University, Taoyuan 333, Taiwan

Yana Luo
State Key Laboratory of Transducer Technology, Institute of Electronics, Chinese Academy of Sciences, Beijing 100190, China

Tzu-Keng Chiu
Department of Chemical and Materials Engineering, Chang Gung University, Taoyuan 333, Taiwan

Junbo Wang
State Key Laboratory of Transducer Technology, Institute of Electronics, Chinese Academy of Sciences, Beijing 100190, China

Jian Chen
State Key Laboratory of Transducer Technology, Institute of Electronics, Chinese Academy of Sciences, Beijing 100190, China

Min-Hsien Wu
Graduate Institute of Biochemical and Biomedical Engineering, Chang Gung University, Taoyuan 333, Taiwan

Elmano Pinto
School of Technology and Management (ESTiG), Polytechnic Institute of Bragança (IPB), Campus de Santa Apolónia, 5300-253 Bragança, Portugal
Transport Phenomena Research Center, Department of Chemical Engineering, Engineering Faculty,
University of Porto, Rua Dr. Roberto Frias, 4200-465 Porto, Portugal

Vera Faustino
School of Technology and Management (ESTiG), Polytechnic Institute of Bragança (IPB), Campus de Santa Apolónia, 5300-253 Bragança, Portugal

Raquel O. Rodrigues
School of Technology and Management (ESTiG), Polytechnic Institute of Bragança (IPB), Campus de Santa Apolónia, 5300-253 Bragança, Portugal
Laboratory of Catalysis and Materials—Associate Laboratory LSRE/LCM, Engineering Faculty,
University of Porto, Rua Dr. Roberto Frias, 4200-465 Porto, Portugal

Diana Pinho
School of Technology and Management (ESTiG), Polytechnic Institute of Bragança (IPB), Campus de Santa Apolónia, 5300-253 Bragança, Portugal
Transport Phenomena Research Center, Department of Chemical Engineering, Engineering Faculty, University of Porto, Rua Dr. Roberto Frias, 4200-465 Porto, Portugal

Valdemar Garcia
School of Technology and Management (ESTiG), Polytechnic Institute of Bragança (IPB), Campus de Santa Apolónia, 5300-253 Bragança, Portugal

João M. Miranda
Transport Phenomena Research Center, Department of Chemical Engineering, Engineering Faculty, University of Porto, Rua Dr. Roberto Frias, 4200-465 Porto, Portugal

Rui Lima
School of Technology and Management (ESTiG), Polytechnic Institute of Bragança (IPB), Campus de Santa Apolónia, 5300-253 Bragança, Portugal
Transport Phenomena Research Center, Department of Chemical Engineering, Engineering Faculty, University of Porto, Rua Dr. Roberto Frias, 4200-465 Porto, Portugal
Department of Mechanical Engineering, Minho University, Campus de Azurém, 4800-058 Guimarães, Portugal

Junji Sone
Faculty of Engineering, Tokyo Polytechnic University, 1583 Iiyama Atsugi, Kanagawa 243-0297, Japan

Katsumi Yamada
Faculty of Engineering, Tokyo Polytechnic University, 1583 Iiyama Atsugi, Kanagawa 243-0297, Japan

Akihisa Asami
Graduation School of Engineering, Tokyo Polytechnic University, 1583 Iiyama Atsugi, Kanagawa 243-0297, Japan

Jun Chen
Faculty of Engineering, Tokyo Polytechnic University, 1583 Iiyama Atsugi, Kanagawa 243-0297, Japan

Florian Thoma
Design of Microsystems, Department of Microsystems Engineering (IMTEK), University of Freiburg, Georges-Koehler-Allee 103, 79110 Freiburg, Germany

Frank Goldschmidtböing
Design of Microsystems, Department of Microsystems Engineering (IMTEK), University of Freiburg, Georges-Koehler-Allee 103, 79110 Freiburg, Germany

Peter Woias
Design of Microsystems, Department of Microsystems Engineering (IMTEK), University of Freiburg, Georges-Koehler-Allee 103, 79110 Freiburg, Germany

Chi-Han Chiou
ITRI South Campus, Industrial Technology Research Institute, Tainan City 70955, Taiwan

Tai-Yen Yeh
Department of Mechanical and Automation Engineering, I-Shou University, Kaohsiung City 84001, Taiwan

Jr-Lung Lin
Department of Mechanical and Automation Engineering, I-Shou University, Kaohsiung City 84001, Taiwan

Eszter L. Tóth
Faculty of Information Technology and Bionics, Pázmány Péter Catholic University, Práter utca 50/a, H-1083 Budapest, Hungary

Research Centre for Natural Sciences, Institute for Technical Physics and Materials Science, Hungarian Academy of Sciences, Konkoly Thege M. út 29-33, H-1121 Budapest, Hungary

Eszter G. Holczer
Research Centre for Natural Sciences, Institute for Technical Physics and Materials Science, Hungarian Academy of Sciences, Konkoly Thege M. út 29-33, H-1121 Budapest, Hungary

Kristóf Iván
Faculty of Information Technology and Bionics, Pázmány Péter Catholic University, Práter utca 50/a, H-1083 Budapest, Hungary

Péter Fürjes
Research Centre for Natural Sciences, Institute for Technical Physics and Materials Science, Hungarian Academy of Sciences, Konkoly Thege M. út 29-33, H-1121 Budapest, Hungary

Enza Fazio
Dipartimento di Fisica e di Scienze della Terra, Università di Messina, v.le F. Stagno d'Alcontres 31, 98166 Messina, Italy

Fortunato Neri
Dipartimento di Fisica e di Scienze della Terra, Università di Messina, v.le F. Stagno d'Alcontres 31, 98166 Messina, Italy

Rosina C. Ponterio
IPCF-CNR, Istituto per i Processi Chimico Fisici, Consiglio Nazionale delle Ricerche, v.le F. Stagno d'Alcontres 37, 98158 Messina, Italy

Sebastiano Trusso
IPCF-CNR, Istituto per i Processi Chimico Fisici, Consiglio Nazionale delle Ricerche, v.le F. Stagno d'Alcontres 37, 98158 Messina, Italy

Matteo Tommasini
Dipartimento di Chimica, Materiali e Ingegneria Chimica, "G. Natta", Politecnico di Milano, P.zza L. da Vinci 32, 20133 Milano, Italy

Paolo Maria Ossi
Dipartimento di Energia & Center for NanoEngineered Materials and Surfaces-NEMAS, Politecnico di Milano, via Ponzio 34-3, 20133 Milano, Italy

Tatsiana Malechka
Institute of Automation (IAT), University Bremen, Otto Hahn Allee NW1, 28359 Bremen, Germany
Friedrich Wilhelm Bessel Institute gGmbH, Otto Hahn Allee NW1, 28359 Bremen, Germany

Tobias Tetzel
Friedrich Wilhelm Bessel Institute gGmbH, Otto Hahn Allee NW1, 28359 Bremen, Germany

Ulrich Krebs
Institute of Automation (IAT), University Bremen, Otto Hahn Allee NW1, 28359 Bremen, Germany

Diana Feuser
Institute of Automation (IAT), University Bremen, Otto Hahn Allee NW1, 28359 Bremen, Germany

Axel Graeser
Institute of Automation (IAT), University Bremen, Otto Hahn Allee NW1, 28359 Bremen, Germany

Harutaka Mekaru
Research Center for Ubiquitous MEMS and Micro Engineering (UMEMSME), National Institute of Advanced Industrial Science and Technology (AIST), 1-2-1 Namiki, Tsukuba, Ibaraki 305-8564, Japan

Florian Preusch
Applied Laser and Phonics Group, University of Applied Sciences Aschaffenburg, Wuerzburger Strasse 45, D-63743 Aschaffenburg, Germany

Benedikt Adelmann
Applied Laser and Phonics Group, University of Applied Sciences Aschaffenburg, Wuerzburger Strasse 45, D-63743 Aschaffenburg, Germany

Ralf Hellmann
Applied Laser and Phonics Group, University of Applied Sciences Aschaffenburg, Wuerzburger Strasse 45, D-63743 Aschaffenburg, Germany

www.ingramcontent.com/pod-product-compliance
Lightning Source LLC
Chambersburg PA
CBHW080639200326
41458CB00013B/4676